ESSENTIALS OF SOCIAL STATISTICS
FOR A DIVERSE SOCIETY

Second Edition

ESSENTIALS OF SOCIAL STATISTICS FOR A DIVERSE SOCIETY

Second Edition

Anna Leon-Guerrero
Pacific Lutheran University

Chava Frankfort-Nachmias
University of Wisconsin–Milwaukee

Los Angeles | London | New Delhi
Singapore | Washington DC

Los Angeles | London | New Delhi
Singapore | Washington DC

FOR INFORMATION:

SAGE Publications, Inc.
2455 Teller Road
Thousand Oaks, California 91320
E-mail: order@sagepub.com

SAGE Publications Ltd.
1 Oliver's Yard
55 City Road
London EC1Y 1SP
United Kingdom

SAGE Publications India Pvt. Ltd.
B 1/I 1 Mohan Cooperative Industrial Area
Mathura Road, New Delhi 110 044
India

SAGE Publications Asia-Pacific Pte. Ltd.
3 Church Street
#10-04 Samsung Hub
Singapore 049483

Acquisitions Editor: Jerry Westby
Publishing Associate: MaryAnn Vail
Editorial Assistant: Laura Kirkhuff
Production Editor: David C. Felts
Copy Editor: Jim Kelly
Typesetter: C&M Digitals (P) Ltd.
Proofreader: Scott Oney
Indexer: Will Ragsdale
Cover Designer: Candice Harman
Marketing Manager: Erica DeLuca

Copyright © 2015 by SAGE Publications, Inc.

Printed in the United States of America

Library of Congress Cataloging-in-Publication Data

Leon-Guerrero, Anna.

Essentials of social statistics for a diverse society / Anna Leon-Guerrero, Pacific Lutheran University, Chava Frankfort-Nachmias, University of Wisconsin–Milwaukee.—Second edition.

pages cm
Brief version of the authors' Social statistics for a diverse society now in its seventh edition.
Includes bibliographical references and index.

ISBN 978-1-4833-5949-6 (pbk.)

1. Social sciences—Statistical methods. 2. Statistics. I. Frankfort-Nachmias, Chava. II. Frankfort-Nachmias, Chava. Social statistics for a diverse society. III. Title.

HA29.L364 2015
519.5—dc23   2014030442

This book is printed on acid-free paper.

14 15 16 17 18 10 9 8 7 6 5 4 3 2 1

BRIEF CONTENTS

DETAILED CONTENTS

PREFACE

S tatistics is not just a part of our lives in the form of news bits or information. And it isn't just numbers either. Throughout this book, we encourage you to move beyond being just a consumer of statistics and begin to recognize and use the many ways that statistics can increase our understanding of our world.

Recently data have been characterized as "big,"[1] referring not only to the amount of available data, but also to the application of such information. Data are used to predict public opinion, health and illness, consumer spending, and even a presidential election. Throughout our text, we emphasize the relevance of statistics in our daily and professional lives. How Americans feel about a variety of political and social topics—the economy, same-sex marriage, gun control, immigration, health care reform, or our president—are measured by surveys and polls and reported daily by the news media. The latest from a health care study on women was just reported on a morning talk show. And that outfit you just purchased—it didn't go unnoticed. The study of consumer trends, specifically focusing on teens and young adults, helps determine commercial programming, product advertising and placement, and, ultimately, consumer spending. President Obama's 2012 reelection campaign victory was attributed in part to a team of data experts who analyzed voter files to predict likely voters.

As social scientists, we have always known that statistics can be a valuable set of tools to help us analyze and understand the differences in our American society and the world. We use statistics to track demographic trends, to assess differences among groups in society, and to make an impact on social policy and social change. Statistics can help us gain insight into real-life problems that affect our lives.

▣ TEACHING AND LEARNING GOALS

The following three teaching and learning goals continue to be the guiding principles of our book.

The first goal is to introduce you to social statistics and demonstrate its value. Although most of you will not use statistics in your own student research, you will be expected to read and interpret statistical information presented by others in professional and scholarly publications, in the workplace, and in the popular media. This book will help you understand the concepts behind the statistics so that you will be able to assess the circumstances in which certain statistics should and should not be used.

Our second goal is to demonstrate that substance and statistical techniques are truly related in social science research. A special quality of this book is its integration of statistical techniques with substantive issues of particular relevance in the social sciences. Your learning will not be limited

---

[1] Big data are described by Viktor Mayer-Schönberger and Kenneth Cukier in *Big Data: A Revolution That Will Transform How We Live, Work, and Think* (2013).

to statistical calculations and formulas. Rather, you will become proficient in statistical techniques while learning about social differences and inequality through numerous substantive examples and real-world data applications. Because the world we live in is characterized by a growing diversity—whereby personal and social realities are increasingly shaped by race, class, gender, and other categories of experience—this book teaches you basic statistics while incorporating social science research related to the dynamic interplay of social variables.

Many of you may lack substantial math background, and some of you may suffer from the "math anxiety syndrome." This anxiety often leads to a less-than-optimal learning environment, with students trying to memorize every detail of a statistical procedure rather than attempting to understand the general concept involved. Hence, our third goal is to address math anxiety by using straightforward prose to explain statistical concepts and by emphasizing intuition, logic, and common sense over rote memorization and derivation of formulas.

▣ DISTINCTIVE AND UPDATED FEATURES OF OUR BOOK

The three learning goals we emphasize are accomplished through a variety of specific and distinctive features throughout this book.

A Close Link Between the Practice of Statistics, Important Social Issues, and Real-World Examples. A special quality of this book is its integration of statistical technique with pressing social issues of particular concern to society and social science. We emphasize how the conduct of social science is the constant interplay between social concerns and methods of inquiry. In addition, the examples throughout the book—mostly taken from news stories, government reports, public opinion polls, scholarly research, and the National Opinion Research Center's General Social Survey (GSS)—are formulated to emphasize to students like you that we live in a world in which statistical arguments are common. Statistical concepts and procedures are illustrated with real data and research, providing a clear sense of how questions about important social issues can be studied with various statistical techniques.

A Focus on Diversity: U.S. and International. A strong emphasis on race, class, and gender as central substantive concepts is mindful of a trend in the social sciences toward integrating issues of diversity in the curriculum. This focus on the richness of social differences within our society and our global neighbors is manifested in the application of statistical tools to examine how race, class, gender, and other categories of experience shape our social world and explain social behavior. There is a special focus on the interplay between local and global concern. Throughout the text, we rely on data from the International Social Survey Programme, and we created a special global data set for this edition to help expand our statistical focus beyond the United States.

Reading the Research Literature. In your student career and in the workplace, you may be expected to read and interpret statistical information presented by others in professional and scholarly publications. The statistical analyses presented in these publications are a good deal more complex than most class and textbook presentations. To guide you in reading and interpreting research reports written by social scientists, most chapters include a section presenting excerpts of published research reports using the statistical concepts under discussion.

Tools to Promote Effective Study. Each chapter concludes with a list of main points and key terms discussed in that chapter. Boxed definitions of the key terms also appear in the body of the chapter, as do learning checks keyed to the most important points. Key terms are also clearly defined and explained in the glossary, another special feature in our book. Answers to all the odd-numbered exercises and Learning Checks in the text are included at the end of the book, as well as on the study site at **edge.sagepub.com/ssdsess2e**. Complete step-by-step solutions are in the manual for instructors, available from the publisher on adoption of the text.

Emphasis on Computing. Real data are used to motivate and make concrete the coverage of statistical topics. These data, from the GSS, Health Information National Trends Survey (HINTS), Monitoring the Future (MTF) survey, and the global data set constructed for this edition, are available on the study site at **edge.sagepub.com/ssdsess2e**.

▣ HIGHLIGHTS OF THE SECOND EDITION

We have made a number of important changes to this book in response to the valuable comments that we have received from the many instructors adopting the first edition and from other interested instructors (and their students).

- *Chapter reorganization:* In this edition, we've created a new chapter on estimation. The text concludes with chapters on analysis of variance and regression and correlation.
- *Real-world examples and exercises:* A hallmark of our text is the extensive use of real data from a variety of sources for chapter illustrations and exercises. Throughout this edition, we have updated the majority of exercises and examples based on the GSS, MTF, and HINTS, our global data set, or U.S. census data.
- *GSS 2010, HINTS 2012, MTF 2011, and Global 2013:* As a companion to the edition's SPSS demonstrations and exercises, we have created five data sets. The GSS10SSDS.SAV contains an expanded selection of variables and cases from the 2010 GSS. Those of you using IBM® SPSS® Statistics Base Integrated Student Edition[2] can work with GSS2010SSDS-A, a subset of 50 GSS variables. The HINTS12SSDS.SAV contains 50 variables from the 2012 HINTS, administered by the National Cancer Institute. HINTS, a nationally representative survey collected in both English and Spanish, aims to monitor changes in the rapidly evolving field of health communication. The MTF11SSDS.SAV contains a selection of variables and cases from the MTF 2011 survey conducted by the University of Michigan Survey Research Center. MTF is a survey of 12th-grade students, and it explores drug use and criminal behavior. GLOBAL13SSDS.SAV includes data measuring the social, economic, and political conditions of 70 countries. SPSS exercises, available on the text's website, use certain variables from all data modules. There is ample opportunity for instructors to develop their own SPSS exercises using these data.
- *Supplemental tools on important topics:* For selected chapters, we have added a new section, Focus on Interpretation. In these sections, we highlight the interpretation of data or specific statistical calculations (some calculated via SPSS). Being statistically literate involves more than just completing a calculation; it also includes learning how to apply and interpret statistical information and being able to say what it means.

---

[2]SPSS is a registered trademark of International Business Machines Corporation.

⑤SAGE edge™

edge.sagepub.com/ssdsess2e

SAGE edge offers a robust online environment featuring an impressive array of tools and resources for review, study, and further exploration, keeping both instructors and students on the cutting edge of teaching and learning. SAGE edge content is open access and available on demand. Learning and teaching has never been easier!

SAGE edge for students provides a personalized approach to help students accomplish their coursework goals in an easy-to-use learning environment.

- Mobile-friendly **eFlashcards** strengthen understanding of key terms and concepts.
- Mobile-friendly practice **quizzes** allow for independent assessment by students of their mastery of course material.
- A customized online **action plan** includes tips and feedback on progress through the course and materials, which allows students to individualize their learning experience.
- **Web exercises** and meaningful web links facilitate student use of internet resources, further exploration of topics, and responses to critical thinking questions.
- EXCLUSIVE! Access to full-text **SAGE journal articles** that have been carefully selected to support and expand on the concepts presented in each chapter.
- Access to five new **data sets** including GSS 2010, HINTS 2012, MTF 2011, and Global 2013.

SAGE edge for instructors supports teaching by making it easy to integrate quality content and create a rich learning environment for students.

- **Test banks** provide a diverse range of prewritten options as well as the opportunity to edit any question and/or insert personalized questions to effectively assess students' progress and understanding.
- **Sample course syllabi** for semester and quarter courses provide suggested models for structuring one's course.
- Editable, chapter-specific **PowerPoint®slides** offer complete flexibility for creating a multimedia presentation for the course.
- EXCLUSIVE! Access to full-text **SAGE journal articles** that have been carefully selected to support and expand on the concepts presented in each chapter to encourage students to think critically.
- **Multimedia content** includes web resources and web exercises that appeal to students with different learning styles.
- **Lecture notes** summarize key concepts by chapter to ease preparation for lectures and class discussions.
- Lively and stimulating **ideas for class activities** that can be used in class to reinforce active learning.
- **Chapter-specific discussion questions** help launch classroom interaction by prompting students to engage with the material and by reinforcing important content.
- A **Course cartridge** provides easy LMS integration.

◫ ACKNOWLEDGMENTS

We are both grateful to Jerry Westby, Series Editor for SAGE Publications, for his commitment to our book and for his invaluable assistance through the production process.

Many manuscript reviewers recruited by SAGE provided invaluable feedback. For their comments, we thank

Teresa Casey, Idaho State University

Bethany Coston, Stony Brook University

Molly George, California Lutheran University

Shirley Jackson, Southern Connecticut State University

Steve B. Lem, Kutztown University

Teri Fair Platt, Suffolk University

Adrienne Trier-Bieniek, Valencia College

Nishanth Visgaratnam, Southern Illinois University Carbondale

Yvonne Vissing, Salem State University

We wish to thank our production editor, David Felts, for guiding our book through the production process. We owe a debt of gratitude to our copy editor, Jim Kelly, for his meticulous review of our manuscript.

Both of us extend our deepest appreciation to Michael Clark for creating the student data sets that accompany the second edition. Additionally, we are grateful to Caleb Schaffner, Ben Gilbertsen, and Kaitlyn Elms for their work on our individual chapters.

Anna Leon-Guerrero expresses her thanks to the following: I wish to thank my PLU statistics students. My passion for and understanding of teaching statistics grow with each semester and class experience. I am grateful for the teaching and learning opportunities that we have shared.

I would like to express my gratitude to friends and colleagues for their encouragement and support throughout this project. My love and thanks to my husband, Brian Sullivan.

Chava Frankfort-Nachmias would like to thank and acknowledge her friends and colleagues for their unending support; she also would like to thank her students: I am grateful to my students at the University of Wisconsin–Milwaukee, who taught me that even the most complex statistical ideas can be simplified. The ideas presented in this book are the products of many years of classroom testing. I thank my students for their patience and contributions.

Finally, I thank my partner, Marlene Stern, for her love and support.

Anna Leon-Guerrero
Pacific Lutheran University

Chava Frankfort-Nachmias
University of Wisconsin–Milwaukee

ABOUT THE AUTHORS

Anna Leon-Guerrero is a Professor of Sociology at Pacific Lutheran University in Washington. She received her PhD in sociology from the University of California, Los Angeles. She teaches courses in statistics, social theory, and social problems. Her areas of research and publications include family business, social welfare policy, and social service program evaluation. She is also the author of *Social Problems: Community, Policy, and Social Action.*

Chava Frankfort-Nachmias is an Emeritus Professor of Sociology at the University of Wisconsin–Milwaukee. She is the coauthor of *Research Methods in the Social Sciences* (with David Nachmias), coeditor of *Sappho in the Holy Land* (with Erella Shadmi) and numerous publications on ethnicity and development, urban revitalization, science and gender, and women in Israel. She was the recipient of the University of Wisconsin System teaching improvement grant on integrating race, ethnicity, and gender into the social statistics and research methods curriculum. She is also the coauthor (with Anna Leon-Guerrero) of *Social Statistics for a Diverse Society.*

Chapter 1

The What and the Why of Statistics

Chapter Learning Objectives

❖ Understanding the research process
❖ Identifying and distinguishing between independent and dependent variables
❖ Identifying and distinguishing between three levels of measurement
❖ Understanding descriptive versus inferential statistical procedures

A re you taking statistics because it is required in your major—not because you find it interesting? If so, you may be feeling intimidated because you know that statistics involves numbers and math. Perhaps you feel intimidated not only because you're uncomfortable with math but also because you suspect that numbers and math don't leave room for human judgment or have any relevance to your own personal experience. In fact, you may even question the relevance of statistics to understanding people, social behavior, or society.

In this book, we will show you that statistics can be a lot more interesting and easy to understand than you may have been led to believe. In fact, as we draw on your previous knowledge and experience and relate materials to interesting and important social issues, you'll begin to see that statistics is not just a course you have to take but a useful tool as well.

There are two reasons why learning statistics may be of value to you. First, you are constantly exposed to statistics every day of your life. Marketing surveys, voting polls, and the social research findings appear daily in newspapers and popular magazines. By learning statistics, you will become a sharper consumer of statistical material. Second, as a major in the social sciences, you may be expected to read and interpret statistical information presented to you in the workplace. Even if conducting research is not a part of your job, you may still be expected to understand and learn from other people's research or to be able to write reports based on statistical analyses.

Just what *is* statistics, anyway? You may associate the word with numbers that indicate birthrates, conviction rates, per capita income, marriage and divorce rates, and so on. But the word

statistics also refers to a set of procedures used by social scientists. They use these procedures to organize, summarize, and communicate information. Only information represented by numbers can be the subject of statistical analysis. Such information is called **data**; researchers use statistical procedures to analyze data to answer research questions and test theories. It is the latter use—answering research questions and testing theories—that this textbook explores.

Statistics   A set of procedures used by social scientists to organize, summarize, and communicate information.

Data   Information represented by numbers, which can be the subject of statistical analysis.

▣ THE RESEARCH PROCESS

To give you a better idea of the role of statistics in social research, let's start by looking at the **research process**. We can think of the research process as a set of activities in which social scientists engage so that they can answer questions, examine ideas, or test theories.

As illustrated in Figure 1.1, the research process consists of five stages:

1. Asking the research question

2. Formulating the hypotheses

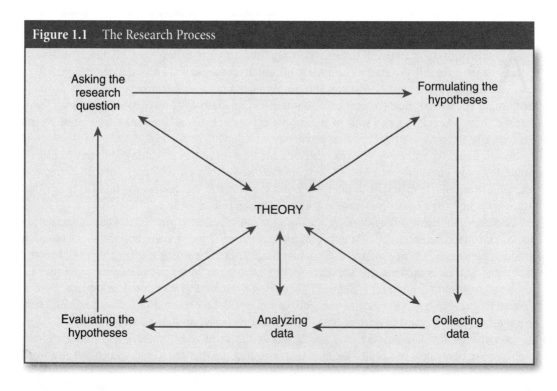

Figure 1.1   The Research Process

3. Collecting data

4. Analyzing data

5. Evaluating the hypotheses

Each stage affects the *theory* and is affected by it as well. Statistics is most closely tied to the data analysis stage of the research process. As we will see in later chapters, statistical analysis of the data helps researchers test the validity and accuracy of their hypotheses.

Research process   A set of activities in which social scientists engage to answer questions, examine ideas, or test theories.

▣ ASKING RESEARCH QUESTIONS

The starting point for most research is asking a *research question*. Consider the following research questions taken from a number of social science journals:

How will the Affordable Care Act influence the quality of health care?

Has support for gay marriage increased during the past decade?

Does race or ethnicity predict voting behavior?

What factors affect the economic mobility of female workers?

These are all questions that can be answered by conducting **empirical research**—research based on information that can be verified by using our direct experience. To answer research questions, we cannot rely on reasoning, speculation, moral judgment, or subjective preference. For example, the questions "Is racial equality good for society?" and "Is an urban lifestyle better than a rural lifestyle?" cannot be answered empirically, because the terms *good* and *better* are concerned with values, beliefs, or subjective preference and, therefore, cannot be independently verified. One way to study these questions is by defining *good* and *better* in terms that can be verified empirically. For example, we can define *good* in terms of economic growth and *better* in terms of psychological well-being. These questions could then be answered by conducting empirical research.

Empirical research   Research based on evidence that can be verified by using our direct experience.

You may wonder how to come up with a research question. The first step is to pick a question that interests you. If you are not sure, look around! Ideas for research problems are all around you, from media sources to personal experience or your own intuition. Talk to other people, write down your own observations and ideas, or learn what other social scientists have written about.

Take, for instance, the issue of gender and work. As a college student about to enter the labor force, you may wonder about the similarities and differences between women's and men's work experiences and about job opportunities when you graduate. Here are some facts and observations based on research reports: In 2012, women who were employed full-time earned about $691 per week on average; men who were employed full-time earned $854 per week on average.[1] Women's and men's work are also very different. Women continue to be the minority in many of the higher ranking and higher salaried positions in professional and managerial occupations. For example, in 2010 women made up 9.7% of civil engineers, 32.3% of physicians, 25.5% of dentists, and 1.5% of electricians. In comparison, among all those employed as preschool and kindergarten teachers, 97% were women. Among all receptionists and information clerks in 2010, 92.7% were women.[2] Another noteworthy development in the history of labor in the United States took place in January 2010: Women outnumbered men for the first time by holding 50.3% of the nonfarm payroll jobs.[3] These observations may prompt us to ask research questions such as the following: How much change has there been in women's work over time? Are women paid, on average, less than men for the same type of work?

✓ *Learning Check*

> *Identify one or two social science questions amenable to empirical research. You can almost bet that you will be required to do a research project sometime in your college career. Get a head start and start thinking about a good research question now.*

▣ THE ROLE OF THEORY

You may have noticed that each preceding research question was expressed in terms of a *relationship*. This relationship may be between two or more attributes of individuals or groups, such as gender and income or gender segregation in the workplace and income disparity. The relationship between attributes or characteristics of individuals and groups lies at the heart of social scientific inquiry.

Most of us use the term *theory* quite casually to explain events and experiences in our daily lives. We may have a "theory" about why our boss has been so nice to us lately or why we didn't do so well on our last history test. In a somewhat similar manner, social scientists attempt to explain the nature of social reality. Whereas our theories about events in our lives are commonsense explanations based on educated guesses and personal experience, to the social scientist, a theory is a more precise explanation that is frequently tested by conducting research.

A **theory** is an explanation of the relationship between two or more observable attributes of individuals or groups. The theory attempts to establish a link between what we observe (the data) and our conceptual understanding of why certain phenomena are related to each other in a particular way. For instance, suppose we wanted to understand the reasons for the income disparity between men and women; we may wonder whether the types of jobs men and women have and the organizations in which they work have something to do with their wages.

Theory   An elaborate explanation of the relationship between two or more observable attributes of individuals or groups.

One explanation for gender inequality in wages is *gender segregation in the workplace*—the fact that American men and women are concentrated in different kinds of jobs and occupations. What is the significance of gender segregation in the workplace? In our society, people's occupations and jobs are closely associated with their levels of prestige, authority, and income. The jobs in which women and men are segregated are not only different but also unequal. Although the proportion of women in the labor force has markedly increased, women are still concentrated in occupations with low pay, low prestige, and few opportunities for promotion. Thus, gender segregation in the workplace is associated with unequal earnings, authority, and status. In particular, women's segregation into different jobs and occupations from those of men is the most immediate cause of the pay gap. Women receive lower pay than men do even when they have the same levels of education, skills, and experience as men in comparable occupations.

▣ FORMULATING THE HYPOTHESES

So far, we have come up with a number of research questions about the income disparity between men and women in the workplace. We have also discussed a possible explanation—a theory—that helps us make sense of gender inequality in wages. Is that enough? Where do we go from here?

Our next step is to test some of the ideas suggested by the gender segregation theory. But this theory, even if it sounds reasonable and logical to us, is too general and does not contain enough specific information to be tested. Instead, theories suggest specific concrete predictions about the way that observable attributes of people or groups are interrelated in real life. These predictions, called **hypotheses**, are tentative answers to research problems. Hypotheses are tentative because they can be verified only after they have been tested empirically.[4] For example, one hypothesis we can derive from the gender segregation theory is that wages in occupations in which the majority of workers are female are lower than the wages in occupations in which the majority of workers are male.

Hypothesis   A tentative answer to a research problem.

Not all hypotheses are derived directly from theories. We can generate hypotheses in many ways—from theories, directly from observations, or from intuition. Probably the greatest source of hypotheses is the professional literature. A critical review of the professional literature will familiarize you with the current state of knowledge and with hypotheses that others have studied.

Let's restate our hypothesis:

Wages in occupations in which the majority of workers are female are lower than the wages in occupations in which the majority of workers are male.

Note that this hypothesis is a statement of a relationship between two characteristics that vary: the *wages* and *gender composition* of occupations. Such characteristics are called variables. A **variable** is a property of people or objects that takes on two or more values. For example, people can be classified into a number of *social class* categories, such as upper class, middle class, or working class. Similarly, people have different levels of education; therefore, *education* is a variable. *Family income* is a variable; it can take on values from zero to hundreds of thousands of dollars or more. *Wages* is a variable, with values from zero to thousands of dollars or more. Similarly, *gender composition* is a variable. The percentage of female (or male) workers in an occupation can vary from 0 to 100. (See Figure 1.2 for examples of some variables and their possible values.)

Variable   A property of people or objects that takes on two or more values.

Each variable must include categories that are both *exhaustive* and *mutually exclusive*. Exhaustiveness means that there should be enough categories composing the variables to classify every observation. For example, the common classification of the variable *marital status* into the categories "married," "single," "divorced," and "widowed" violates the requirement of exhaustiveness. As defined, it does not allow us to classify same-sex couples or heterosexual couples who are not legally married. (We can make every variable exhaustive by adding the category "other" to the list of categories. However, this practice is not recommended if it leads to the exclusion of categories that have theoretical significance or a substantial number of observations.)

Figure 1.2   Variables and Value Categories

Variable	Categories
Social class	Upper class Middle class Working class
Religion	Christian Jewish Muslim
Monthly income	$1,000 $2,500 $10,000 $15,000
Gender	Male Female

Mutual exclusiveness means that there is only one category suitable for each observation. For example, we need to define *religion* in such a way that no one would be classified into more than one category. For instance, the categories "Protestant" and "Methodist" are not mutually exclusive because Methodists are also considered Protestant and, therefore, could be classified into both categories.

✓ *Learning*
Check

Review the definitions of exhaustive and mutually exclusive. Now look at Figure 1.2. What other categories could be added to the variable religion to make it exhaustive and mutually exclusive? What other categories could be added to social class? To income?

Social scientists can choose which level of social life to focus their research on. They can focus on individuals or on groups of people such as families, organizations, and nations. These distinctions are referred to as **units of analysis**. A variable is a property of whatever the unit of analysis is for the study. Variables can be properties of individuals, of groups (e.g., the family or a social group), of organizations (e.g., a hospital or university), or of societies (e.g., a country or a nation). For example, in a study that looks at the relationship between individuals' levels of education and their incomes, the variable *income* refers to the income level of an individual. On the other hand, a study that compares how differences in corporations' revenues relate to differences in the fringe benefits they provide to their employees uses the variable *revenue* as a characteristic of an organization (the corporation). The variables *wages* and *gender composition* in our example are characteristics of occupations. Figure 1.3 illustrates different units of analysis frequently employed by social scientists.

Unit of analysis The level of social life on which social scientists focus. Examples of different levels are individuals and groups.

✓ *Learning*
Check

Remember that research question you came up with? Can you formulate a hypothesis you could test? Remember that the variables must take on two or more values and you must determine the unit of analysis.

Independent and Dependent Variables: Causality

Hypotheses are usually stated in terms of a relationship between an *independent* and a *dependent variable.* The distinction between an independent and a dependent variable is important in the language of research. Social theories often intend to provide an explanation for social patterns or causal relations between variables. For example, according to the gender segregation theory,

Figure 1.3 Examples of Units of Analysis

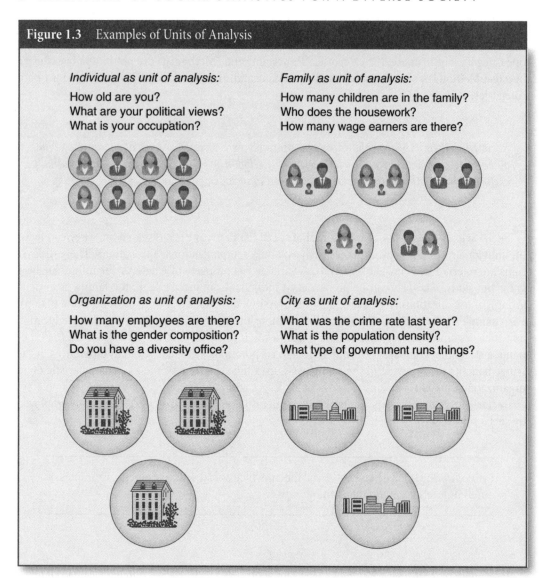

gender segregation in the workplace is the primary explanation (although certainly not the only one) of the male-female earning gap. Why should jobs where the majority of workers are women pay less than jobs that employ mostly men? One explanation is that

societies undervalue the work women do, regardless of what those tasks are, because women do them. . . . For example, our culture tends to devalue caring or nurturant work at least partly because it is done by women. This tendency accounts for child care workers' low rank in the pay hierarchy.[5]

In the language of research, the variable the researcher wants to explain (the "effect") is called the **dependent variable**. The variable that is expected to "cause" or account for the dependent variable is called the **independent variable**. Therefore, in our example, *gender composition of occupations* is the independent variable, and *wages* is the dependent variable.

Dependent variable   The variable to be explained (the "effect").

Independent variable   The variable expected to account for (the "cause" of) the dependent variable.

Cause-and-effect relationships between variables are *not* easy to infer in the social sciences. To establish that two variables are causally related, you need to meet three conditions: (1) The cause has to precede the effect in time, (2) there has to be an empirical relationship between the cause and the effect, and (3) this relationship cannot be explained by other factors.

Independent and Dependent Variables: Guidelines

Because of the limitations in inferring cause-and-effect relationships in the social sciences, be cautious about using the terms *cause* and *effect* when examining relationships between variables. However, using the terms *independent variable* and *dependent variable* is still appropriate, even when this relationship is not articulated in terms of direct cause and effect. Here are a few guidelines that may help you identify the independent and dependent variables:

1. The dependent variable is always the property that you are trying to explain; it is always the object of the research.

2. The independent variable usually occurs earlier in time than the dependent variable.

3. The independent variable is often seen as influencing, directly or indirectly, the dependent variable.

The purpose of the research should help determine which is the independent variable and which is the dependent variable. In the real world, variables are neither dependent nor independent; they can be switched around depending on the research problem. A variable defined as independent in one research investigation may be a dependent variable in another.[6] For instance, *educational attainment* may be an independent variable in a study attempting to explain how education influences political attitudes. However, in an investigation of whether a person's level of education is influenced by the social status of his or her family of origin, *educational attainment* is the dependent variable. Some variables, such as race, age, and ethnicity, because they are primordial characteristics that cannot be explained by social scientists, are never considered dependent variables in a social science analysis.

Identify the independent and dependent variables in the following hypotheses:

- *Younger Americans are more likely to support stricter gun control laws than older Americans.*
- *People who attend church regularly are more likely to oppose abortion than people who do not attend church regularly.*
- *Elderly women are more likely to live alone than elderly men.*
- *Individuals with postgraduate education are likely to have fewer children than those with less education.*

What are the independent and dependent variables in your hypothesis?

▣ COLLECTING DATA

Once we have decided on the research question, the hypothesis, and the variables to be included in the study, we proceed to the next stage in the research cycle. This step includes measuring our variables and collecting the data. As researchers, we must decide how to measure the variables of interest to us, how to select the cases for our research, and what kind of data collection techniques we will be using. A wide variety of data collection techniques are available to us, from direct observations to survey research, experiments, or secondary sources. Similarly, we can construct numerous measuring instruments. These instruments can be as simple as a single question included in a questionnaire or as complex as a composite measure constructed through the combination of two or more questionnaire items. The choice of a particular data collection method or instrument to measure our variables depends on the study objective. For instance, suppose we decide to study how social class position is related to attitudes about abortion. Since attitudes about abortion are not directly observable, we need to collect data by asking a group of people questions about their attitudes and opinions. A suitable method of data collection for this project would be a *survey* that uses some kind of questionnaire or interview guide to elicit verbal reports from respondents. The questionnaire could include numerous questions designed to measure attitudes toward abortion, social class, and other variables relevant to the study.

How would we go about collecting data to test the hypothesis relating the gender composition of occupations to wages? We want to gather information on the proportions of men and women in different occupations and the average earnings for these occupations. This kind of information is routinely collected by the government and published in sources such as bulletins distributed by the U.S. Department of Labor's Bureau of Labor Statistics and the U.S. Census Bureau's *Statistical Abstract of the United States.* The data obtained from these sources could then be analyzed and used to test our hypothesis.

Levels of Measurement

The statistical analysis of data involves many mathematical operations, from simple counting to addition and multiplication. However, not every operation can be used with every variable. The types of statistical operations we employ depend on how our variables are measured. For example, for the variable *gender*, we can use the number 1 to represent female and the number 2 to represent male. Similarly, 1 can also be used as a numerical code for the category "one child" in the variable *number of children*. Clearly, in the first example, the number is an arbitrary symbol that does not correspond to the property "female," whereas in the second example the number 1 has a distinct numerical meaning that does correspond to the property "one child." The correspondence between the properties we measure and the numbers representing these properties determines the types of statistical operations we can use. The degree of correspondence also leads to different ways of measuring—that is, to distinct *levels of measurement*. In this section, we will discuss three levels of measurement: *nominal, ordinal*, and *interval-ratio*.

Nominal Level of Measurement

At the **nominal** level of measurement, numbers or other symbols are assigned a set of categories for the purpose of naming, labeling, or classifying the observations. *Gender* is an example of a nominal-level variable. Using the numbers 1 and 2, for instance, we can classify our observations into the categories "female" and "male," with 1 representing female and 2 representing male. We could use any of a variety of symbols to represent the different categories of a nominal variable; however, when numbers are used to represent the different categories, we do not imply anything about the magnitude or quantitative difference between the categories. Because the different categories (e.g., male vs. female) vary in the quality inherent in each but not in quantity, nominal variables are often called *qualitative*. Other examples of nominal-level variables are political party, religion, and race.

Nominal measurement   Numbers or other symbols are assigned to a set of categories for the purpose of naming, labeling, or classifying the observations.

Ordinal Level of Measurement

Whenever we assign numbers to rank-ordered categories ranging from low to high, we have an **ordinal** level of measurement. *Social class* is an example of an ordinal variable. We might classify individuals with respect to their social class status as "upper class," "middle class," or "working class." We can say that a person in the category "upper class" has a higher class position than a person in a "middle-class" category (or that a "middle-class" position is higher than a "working-class" position), but we do not know the magnitude of the differences between the categories—that is, we don't know how much higher "upper class" is compared with the "middle class."

Many attitudes that we measure in the social sciences are ordinal-level variables. Take, for instance, the following statement used to measure attitudes toward gun control: "There should be background checks for private and gun show sales." Respondents are asked to mark the number representing their degree of agreement or disagreement with this statement. One form in which numbers might be made to correspond with the answers can be seen in Table 1.1. Although the differences between these numbers represent higher or lower degrees of agreement with background checks, the distance between any two of those numbers does not have a precise numerical meaning.

Table 1.1 Ordinal Ranking Scale

Rank	Value
1	Strongly agree
2	Agree
3	Neither agree nor disagree
4	Disagree
5	Strongly disagree

Ordinal measurement Numbers are assigned to rank-ordered categories ranging from low to high.

Interval-Ratio Level of Measurement

If the categories (or values) of a variable can be rank-ordered, and if the measurements for all the cases are expressed in the same units, then an **interval-ratio** level of measurement has been achieved. Examples of variables measured at the interval-ratio level are *age*, *income*, and *SAT score*. With all these variables, we can compare values not only in terms of which is larger or smaller but also in terms of *how much* larger or smaller one is compared with another. In some discussions of levels of measurement, you will see a distinction made between interval-ratio variables that have natural zero points (where zero means the absence of the property) and those variables that have zero as an arbitrary point. For example, weight and length have natural zero points, whereas temperature has an arbitrary zero point. Variables with natural zero points are also called *ratio variables*. In statistical practice, however, ratio variables are subjected to operations that treat them as intervals and ignore their ratio properties. Therefore, no distinction between these two types is made in this text.

Interval-ratio measurement Measurements for all cases are expressed in the same units.

Cumulative Property of Levels of Measurement

Variables that can be measured at the interval-ratio level of measurement can also be measured at the ordinal and nominal levels. As a rule, properties that can be measured at a higher level (interval-ratio is the highest) can also be measured at lower levels, but not vice versa. Let's take, for example, *gender composition of occupations*, the independent variable in our research example. Table 1.2 shows the percentages of women in five major occupational groups as reported in the *Statistical Abstract of the United States: 2012.*

The variable *gender composition* (measured as the percentage of women in the occupational group) is an interval-ratio variable and, therefore, has the properties of nominal, ordinal, and interval-ratio measures. For example, we can say that the management group differs from the natural resources group (a nominal comparison), that service occupations have more women than the other occupational categories (an ordinal comparison), and that service occupations have 35.6 percentage points more women (56.8 − 21.2) than production occupations (an interval-ratio comparison).

The types of comparisons possible at each level of measurement are summarized in Table 1.3 and Figure 1.4. Note that differences can be established at each of the three levels, but only at the interval-ratio level can we establish the magnitude of the difference.

Levels of Measurement of Dichotomous Variables

A variable that has only two values is called a **dichotomous variable**. Several key social factors, such as gender, employment status, and marital status, are dichotomies—that is, you are male or female, employed or unemployed, married or not married. Such variables may seem to be

Table 1.2　Gender Composition of Five Major Occupational Groups

Occupational Group	Women in Occupation (%)
Management, professional, and related occupations	51.5
Service occupations	56.8
Production, transportation, and materials occupations	21.2
Sales and office occupations	62.9
Natural resources, construction, and maintenance occupations	4.6

Source: U.S. Census Bureau, *Statistical Abstract of the United States: 2012,* Table 616.

✓ *Learning Check*

Make sure you understand these levels of measurement. As the course progresses, your instructor is likely to ask you what statistical procedure you would use to describe or analyze a set of data. To make the proper choice, you must know the level of measurement of the data.

Table 1.3　Levels of Measurement and Possible Comparisons

Level	Different or Equivalent	Higher or Lower	How Much Higher
Nominal	Yes	No	No
Ordinal	Yes	Yes	No
Interval-ratio	Yes	Yes	Yes

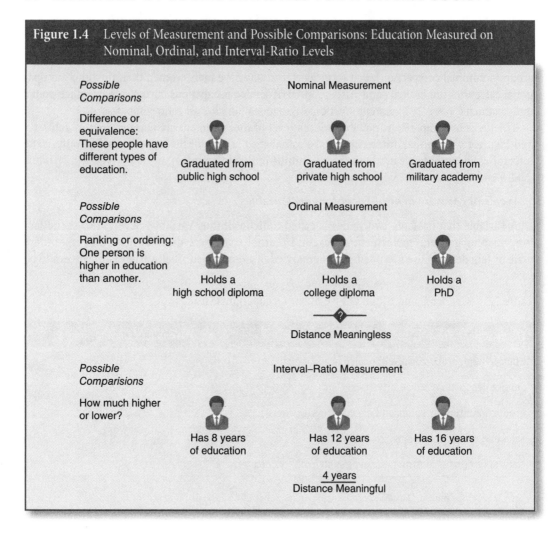

Figure 1.4   Levels of Measurement and Possible Comparisons: Education Measured on Nominal, Ordinal, and Interval-Ratio Levels

measured at the nominal level: You fit in either one category or the other. No category is naturally higher or lower than the other, so they can't be ordered.

***

Dichotomous variable   A variable that has only two values.

***

However, because there are only two possible values for a dichotomy, we can measure it at the ordinal or the interval-ratio level. For example, we can think of "femaleness" as the ordering principle for gender, so that "female" is higher and "male" is lower. Using "maleness" as the ordering principle, "female" is lower and "male" is higher. In either case, with only two classes, there is no way to get them out of order; therefore, gender could be considered at the ordinal level.

Dichotomous variables can also be considered to be interval-ratio level. Why is this? In measuring interval-ratio data, the size of the interval between the categories is *meaningful:* The distance between 4 and 7, for example, is the same as the distance between 11 and 14. But with a dichotomy, there is only one interval. Therefore, there is really no other distance with which we can compare it.

Mathematically, this gives the dichotomy more power than other nominal-level variables (as you will notice later in the text).

For this reason, researchers often dichotomize some of their variables, turning a multicategory nominal variable into a dichotomy. For example, you may see race (originally divided into many categories) dichotomized into "white" and "nonwhite." Though this is substantively suspect, it may be the most logical statistical step to take.

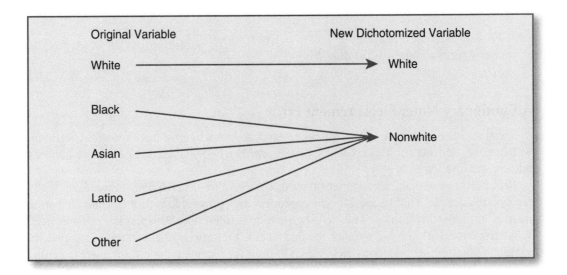

When you dichotomize a variable, be sure that the two categories capture a distinction that is important to your research question (e.g., a comparison of the number of white vs. nonwhite U.S. senators).

Discrete and Continuous Variables

The statistical operations we can perform are also determined by whether the variables are continuous or discrete. *Discrete* variables have minimum-sized units of measurement, which cannot be subdivided. The number of children per family is an example of a discrete variable because the minimum unit is 1 child. A family may have 2 or 3 children, but not 2.5 children. The variable *wages* in our research example is a discrete variable because currency has a minimum unit (1 cent), which cannot be subdivided. One can have $101.21 or $101.22 but not $101.21843. Wages cannot differ by less than 1 cent—the minimum-sized unit.

Unlike discrete variables, continuous variables do not have minimum-sized units of measurement; their ranges of values can be subdivided into increasingly smaller fractional values. *Length* is an example of a continuous variable because there is no minimum unit of length. A particular object may be 12 in. long, it may be 12.5 in. long, or it may be 12.532011 in. long. Although we cannot always measure all possible length values with absolute accuracy, it is possible for objects to exist at an infinite number of lengths.[7] In principle, we can speak of a tenth of an inch, a ten thousandth of an inch, or a ten trillionth of an inch. The variable *gender composition of occupations* is a continuous variable because it is measured in proportions or percentages (e.g., the percentage of women in medicine), which can be subdivided into smaller and smaller fractions.

This attribute of variables—whether they are continuous or discrete—affects subsequent research operations, particularly measurement procedures, data analysis, and methods of inference and generalization. However, keep in mind that, in practice, some discrete variables can be treated as if they were continuous, and vice versa.

✓ *Learning*
Check

Name three continuous and three discrete variables. Determine whether each of the variables in your hypothesis is continuous or discrete.

A Cautionary Note: Measurement Error

Social scientists attempt to ensure that the research process is as error free as possible, beginning with how we construct our measurements. We pay attention to two characteristics of measurement: reliability and validity.

Reliability means that the measurement yields consistent results each time it is used. For example, asking a sample of individuals "Do you approve or disapprove of President Obama's job performance?" is more reliable than asking "What do you think of President Obama's job performance?" While responses to the second question are meaningful, the answers might be vague and could be subject to different interpretations. Researchers look for the consistency of measurement over time, in relationship with other related measures, or in measurements or observations made by two or more researchers. Reliability is a prerequisite for validity: We cannot measure a phenomenon if the measure we are using gives us inconsistent results.

Validity refers to the extent to which measures indicate what they are intended to measure. While standardized IQ tests are reliable, it is still debated whether such tests measure intelligence or one's test-taking ability. A measure may not be valid because of individual error (individuals may want to provide socially desirable responses) or method error (questions may be unclear or poorly written).

Specific techniques and practices for determining and improving measurement reliability and validity are the subject of research methods courses.

▣ ANALYZING DATA AND EVALUATING THE HYPOTHESES

Following the data collection stage, researchers analyze their data and evaluate the hypotheses of the study. The data consist of codes and numbers used to represent our observations. In our example, each occupational group would be represented by two scores: (1) the percentage of women and (2) the average wage. If we had collected information on 100 occupations, we would end up with 200 scores, 2 per occupational group. However, the typical research project includes more variables; therefore, the amount of data the researcher confronts is considerably larger. We now must find a systematic way to organize these data, analyze them, and use some set of procedures to decide what they mean. These last steps make up the *statistical analysis* stage, which is the main topic of this textbook. It is also at this point in the research cycle that statistical procedures will help us *evaluate* our research hypothesis and assess the theory from which the hypothesis was derived.

Descriptive and Inferential Statistics

Statistical procedures can be divided into two major categories: *descriptive statistics* and *inferential statistics*. Before we can discuss the difference between these two types of statistics, we need to understand the terms *population* and *sample*. A **population** is the total set of individuals, objects, groups, or events in which the researcher is interested. For example, if we were interested in looking at voting behavior in the last presidential election, we would probably define our population as all citizens who voted in the election. If we wanted to understand the employment patterns of Latinas in our state, we would include in our population all Latinas in our state who are in the labor force.

Population   The total set of individuals, objects, groups, or events in which the researcher is interested.

Although we are usually interested in a population, quite often, because of limited time and resources, it is impossible to study the entire population. Imagine interviewing all the citizens of the United States who voted in the last election or even all the Latinas who are in the labor force in our state. Not only would that be very expensive and time-consuming, but we would also probably have a very hard time locating everyone! Fortunately, we can learn a lot about a population if we carefully select a subset from that population. A subset selected from a population is called a **sample**. Researchers usually collect their data from a sample and then generalize their observations to the larger population.

Sample   A relatively small subset selected from a population.

Descriptive statistics includes procedures that help us organize and describe data collected from either a sample or a population. Occasionally data are collected on an entire population, as in a census. **Inferential statistics**, on the other hand, is concerned with making predictions or inferences about a population from observations and analyses of a sample. For instance, the General Social Survey (GSS), from which numerous examples presented in this book are drawn, is conducted every other year by the National Opinion Research Center (NORC) on a representative sample of several thousands of respondents (e.g., a total of 4,901 cases were included in the GSS 2010: 2,044 new cases and reinterviews with 2,857 respondents from the 2006 and 2008 GSSs). The survey, which includes several hundred questions, is designed to provide social science researchers with a readily accessible database of socially relevant attitudes, behaviors, and attributes of a cross section of the U.S. adult population. NORC has verified that the composition of the GSS samples closely resembles census data. But because the data are based on a sample rather than on the entire population, the average of the sample does not equal the average of the population as a whole. For example, in the 2010 GSS, men and women were asked to report their total years of education. GSS researchers found the average to be 13.47 years, a little more than a year beyond a high school degree. This average probably differs from the average of the population from which the GSS sample was drawn. The tools of statistical inference help determine the accuracy of the sample average obtained by the researchers.

Descriptive statistics   Procedures that help us organize and describe data collected from either a sample or a population.

Inferential statistics   The logic and procedures concerned with making predictions or inferences about a population from observations and analyses of a sample.

Evaluating the Hypotheses

At the completion of these descriptive and inferential procedures, we can move to the next stage of the research process: the assessment and evaluation of our hypotheses and theories in light of the analyzed data. At this next stage, new questions might be raised about unexpected trends in the data and about other variables that may have to be considered in addition to our original variables. For example, we may have found that the relationship between the gender composition of occupations and earnings can be observed with respect to some groups of occupations but not others. Similarly, the relationship between these variables may apply for some racial/ethnic groups but not for others.

These findings provide evidence to help us decide how our data relate to the theoretical framework that guided our research. We may decide to revise our theory and hypothesis to take account of these later findings. Recent studies are modifying what we know about gender segregation in the workplace. These studies suggest that race as well as gender shapes the occupational structure in the United States and helps explain disparities in income. This reformulation of the theory calls for a modified hypothesis and new research, which starts the circular process of research all over again.

Statistics provides an important link between theory and research. As our example on gender segregation demonstrates, the application of statistical techniques is an indispensable part of the research process. The results of statistical analyses help us evaluate our hypotheses and theories, help us discover unanticipated patterns and trends, and provide the impetus for shaping and reformulating our theories. Nevertheless, the importance of statistics should not diminish the significance of the preceding phases of the research process. Nor does the use of statistics lessen the importance of our own judgment in the entire process. Statistical analysis is a relatively small part of the research process, and even the most rigorous statistical procedures cannot speak for themselves. If our research questions are poorly conceived or our data are flawed because of errors in our design and measurement procedures, our results will be useless.

▣ LOOKING AT SOCIAL DIFFERENCES

By the middle of this century, if current trends continue unchanged, the United States will no longer be a predominantly European society. Mostly because of renewed immigration and higher birthrates, the United States is being transformed into a "global society" in which nearly half the population will be of African, Asian, Latino, or Native American ancestry. Is the increasing diversity of American society relevant to social scientists? What impact will such diversity have on the research methodologies we employ?

In a diverse society stratified by race, ethnicity, class, and gender, less partial and distorted explanations of social relations tend to result when researchers, research participants, and the research process itself reflect that diversity. Such diversity shapes the research questions we ask, how we observe and interpret our findings, and the conclusions we draw.

How does a consciousness of social differences inform social statistics? How can issues of race, class, gender, and other demographic categories shape the way we approach statistics? A statistical approach that focuses on social differences uses statistical tools to examine how variables such as race, class, and gender as well as other demographic categories such as age, religion, and sexual orientation shape our social world and explain our social behavior. Numerous statistical procedures can be applied to describe these processes, and we will begin to look at some of those options in the next chapter.

Whichever model of social research you use—whether you follow a traditional one or integrate your analysis with qualitative data, whether you focus on social differences or any other aspect of social behavior—remember that any application of statistical procedures requires a basic understanding of the statistical concepts and techniques. This introductory text is intended to familiarize you with the range of descriptive and inferential statistics widely applied in the social sciences. Our emphasis on statistical techniques should not diminish the importance of human judgment and your awareness of the person-made quality of statistics. Only with this awareness can statistics become a useful tool for viewing social life.

MAIN POINTS

- Statistics is a set of procedures used by social scientists to organize, summarize, and communicate information. Only information represented by numbers can be the subject of statistical analysis.

- The research process is a set of activities in which social scientists engage to answer questions, examine ideas, or test theories. It consists of the following stages: asking the research question, formulating the hypotheses, collecting data, analyzing data, and evaluating the hypotheses.

- A theory is an elaborate explanation of the relationship between two or more observable attributes of individuals or groups.

- Theories offer specific concrete predictions about the way observable attributes of people or groups would be interrelated in real life. These predictions, called hypotheses, are tentative answers to research problems.

- A variable is a property of people or objects that takes on two or more values. The variable that the researcher wants to explain (the "effect") is called the dependent variable. The variable that is expected to "cause" or account for the dependent variable is called the independent variable.

- Three conditions are required to establish causal relations: (1) The cause has to precede the effect in time, (2) there has to be an empirical relationship between the cause and the effect, and (3) this relationship cannot be explained by other factors.

- At the nominal level of measurement, numbers or other symbols are assigned to a set of categories to name, label, or classify the observations. At the ordinal level of measurement, categories can be rank ordered from low to high (or vice versa). At the interval-ratio level of measurement, measurements for all cases are expressed in the same unit.

- A population is the total set of individuals, objects, groups, or events in which the researcher is interested. A sample is a relatively small subset selected from a population.

- Descriptive statistics includes procedures that help us organize and describe data collected from either a sample or a population. Inferential statistics is concerned with making predictions or inferences about a population from observations and analyses of a sample.

KEY TERMS

data	inferential statistics	sample
dependent variable	interval-ratio	statistics
descriptive statistics	measurement	theory
dichotomous variable	nominal measurement	unit of analysis
empirical research	ordinal measurement	variable
hypothesis	population	
independent variable	research process	

⑤SAGE edge™

Sharpen your skills with SAGE edge at **edge.sagepub.com/ssdsess2e**. **SAGE edge for students** provides a personalized approach to help you accomplish your coursework goals in an easy-to-use learning environment.

CHAPTER EXERCISES

1. In your own words, explain the relationship of data (collecting and analyzing) to the research process. (Refer to Figure 1.1.)

2. Construct potential hypotheses or research questions to relate the variables in each of the following examples. Also, write a brief statement explaining why you believe there is a relationship between the variables as specified in your hypotheses.
   a. Gender and educational level
   b. Income and race
   c. The crime rate and the number of police in a city
   d. Life satisfaction and marital status
   e. A nation's military expenditures as a percentage of its gross domestic product and that nation's overall level of security
   f. Care of elderly parents and ethnicity

3. Determine the level of measurement for each of the following variables:
   a. The number of people in your family
   b. Place of residence classified as urban, suburban, or rural
   c. The percentage of university students who attended public high school
   d. The rating of the overall quality of a textbook, on a scale from "excellent" to "poor"
   e. The type of transportation a person takes to work (e.g., bus, walking, car)
   f. Your annual income
   g. The U.S. unemployment rate
   h. The presidential candidate the respondent voted for in 2012

4. For each of the variables in Exercise 3 that you classified as interval-ratio, identify whether it is discrete or continuous.

5. Why do you think men and women, on average, do not earn the same amount of money? Develop your own theory to explain the difference. Use three independent variables in your theory, with annual income as your dependent variable. Construct hypotheses to link each independent variable with your dependent variable.

6. For each of the following examples, indicate whether it involves the use of descriptive or inferential statistics. Justify your answer.
   a. The number of unemployed people in the United States
   b. Determining students' opinions about the quality of food at the cafeteria on the basis of a sample of 100 students
   c. The national incidence of breast cancer among Asian women
   d. Conducting a study to determine the rating of the quality of a new smartphone, gathered from 1,000 new buyers
   e. The average grade point average of various majors (e.g., sociology, psychology, English) at your university
   f. The change in the number of immigrants coming to the United States from Southeast Asian countries between 2005 and 2010

7. Identify three social problems or issues that can be investigated with statistics. (One example of a social problem is criminal acts, such as murder.) Which one of the three issues would be the most difficult to study? Which would be the easiest? Why?

8. Construct measures of political participation at the nominal, ordinal, and interval-ratio levels. (*Hint:* You can use behaviors such as voting frequency or political party membership.) Discuss the advantages and disadvantages of each.

9. Variables can be measured according to more than one level of measurement. For the following variables, identify at least two levels of measurement. Is one level of measurement better than another? Explain.
   a. Individual age
   b. Annual income
   c. Religiosity
   d. Student performance
   e. Social class
   f. Attitude toward gun control

Chapter 2

The Organization and Graphic Presentation of Data

Chapter Learning Objectives

❖ Understanding how to construct and analyze frequency, percentage, and cumulative distributions

❖ Understanding how to calculate proportions and percentages

❖ Recognizing the differences in frequency distributions for nominal, ordinal, and interval-ratio variables

❖ Reading statistical tables in research literature

❖ Constructing and interpreting a pie chart, bar graph, histogram, line graph, and time-series chart

❖ Analyzing and interpreting charts and graphs in the literature

As social researchers, we often have to deal with very large amounts of data. For example, in a typical survey, by the completion of your data collection phase you will have accumulated thousands of individual responses, represented by a jumble of numbers. To make sense out of these data, you will have to organize and summarize them in some systematic fashion. In this chapter, we review two methods used by social scientists: the creation of frequency distributions and graphics.

The most basic method for organizing data is to classify the observations into a frequency distribution. A **frequency distribution** is a table that reports the number of observations that fall into each category of the variable we are analyzing. Constructing a frequency distribution is usually the first step in the statistical analysis of data.

Frequency distribution   A table reporting the number of observations falling into each category of the variable.

▣ FREQUENCY DISTRIBUTIONS

Immigration has been described as "remaking America with political, economic, and cultural ramifications."[1] Globalization has fueled migration, particularly since the beginning of the 21st century. Workers migrate because of the promise of employment and higher standards of living than in their home countries. Data reveal that the United States is the destination for many migrants.[2] The Census Bureau uses the term *foreign born* to refer to those who are not U.S. citizens at birth. The Census Bureau estimates that nearly 12.9% of the U.S. population, or approximately 40 million people, are foreign born.[3] Immigrants are not one homogeneous group but many diverse groups. Table 2.1 shows the frequency distribution of the world region of birth for the foreign-born population.

Table 2.1   Frequency Distribution for Categories of World Region of Birth for Foreign-Born Population, 2010

World Region of Birth	Frequency (f)
Africa	1,607,000
Asia	11,284,000
Europe	4,817,000
Latin America and the Caribbean	21,224,000
Northern America	807,000
Oceania	217,000
Total (N)	39,956,000

Source: Elizabeth Grieco, Yesenia Acosta, C. Patricia de la Cruz, Christine Gambino, Thomas Gryn, Luke Larsen, Edward Trevelyan, and Nathan Walters. *The Foreign-Born Population in the United States: 2010* (ACS-19; Washington, DC: U.S. Census Bureau), 2012.

Note that the frequency distribution is organized in a table, which has a number (2.1) and a descriptive title. The title indicates the kind of data presented: "Categories of World Region of Birth for Foreign-Born Population." The table consists of two columns. The first column identifies the variable (world region of birth) and its categories. The second column, headed "Frequency (f)," tells the number of cases in each category as well as the total number of cases ($N = 39,956,000$). Note also that the source of the table is clearly identified in a source note. It tells us that the data are from a 2012 Census Bureau report (though the information is based on 2010 data). In general, the source of data for a table should appear as a source note unless it is clear from the general discussion of the data.

What can you learn from the information presented in Table 2.1? The table shows that as of 2010, approximately 40 million people were classified as foreign born. Out of this group, the majority, about 21.2 million people, were from Latin America, 11.3 million were from Asia, followed by 4.8 million from Europe.

▣ PROPORTIONS AND PERCENTAGES

Frequency distributions are helpful in presenting information in a compact form. However, when the number of cases is large, the frequencies may be difficult to grasp. To standardize these raw frequencies, we can translate them into relative frequencies—that is, proportions or percentages.

A **proportion** is a relative frequency obtained by dividing the frequency in each category by the total number of cases. To find a proportion (p), divide the frequency (f) in each category by the total number of cases (N):

$$p = \frac{f}{N} \tag{2.1}$$

where

f = frequency

N = total number of cases

We've calculated the proportions for the three largest groups of foreign born. First, the proportion of foreign born originally from Latin America is

$$\frac{21,224,000}{39,956,000} = .53$$

The proportion of foreign born who were originally from Asia is

$$\frac{11,284,000}{39,956,000} = .28$$

The proportion of foreign born who were originally from Europe is

$$\frac{4,817,000}{39,956,000} = .12$$

The proportion of foreign born who were originally from other reported areas is

$$\frac{2,631,000}{39,956,000} = .07$$

We rounded this last proportion from .065 to .07. Confirm these calculations on your own. You can also calculate the proportions individually for Africa, Northern America, and Oceania.

Proportions should always sum to 1.00 (allowing for some rounding errors). Thus, in our example the sum of the six proportions is

$$.53 + .28 + .12 + .07 = 1.0$$

To determine a frequency from a proportion, we simply multiply the proportion by the total N:

$$f = p(N) \tag{2.2}$$

Thus, the frequency of foreign born from Asia can be calculated as

$$0.28(39,956,000) = 11,187,680$$

Note that the obtained frequency differs somewhat from the actual frequency of 11,284,000. This difference is due to rounding off of the proportion. If we use the actual proportion instead of the rounded proportion, we obtain the correct frequency:

$$0.282410651(39,956,000) = 11,284,000$$

Proportion   A relative frequency obtained by dividing the frequency in each category by the total number of cases.

We can also express frequencies as percentages. A **percentage** is a relative frequency obtained by dividing the frequency in each category by the total number of cases and multiplying by 100. In most statistical reports, frequencies are presented as percentages rather than proportions. Percentages express the size of the frequencies as if there were a total of 100 cases.

To calculate a percentage, simply multiply the proportion by 100:

$$\text{Percentage } (\%) = \frac{f}{N}(100) \tag{2.3}$$

or

$$\text{Percentage } (\%) = p(100) \tag{2.4}$$

Thus, the percentage of respondents who were originally from Asia is

$$0.28(100) = 28\%$$

The percentage of respondents who were originally from Latin America is

$$0.53(100) = 53\%$$

Percentage   A relative frequency obtained by dividing the frequency in each category by the total number of cases and multiplying by 100.

✓ *Learning*
Check

Calculate the proportion of male and female students in your statistics class. What proportion is female?

▣ PERCENTAGE DISTRIBUTIONS

Percentages are usually displayed as percentage distributions. A **percentage distribution** is a table showing the percentage of observations falling into each category of the variable. For example, Table 2.2 presents the frequency distribution of categories of places of origin (Table 2.1) along with the corresponding percentage distribution. Percentage distributions (or proportions) should always show the base (N) on which they were computed. Thus, in Table 2.2 the base on which the percentages were computed is $N = 39{,}956{,}000$. While in most cases the total percentages should equal 100%, in Table 2.2, the total is 99.9%.

Percentage distribution   A table showing the percentage of observations falling into each category of the variable.

▣ COMPARISONS

In Table 2.2, we illustrated that there are six primary places of origin for foreign born in the United States. These distinctions help us understand the specific characteristics and backgrounds of each group. We can resist the temptation to group all foreign born into one category and ask, for

Table 2.2   Frequency and Percentage Distributions for Categories of World Region of Birth for Foreign Born, 2010

World Region of Birth	Frequency (f)	Percentage (%)
Africa	1,607,000	4.0
Asia	11,284,000	28.2
Europe	4,817,000	12.1
Latin America and the Caribbean	21,224,000	53.1
Northern America	807,000	2.0
Oceania	217,000	0.5
Total (N)	39,956,000	99.9%

Source: Elizabeth Grieco, Yesenia Acosta, C. Patricia de la Cruz, Christine Gambino, Thomas Gryn, Luke Larsen, Edward Trevelyan, and Nathan Walters. *The Foreign-Born Population in the United States: 2010* (ACS-19; Washington, DC: U.S. Census Bureau), 2012.

instance, is one group more educated than another? Is one group younger than the other groups? As students, as social scientists, and even as consumers, we are frequently faced with problems that call for some way to make clear and valid comparisons.

The decision to consider these groups separately or to pool them depends to a large extent on our research question. For instance, we know that in 2010, 19% of the foreign-born population were living in poverty.[4] Among the foreign born, poverty rates were highest among those from Latin America (24%) and lowest among those from Northern America (9.1%). What do these figures tell us about the demographic characteristics of the foreign born? Of the Latin American foreign born? Of the Northern American foreign born? To answer these questions and determine whether the two categories of region of birth have markedly different social characteristics, we need to *compare* them.

Several types of comparisons are quite common in the social sciences. One type is the comparison between groups that have different characteristics—for example, comparisons between older and younger Americans, white and Asian Pacific Islander, or, as in our chapter example, between different categories of foreign-born individuals. Sometimes, we may be interested in looking at regional differences among groups or in comparing groups from different segments of society. You may have read news stories about contrasts in voting patterns between gun owners and non–gun owners or between liberals and conservatives. Also, we may be interested in comparing changes in the same group over time, such as the percentage change in foreign-born residents in the United States over the past decade or how the population has shifted from the cities toward the suburbs.

▣ STATISTICS IN PRACTICE: LABOR FORCE PARTICIPATION AMONG THE FOREIGN BORN

Very often, we are interested in comparing two or more groups that differ in size. Percentages are especially useful for making such comparisons. For example, we know that differences in socioeconomic status mark divisions between populations, indicating differential access to economic opportunities. Labor participation (either employed or seeking employment) is an important indicator of access to economic opportunities and is strongly associated with socioeconomic status. Table 2.3 shows the raw frequency distributions for the variable *labor force participation* for foreign-born individuals by race and Hispanic ethnicity. Notice that the title of the table includes the notation "(numbers in thousands)." The Census Bureau frequently presents figures this way, eliminating the set of three zeroes at the end. The numbers can be converted to millions, by multiplying each by 1,000 (e.g., 4,138 × 1,000 = 4,138,000).

Table 2.3   Employment Status of the Foreign-Born Population, by Race and Hispanic Ethnicity, 2010 (numbers in thousands)

Employment Status	White Non-Hispanic	Black Non-Hispanic	Asian Non-Hispanic	Hispanic
Employed	4,138	1,893	4,928	10,776
Unemployed	332	269	386	1,376
Not in labor force	2,893	736	2,758	5,010
Total (*N*)	7,363	2,898	8,072	17,162

Source: U.S. Census Bureau, *Statistical Abstract of the United States: 2012*, Table 589.

Which group has the highest relative number of persons who are not in the labor force? Because of the differences in the population sizes of the four groups, this is a difficult question to answer based on only the raw frequencies. To make a valid comparison, we have to compare the percentage distributions for all the three groups. These are presented in Table 2.4. Note that the percentage distributions make it easier to identify differences between the groups. The base *N*'s are reported for each racial/ethnic group. Compared with the other groups, black non-Hispanic foreign-born individuals have the highest percentage employed in the labor force (1,893/2,898 = 65.3%). This group also has the highest percentage of unemployed (269/2,898 = 9.3%), followed by Hispanics (1,376/17,162 = 8.0%).

✓ *Learning Check*

Examine Table 2.4 and answer the following questions: What is the percentage of white non-Hispanics who are employed? What is the base (N) for this percentage? What is the percentage of Hispanics who are not in the labor force? What is the base (N) for this percentage?

Table 2.4   Employment Status of the Foreign-Born Population by Race and Hispanic Ethnicity, 2010 (percentages)

Employment Status	White Non-Hispanic N = 7,363	Black Non-Hispanic N = 2,898	Asian Non-Hispanic N = 8,072	Hispanic N = 17,162
Employed	56.2	65.3	61.1	62.8
Unemployed	4.5	9.3	4.8	8.0
Not in labor force	39.3	25.4	34.2	29.2
Total	100	100	100.1	100

Source: U.S. Census Bureau, *Statistical Abstract of the United States: 2012*, Table 589.

Whenever one group is compared with another, the most meaningful conclusions can usually be drawn based on comparison of the relative frequency distributions. In fact, we are seldom interested in a single distribution. Most interesting questions in the social sciences are about differences between two or more groups.[5] The finding that the foreign-born population labor force participation patterns vary among different race and ethnic groups raises doubt about whether the foreign born can be legitimately regarded as a single, relatively homogeneous group. Further analyses could examine *why* these differences exist. Other variables that explain these differences could be identified (such as educational attainment, age, or previous work experience). These kinds of questions can be answered using more complex multivariate statistical techniques that involve more than two variables. The comparison of percentage distributions is an important foundation for these more complex techniques.

▣ THE CONSTRUCTION OF FREQUENCY DISTRIBUTIONS

In this section, you will learn how to construct frequency distributions. Most often, this can be done by your computer, but it is important to go through the process to understand how frequency distributions are actually put together.

For nominal and ordinal variables, constructing a frequency distribution is quite simple. Count and report the number of cases that fall into each category of the variable along with the total number of cases (*N*). For the purpose of illustration, let's take a small random sample of 40 cases from a General Social Survey (GSS) sample and record their scores on the following variables: gender, a nominal-level variable; degree, an ordinal measurement of education; and age and number of children, both interval-ratio variables. The use of "male" and "female" in parts of this book is in keeping with the GSS categories for the variable "sex" (respondent's sex).

You can see that it is going to be difficult to make sense of these data just by eyeballing Table 2.5. How many of these 40 respondents are male? How many said that they had graduate degrees? How many were older than 50 years of age? To answer these questions, we construct the frequency distributions for all four variables.

Table 2.5   A GSS Subsample of 40 Respondents

Gender of Respondent	Degree	Number of Children	Age (years)
M	Bachelor	1	43
F	High school	2	71
F	High school	0	71
M	High school	0	37
M	High school	0	28
F	High school	6	34
F	High school	4	69
F	Graduate	0	51
F	Bachelor	0	76
M	Graduate	2	48
M	Graduate	0	49
M	Less than high school	3	62
F	Less than high school	8	71
F	High school	1	32
F	High school	1	59
F	High school	1	71
M	High school	0	34
M	Bachelor	0	39
F	Bachelor	2	50
M	High school	3	82
F	High school	1	45
M	High school	0	22
M	High school	2	40
F	High school	2	46
M	High school	0	29
F	High school	1	75
F	High school	0	23
M	Bachelor	2	35
M	Bachelor	3	44
F	High school	3	47
M	High school	1	84
F	Graduate	1	45
F	Less than high school	3	24

(Continued)

Table 2.5   (Continued)

Gender of Respondent	Degree	Number of Children	Age (years)
F	Graduate	0	47
F	Less than high school	5	67
F	High school	1	21
F	High school	0	24
F	High school	3	49
F	High school	3	45
F	Graduate	3	37

Note: M, male; F, female.

Frequency Distributions for Nominal Variables

Let's begin with the nominal variable, *gender.* First, we tally the number of male respondents, then the number of female respondents (the column of tallies has been included in Table 2.6 for the purpose of illustration). The tally results are then used to construct the frequency distribution presented in Table 2.6. The table has a title describing its content ("Frequency Distribution of the Variable Gender: GSS Subsample"). Its categories (male and female) and their associated frequencies are clearly listed; in addition, the total number of cases (N) is also reported. The Percentage column is the percentage distribution for this variable. To convert the Frequency column to percentages, simply divide each frequency by the total number of cases and multiply by 100. Percentage distributions are routinely added to almost any frequency table and are especially important if comparisons with other groups are to be considered. Immediately, we can see that it is easier to read the information. There are 25 female and 15 male respondents in this sample. Based on this frequency distribution, we can also conclude that the majority of sample respondents are female.

✓ *Learning*
Check

> *Construct a frequency and percentage distribution for male and female students in your statistics class.*

Table 2.6   Frequency Distribution of the Variable Gender: GSS Subsample

Gender	Tally	Frequency (f)	Percentage
Male	⊔⊓ ⊔⊓ ⊔⊓	15	37.5
Female	⊔⊓ ⊔⊓ ⊔⊓ ⊔⊓ ⊔⊓	25	62.5
Total (N)		40	100.0

Frequency Distributions for Ordinal Variables

To construct a frequency distribution for ordinal-level variables, follow the same procedures outlined for nominal-level variables. Table 2.7 presents the frequency distribution for the variable degree. The table shows that 60.0%, a majority, indicated that their highest degree was a high school degree.

Table 2.7 Frequency Distribution of the Variable Degree: GSS Subsample

Degree	Tally	Frequency (f)	Percentage
Less than high school	llll	4	10.0
High school	llll llll llll llll llll	24	60.0
Bachelor	llll l	6	15.0
Graduate	llll l	6	15.0
Total (N)		40	100.0

The major difference between frequency distributions for nominal and ordinal variables is the order in which the categories are listed. The categories for nominal-level variables do not have to be listed in any particular order. For example, we could list female respondents first and male respondents second without changing the nature of the distribution. Because the categories or values of ordinal variables are rank ordered, however, they must be listed in a way that reflects their rank—from the lowest to the highest or from the highest to the lowest. Thus, the data on degree in Table 2.7 are presented in declining order from "less than high school" (the lowest educational category) to "graduate" (the highest educational category).

✓ Learning Check

Figures 2.1, 2.2, and 2.3 illustrate the gender and degree data in stages as presented in Tables 2.5, 2.6, and 2.7. To convince yourself that classifying the respondents by gender (Figure 2.2) and by degree (Figure 2.3) makes the job of counting much easier, turn to Figure 2.1 and answer these questions: How many men are in the group? How many women? How many said that they completed a bachelor's degree? Now turn to Figure 2.2: How many men are in the group? How many women? Finally, examine Figure 2.3: How many said that they completed a bachelor's degree?

Figure 2.1   Forty Respondents From the GSS Subsample, Their Gender, and Their Levels of Degree (see Table 2.5)

Figure 2.2   Forty Respondents From the GSS Subsample, Classified by Gender (see Table 2.6)

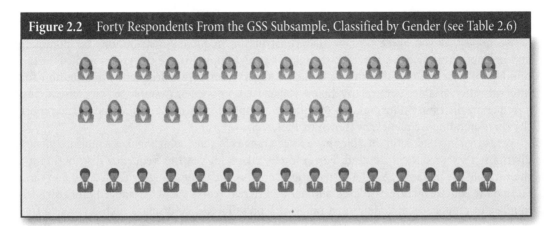

Figure 2.3   Forty Respondents From the GSS Subsample, Classified by Gender and Degree

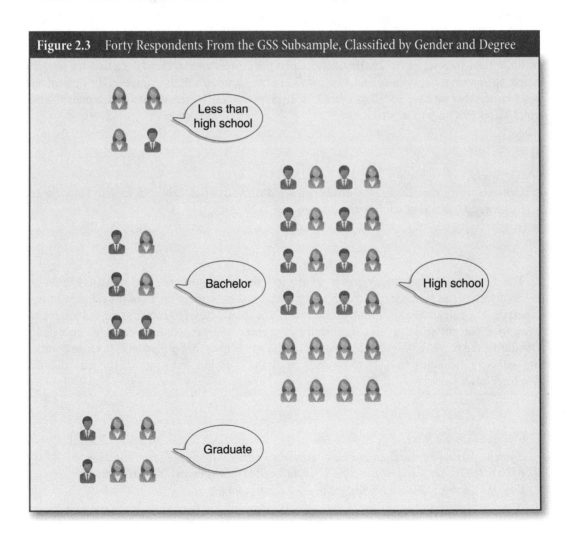

Frequency Distributions for Interval-Ratio Variables

We hope that you agree by now that constructing frequency distributions for nominal- and ordinal-level variables is rather straightforward. Simply list the categories and count the number of observations that fall into each category. Building a frequency distribution for interval-ratio variables with relatively few values is also easy. For example, when constructing a frequency distribution for number of children, simply list the number of children and report the corresponding frequency, as shown in Table 2.8.

Very often interval-ratio variables have a wide range of values, which makes simple frequency distributions very difficult to read. For example, take a look at the frequency distribution for the variable *age* in Table 2.9. The distribution contains age values ranging from 21 to 84 years. For a more concise picture, the large number of different scores could be reduced into a smaller number of groups, each containing a range of scores. Table 2.10 displays such a grouped frequency distribution of the data in Table 2.9. Each group, known as a *class interval*, now contains 10 possible scores instead of 1. Thus, the ages of 21, 22, 23, 24, 28, and 29 all fall into a single class interval of 20–29. The second column of Table 2.10, Frequency, tells us the number of respondents who fall into each of the intervals—for the example, that seven respondents fall into the class interval of 20–29. Having grouped the scores, we can clearly see that the biggest single age-group is between 40 and 49 years (12 out of 40, or 30% of the sample). The percentage distribution that we have added to Table 2.10 displays the relative frequency of each interval and emphasizes this pattern as well.

✓ *Learning*
  Check

> *Can you verify that Table 2.10 was constructed correctly? Use Table 2.9 to determine the frequency of cases that fall into the categories of Table 2.10.*

The decision as to how many groups to use and, therefore, how wide the intervals should be is usually up to the researcher and depends on what makes sense in terms of the purpose of the research. The rule of thumb is that an interval width should be large enough to avoid too many categories but not so large that significant differences between observations are concealed.[6] Obviously, the number of intervals depends on the width of each. For instance, if you are working with scores ranging from 10 to 60 and you establish an interval width of 10, you will have five intervals.

✓ *Learning*
  Check

> *If you are having trouble distinguishing between nominal, ordinal, and interval-ratio variables, go back to Chapter 1 and review the section on levels of measurement. The distinction between these three levels of measurement is important throughout the book.*

Table 2.8   Frequency Distribution of Variable Number of
            Children: GSS Subsample

Number of Children	Frequency (f)	Percentage
0	13	32.5
1	9	22.5
2	6	15.0
3	8	20.0
4	1	2.5
5	1	2.5
6	1	2.5
7+	1	2.5
Total (N)	40	100.0

Table 2.9   Frequency Distribution of the Variable Age: GSS Subsample

Age of Respondent	Frequency (f)	Age of Respondent (years)	Frequency (f)
21	1	59	1
22	1	62	1
23	1	67	1
24	2	69	1
28	1	71	4
29	1	75	1
32	1	76	1
34	2	82	1
35	1	84	1
37	2		
39	1		
40	1		
43	1		
44	1		
45	3		
46	1		
47	2		
48	1		
49	2		
50	1		
51	1		

Table 2.10 Grouped Frequency Distribution of the Variable
Age: GSS Subsample

Age Category (years)	Frequency (f)	Percentage
20–29	7	17.5
30–39	7	17.5
40–49	12	30.0
50–59	3	7.5
60–69	3	7.5
70–79	6	15.0
80–89	2	5.0
Total (N)	40	100.0

▣ CUMULATIVE DISTRIBUTIONS

Sometimes, we may be interested in locating the relative position of a given score in a distribution. For example, we may be interested in finding out how many or what percentage of our sample was younger than 40 or older than 60. Frequency distributions can be presented in a cumulative fashion to answer such questions. A **cumulative frequency distribution** shows the frequencies at or below each category of the variable.

Cumulative frequency distribution A distribution showing the frequency at or below each category (class interval or score) of the variable.

Cumulative frequencies are appropriate only for variables that are measured at an ordinal level or higher. They are obtained by adding to the frequency in each category the frequencies of all the categories below it.

Let's look at Table 2.11. It shows the cumulative frequencies based on the frequency distribution from Table 2.10. The cumulative frequency column, denoted by *Cf*, shows the number of persons at or below each interval. For example, you can see that 14 of the 40 respondents were 39 years old or younger, and 29 respondents were 59 years old or younger.

To construct a cumulative frequency distribution, start with the frequency in the lowest class interval (or with the lowest score, if the data are ungrouped), and add to it the frequencies in the next highest class interval. Continue adding the frequencies until you reach the last class interval. The cumulative frequency in the last class interval will be equal to the total number of cases (*N*). In Table 2.11, the frequency associated with the first class interval (20–29) is 7. The cumulative frequency associated with this interval is also 7, since there are no cases below this class interval.

Table 2.11   Grouped Frequency Distribution and Cumulative
Frequency for the Variable Age: GSS Subsample

Age Category (years)	Frequency (f)	Cf
20–29	7	7
30–39	7	14
40–49	12	26
50–59	3	29
60–69	3	32
70–79	6	38
80–89	2	40
Total (N)	40	

The frequency for the second class interval is 7. The cumulative frequency for this interval is $7 + 7 = 14$. To obtain the cumulative frequency of 26 for the third interval, we add its frequency (12) to the cumulative frequency associated with the second class interval (14). Continue this process until you reach the last class interval. Therefore, the cumulative frequency for the last interval is equal to 40, the total number of cases (N).

We can also construct a cumulative percentage distribution (C%), which has wider applications than the cumulative frequency distribution (Cf). A **cumulative percentage distribution** shows the percentage at or below each category (class interval or score) of the variable. A cumulative percentage distribution is constructed using the same procedure as for a cumulative frequency distribution except that the percentages—rather than the raw frequencies—for each category are added to the total percentages for all the previous categories.

Cumulative percentage distribution   A distribution showing the percentage at or below each category (class interval or score) of the variable.

In Table 2.12, we have added the cumulative percentage distribution to the frequency and percentage distributions shown in Table 2.10. The cumulative percentage distribution shows, for example, that 35% of the sample was younger than 40 years of age—that is, 39 years or younger.

Like the percentage distributions described earlier, cumulative percentage distributions are especially useful when you want to compare differences between groups. For an example of how cumulative percentages are used in a comparison, we used the 2010 GSS data to contrast the opinions of whites and blacks about whether they believe immigrants take jobs away from native-born Americans. Respondents were asked the following: "How much do you agree or disagree with the following statement? Immigrants take jobs away from people who were born in America."

Table 2.12   Grouped Frequency Distribution and Cumulative
Percentages for the Variable Age: GSS Subsample

Age Category (years)	Frequency (f)	Percentage	C%
20–29	7	17.5	17.5
30–39	7	17.5	35.0
40–49	12	30.0	65.0
50–59	3	7.5	72.5
60–69	3	7.5	80.0
70–79	6	15.0	95.0
80–89	2	5.0	100.0
Total (N)	40	100.0	

Table 2.13   Immigrants Take Jobs Away: White Versus Black
Respondents

	Whites		Blacks	
	%	C%	%	C%
Strongly agree	8.9	8.9	13.5	13.5
Agree	32.5	41.4	37.8	51.3
Neither	22.2	63.6	17.6	68.5
Disagree	28.5	92.1	24.3	93.2
Strongly disagree	7.9	100	6.8	100.0
Total	100.0		100.0	
(N)	369		74	

Source: Author created based on data from GSS, 2010.

The percentage distributions and the cumulative percentage distributions for whites and blacks are shown in Table 2.13. The cumulative percentage distributions suggest that a higher percentage of blacks agree with the statement that immigrants take away jobs. The two groups are separated by 9.9 percentage points—51.3% of black respondents indicated that they either strongly agreed or agreed with the statement, while 41.4% of white respondents said the same. (Note that a higher percentage of whites disagreed with the statement than blacks.) What might explain these differences? These data prompt many other questions about the role that race or other variables may play in attitudes about legal and unauthorized immigration. For instance, what would the differences be if we compared men with women? Whites with Latinos?

▣ RATES

Terms such as *birthrate, unemployment rate*, and *marriage rate* are often used by social scientists and demographers and then quoted in the popular media to describe population trends. But what exactly are rates, and how are they constructed? A **rate** is a number obtained by dividing the number of actual occurrences in a given time period by the number of possible occurrences. For example, to determine the poverty rate for 2011, the Census Bureau took the number of men and women in poverty in 2011 (actual occurrences) and divided it by the total population in 2011 (possible occurrences). The rate for 2011 can be expressed as

Poverty rate, 2011 = Number of people in poverty in 2011/Total population in 2011

Since 46,247,000 people were poor in 2011 and the number for the total population was 308,456,000, the poverty rate for 2011 can be expressed as

Poverty rate, 2011 = 46,247,000/308,456,000 = .15

The poverty rate in 2011 as reported by the U.S. Census Bureau was 15% (0.15 × 100). This means that for every 1,000 people, 150 were poor according to the U.S. Census Bureau definition. Rates are often expressed as rates per thousand or hundred thousand to eliminate decimal points and make the number easier to interpret.

The preceding poverty rate can be referred to as a *crude rate* because it is based on the total population. Rates can be calculated on the general population or on a more narrowly defined select group. For instance, poverty rates are often given for the number of people who are under 18 years—highlighting how our young are vulnerable to poverty. The poverty rate for those under 18 years is as follows:

Poverty rate for those 18 years or younger, 2011 = 16,134,000/73,737,000 = .22

We could even take a look at the poverty rate for older Americans:

Poverty rate for those 65 years of age or older, 2011 = 3,620,000/41,507,000 = .09

Rate   A number obtained by dividing the number of actual occurrences in a given time period by the number of possible occurrences.

Law enforcement agencies routinely record crime rates (the number of crimes committed relative to the size of a population), arrest rates (the number of arrests made relative to the number of crimes reported), and conviction rates (the number of convictions relative to the number of cases tried). Can you think of some other variables that could be expressed as rates?

▣ READING THE RESEARCH LITERATURE: STATISTICAL TABLES[7]

Statistical tables that display frequency distributions or other kinds of statistical information are found in virtually every book, article, or newspaper report that makes any use of statistics. However, the inclusion of statistical tables in a report or an article doesn't necessarily mean that the research is more scientific or convincing. You will always have to ask what the tables are saying and judge whether the information is relevant or accurately presented and analyzed. Most statistical tables presented in the social science literature are a good deal more complex than those we describe in this chapter. The same information can sometimes be organized in many different ways, and because of space limitations the researcher may present the information with minimum detail.

In this section, we present some guidelines for how to read and interpret statistical tables displaying frequency distributions. The purpose is to help you see that some of the techniques described in this chapter are actually used in a meaningful way. Remember that it takes time and practice to develop the skill of reading tables. Even experienced researchers sometimes make mistakes when interpreting tables. So take the time to study the tables presented here, do the chapter exercises, and you will find that reading, interpreting, and understanding tables will become easier in time.

Basic Principles

The first step in reading any statistical table is to understand what the researcher is trying to tell you. There must be a reason for including the information, and usually the researcher tells you what it is. Begin your inspection of the table by reading its title. It usually describes the central contents of the table. Check for any source notes to the table. These tell the source of the data or the table and any additional information that the author considers important. Next, examine the column and row headings and subheadings. These identify the variables, their categories, and the kind of statistics presented, such as raw frequencies or percentages. The main body of the table includes the appropriate statistics (frequencies, percentages, rates, etc.) for each variable or group as defined by each heading and subheading.

Table 2.14 was taken from an article written by Yolanda Padilla and her colleagues (2006) about the disadvantages faced by the young children of Mexican immigrants in unmarried families. For their analysis, the researchers relied on data from the Fragile Families and Child Wellbeing Study, a nationally representative, longitudinal survey that follows a cohort of new parents and their children for five years. They compared parental demographic and socioeconomic characteristics, formal and informal support, and child well-being indicators for Mexican immigrant and U.S.-born unmarried

Table 2.14   Percentage Distribution of Access to Public Benefits Among Mexican Immigrant and U.S.-Born Unmarried Mothers

Variable	Mexican Immigrant	U.S. Born			
		Mexican American	Black Non-Hispanic	White Non-Hispanic	Total U.S.-Born Population
Prenatal care in first 3 months of pregnancy	79	72.9	78.4	81.8	78.1
Health insurance					
Medicaid	77	79	70.2	70	69.1
Private	16.3	16.2	22.9	23.8	24.8
Other	6.7	4.8	6.9	5.2	6.1
TANF receipt*	12.2	20.4	38.9	14.8	29.6
Food stamp receipt	21.3	40.3	53.5	30.4	45.8
Rent assistance	7.2	22.3	24.4	99	21.0
Head Start	5.2	5.4	3.5	3.8	4.0
WIC receipt**	84.6	88.1	86.7	76	84.6

*TANF, Temporary Assistance for Needy Families

**WIC, Special Supplemental Nutrition Program for Women, Infants and Children.

Source: Adapted from Yolanda Padilla, Melissa Dalton Radey, Robert Hummer, and Eunjeong Kim, "The Living Conditions of U.S.-Born Children of Mexican Immigrants in Unmarried Families," *Hispanic Journal of Behavioral Sciences* 28, no. 3 (2006), p. 343.

mothers. Table 2.14 summarizes the utilization of public benefits and programs by immigrant and U.S.-born groups. Note that the columns or rows do not add up to 100%.

Note that the frequency (f) for each category is not reported in Table 2.14. Although the table is quite simple, it is important to examine it carefully, including its title and headings, to make sure that you understand what the information means.

✓ *Learning Check*

Inspect Table 2.14 and answer the following questions:

- *What is the source of this table?*
- *How many variables are presented? What are their names?*
- *What is represented by the numbers presented in the second column? In the last row of the table?*

What do the authors tell us about the table?

Among unmarried mothers, there is no significant difference in access to prenatal care or infant health care (well-child visits) based on immigrant status. Differences in access to health insurance are evident only between Mexican immigrant mothers and non-Hispanic White mothers, who tend to have lower rates of Medicaid and higher rates of private insurance.

Immigrant mothers are significantly less likely to receive welfare assistance in the form of Temporary Assistance for Needy Families (TANF) than are U.S.-born mothers. Only about 12.2% of Mexican immigrant mothers receive TANF compared with 20.4% of U.S.-born Mexican mothers and 38.9% of non-Hispanic black mothers. Unmarried Mexican immigrant mothers do not differ significantly from non-Hispanic white mothers in this measure. The same pattern is observed for food stamps and rent assistance. Only 21.3% of unmarried Mexican immigrant mothers receive food stamps, and only 7.2% receive rent assistance. In terms of assistance from Head Start/Early Head Start, we found no significant difference between Mexican immigrants and natives. Finally, rates of receipt of Women, Infants and Children (WIC) Program benefits are similar across all groups, although Mexican immigrant mothers are slightly more likely to receive WIC benefits than are non-Hispanic White mothers.[8]

They conclude that despite having fewer resources, immigrant mothers are less likely than U.S.-born mothers to receive formal support (which includes access to public assistance and private health insurance).

For a more detailed analysis of the relationships between these variables, you need to consider some of the more complex techniques of bivariate (two variable) analysis and statistical inference. We consider these more advanced techniques beginning with Chapter 8.

▣ GRAPHIC PRESENTATION OF DATA

You have probably heard that "a picture is worth a thousand words." The same can be said about statistical graphs because they summarize hundreds or thousands of numbers. Graphs tell a story in pictures rather than in words or numbers and are utilized in news stories, research reports, and government documents. Many are intimidated by statistical information presented in frequency distributions or in other tabular forms, but find the same information to be readable and understandable when presented graphically.

In this section, you will learn about some of the most commonly used graphical techniques. We concentrate less on the technical details of how to create graphs and more on how to choose the appropriate graphs to make statistical information coherent. We also focus on how to interpret information presented graphically.

As we introduce the various graphical techniques, we also show you how to use graphs to tell a story. The particular story we tell is that of the elderly in the United States. Demographers predict that over the next several decades, our nation's overall population growth will be among middle-aged and older Americans, which has been referred to as the graying of America. "Population aging is a long-range trend that will characterize our society as we continue into the 21st century. It is a force we all will cope with for the rest of our lives," warns gerontologist Harry Moody.[9]

The different types of graphs introduced in this section demonstrate the many facets and challenges of our aging society. People have tended to talk about seniors as if they were a homogeneous group, but the different graphical techniques we illustrate here dramatize the wide variations in economic characteristics, living arrangements, and family status among people aged 65 and older. Most of the statistical information presented in this section is based on reports prepared by statisticians from the U.S. Census Bureau and other government agencies that gather information about the elderly in the United States and internationally.

Numerous graphing techniques are available to you, but here we focus on just a few of the most widely used ones in the social sciences. The first two, the pie chart and bar graph, are appropriate for nominal and ordinal variables. The next two, histograms and line graphs, are used with interval-ratio variables. We also discuss statistical maps and time-series charts. The statistical map is most often used with interval-ratio data. Finally, time-series charts are used to show how some variables change over time.

The Pie Chart: Race and Ethnicity of the Elderly

The elderly population of the United States is racially heterogeneous. As the data in Table 2.15 show, of the total 40,489,000 elderly (defined as persons 65 years and older) in 2009–2011, the two largest racial groups were whites (34.4 million)[10] and blacks (3.4 million).

A **pie chart** shows the differences in frequencies or percentages among the categories of a nominal or an ordinal variable. The categories are displayed as segments of a circle whose pieces add up to 100% of the total frequencies. The pie chart shown in Figure 2.4 displays the same information that Table 2.15 presents. Although you can inspect these data in Table 2.15, you can interpret the information more easily by seeing it presented in the pie chart in Figure 2.4. It shows that the elderly population is predominantly white (85%), followed by black (8.5%) Americans.

✓ *Learning*
Check

Notice that the pie chart contains all the information presented in the frequency distribution. Like the frequency distribution, a chart has an identifying number, a title that describes the content of the figure, and a reference to a source. The frequency or percentage is represented both visually and in numbers.

Pie chart   A graph showing the differences in frequencies or percentages among the categories of a nominal or an ordinal variable. The categories are displayed as segments of a circle whose pieces add up to 100% of the total frequencies.

Note that the percentages for several of the racial groups are about 3.5% or less. It might be better to combine categories—American Indian, Asian, Native Hawaiian—into an "other races"

Table 2.15 Three-Year Estimates of the U.S. Population 65 Years and Older by Race, 2009–2011

Race	Frequency (f)	Percentage (%)
White alone	34,415,650	85.0
Black alone	3,441,565	8.5
American Indian alone	202,445	0.5
Asian alone	1,417,115	3.5
Native Hawaiian or Pacific Islander alone	40,489	0.1
Some other race alone	607,335	1.5
Two or more races combined	364,401	0.9
Total	40,489,000	100.0

Source: U.S. Census Bureau, *American Fact Finder*, 2011, Table S0103.

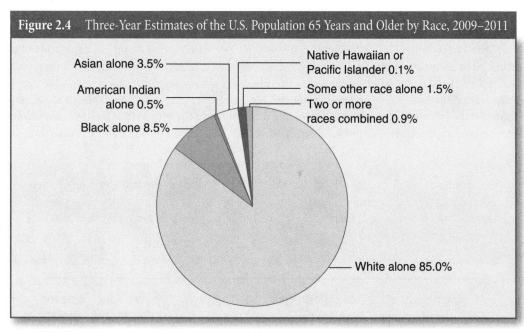

Figure 2.4 Three-Year Estimates of the U.S. Population 65 Years and Older by Race, 2009–2011

Source: U.S. Census Bureau, *American Fact Finder*, 2011, Table S01013.

category. This will leave us with three distinct categories: white, black, and other and two or more races. The revised pie chart is presented in Figure 2.5. Confirm for yourself how the percentages are derived from Table 2.15. We can highlight the diversity of the elderly population by "exploding"

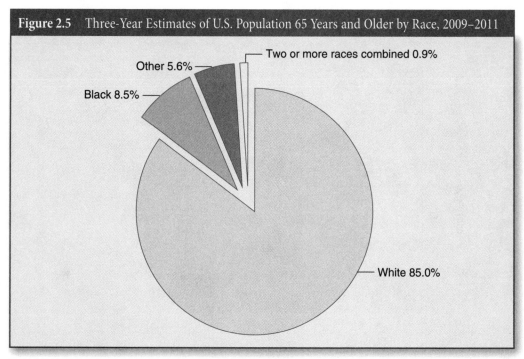

Figure 2.5   Three-Year Estimates of U.S. Population 65 Years and Older by Race, 2009–2011

Source: U.S. Census Bureau, *American Fact Finder*, 2011, Table S01013.

the pie chart, moving the segments representing these groups slightly outward to draw them to the viewer's attention.

The Bar Graph: Marital Status of the Elderly

The **bar graph** provides an alternative way to present nominal or ordinal data graphically. It shows the differences in frequencies or percentages among categories of a nominal or an ordinal variable. The categories are displayed as rectangles of equal width with their heights proportional to the frequency or percentage of the category.

Bar graph   A graph showing the differences in frequencies or percentages among the categories of a nominal or an ordinal variable. The categories are displayed as rectangles of equal width with their heights proportional to the frequency or percentage of the category.

Let's illustrate the bar graph with an overview of the marital status of the elderly. Figure 2.6 is a bar graph displaying the percentage distribution of persons 65 years and older by marital status in 2010. This chart is interpreted similarly to a pie chart except that the categories of the variable are arrayed along the horizontal axis (sometimes referred to as the *X*-axis) and the

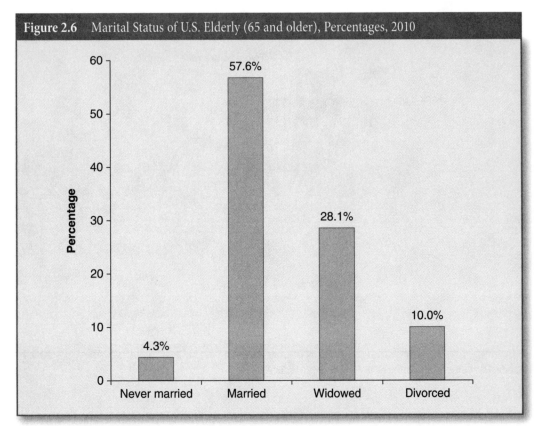

Figure 2.6   Marital Status of U.S. Elderly (65 and older), Percentages, 2010

Source: U.S. Census Bureau, *Statistical Abstract of the United States: 2012*, Table 34.

percentages along the vertical axis (sometimes referred to as the *Y*-axis). This bar graph is easily interpreted: It shows that in 2010, 57.6% were married, 28.1% widowed, 10% divorced, and 4.3% never married.

Construct a bar graph by first labeling the categories of the variables along the horizontal axis. For these categories, construct rectangles of equal width, with the height of each proportional to the frequency or percentage of the category. Note that a space separates each of the categories to make clear that they are nominal categories.

Bar graphs are often used to compare one or more categories of a variable among different groups. For example, the longevity of women is the major factor in the gender differences in marital and living arrangements.[11] In addition, elderly widowed men are more likely to remarry than elderly widowed women.

Suppose we want to show how the patterns in marital status differ between men and women. Figure 2.7 compares the marital status for women and men 65 years and older in 2010. We can also construct bar graphs horizontally, with the categories of the variable arrayed along the vertical axis and the percentages or frequencies displayed on the horizontal axis, as displayed in Figure 2.7. It clearly shows that elderly women are more likely than elderly men to be widowed, and elderly men are more likely to be married than elderly women.

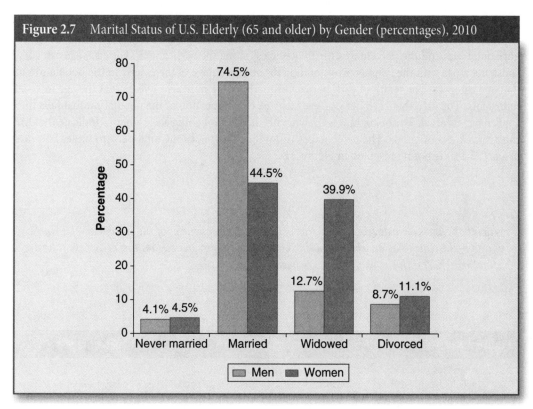

Figure 2.7 Marital Status of U.S. Elderly (65 and older) by Gender (percentages), 2010

Source: U.S. Census Bureau, *Statistical Abstract of the United States: 2012*, Table 34.

The Histogram

The **histogram** is used to show the differences in frequencies or percentages among categories of an interval-ratio variable. The categories are displayed as contiguous bars, each with width proportional to the width of the category and height proportional to the frequency or percentage of that category. A histogram looks very similar to a bar chart except that the bars are contiguous with each other (touching) and may not be of equal width. In a bar chart, the spaces between the bars visually indicate that the categories are separate. Examples of variables with separate categories are *marital status* (married, single), *gender* (male, female), and *employment status* (employed, unemployed). In a histogram, the touching bars indicate that the categories or intervals are ordered from low to high in a meaningful way. For example, the categories of the variables *hours spent studying*, *age*, and *years of school completed* are contiguous, ordered intervals.

Histogram   A graph showing the differences in frequencies or percentages among the categories of an interval-ratio variable. The categories are displayed as contiguous bars, each with width proportional to the width of the category and height proportional to the frequency or percentage of that category.

Figure 2.8 is a histogram displaying the percentage distribution of the population 55 years and older by age. To construct the histogram of Figure 2.8, arrange the age intervals along the horizontal axis and the percentages (or frequencies) along the vertical axis. For each age category, construct a bar with the height corresponding to the percentage of the elderly in the population in that age category. The width of each bar corresponds to the number of years that the age interval represents. The area that each bar occupies tells us the proportion of the population that falls into a given age interval. The histogram is drawn with the bars touching each other to indicate that the categories are contiguous. The percentages will not equal 100, as the total percent under 55 years of age (75.1%) is not represented in Figure 2.8.

✓ *Learning Check*

When bar charts or histograms are used to display the frequencies of the categories of a single variable, the categories are shown on the X-axis and the frequencies on the Y-axis. In a horizontal bar chart or histogram, this is reversed.

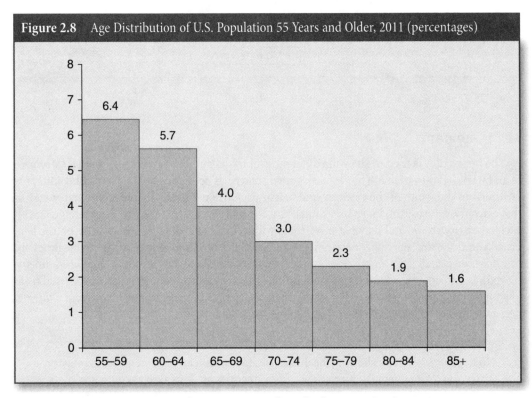

Figure 2.8 Age Distribution of U.S. Population 55 Years and Older, 2011 (percentages)

Source: U.S. Census Bureau, Current Population Survey, Annual Social and Economic Supplement, 2011.

The Line Graph

Numerical growth of the elderly population is taking place worldwide, occurring in both developed and developing countries. In 1994, 30 nations had elderly populations of at least 2 million; demographic projections indicate that there will be 55 such nations by 2020. Japan is one of the nations experiencing dramatic growth of its elderly population. Figure 2.9 is a line graph displaying the elderly population of Japan by age.

The **line graph** is another way to display interval-ratio distributions; it shows the differences in frequencies or percentages among categories of an interval-ratio variable. Points representing the frequencies of the categories are placed above their midpoints and are joined by a straight line. Notice that in Figure 2.9 the age intervals are arranged on the horizontal axis and the frequencies along the vertical axis. Instead of using bars to represent the frequencies, however, points representing the frequencies of each interval are placed above the midpoints of the intervals. Adjacent points are then joined by straight lines.

Figure 2.9   Population of Japan, Age 55 and Older, 2009

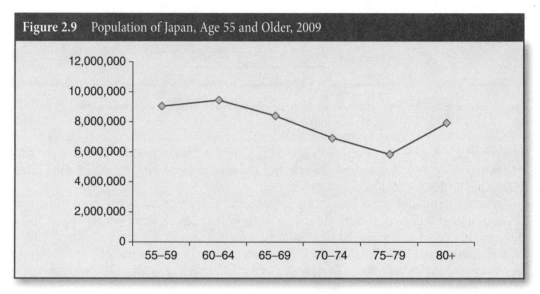

Source: Adapted from Ministry of Internal Affairs and Communications of Japan, Statistics Bureau, *Monthly Report April 2010*, Population Estimates.

Line graph   A graph showing the differences in frequencies or percentages among categories of an interval-ratio variable. Points representing the frequencies of each category are placed above the midpoints of the categories and are joined by a straight line.

Both the histogram and the line graph can be used to depict distributions and trends of interval-ratio variables. How do you choose which one to use? To some extent, the choice is a matter of individual preference, but in general, line graphs are better suited for comparing

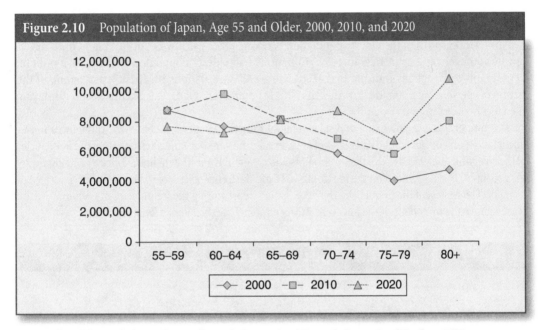

Figure 2.10 Population of Japan, Age 55 and Older, 2000, 2010, and 2020

Source: Adapted from U.S. Census Bureau, Center for International Research, International Database, 2007.

how a variable is distributed across two or more groups or across two or more time periods. For example, Figure 2.10 compares the elderly population in Japan for 2000 with the projected elderly population for 2010 and 2020.

Let's examine this line graph. It shows that Japan's population age 65 and older is expected to grow dramatically in the coming decades. According to projections, Japan's oldest-old population, those 80 years or older, is also projected to grow rapidly, from about 4.8 million (less than 4% of the total population) to 10.8 million (8.9%) by 2020. This projected rise has already led to a reduction in retirement benefits and other adjustments to prepare for the economic and social impact of a rapidly aging society.[12]

✓ *Learning Check*

Look closely at the line graph shown in Figure 2.10, comparing 2010 and 2020 data. How would you characterize the population increase among the Japanese elderly?

Time-Series Charts

We are often interested in examining how some variables change over time. For example, we may be interested in showing changes in the labor force participation of Latinas over the past decade, changes in the public's attitude toward abortion rights, or changes in divorce and marriage rates. A **time-series chart** displays changes in a variable at different points in time. It involves two variables: (1) *time*, which is labeled across the horizontal axis, and (2) another variable of interest whose values (frequencies, percentages, or rates) are labeled along the vertical axis. To construct a time-series chart, use a series of dots to mark the value of the variable at each time interval, and then join the dots by a series of straight lines.

Time-series chart　A graph displaying changes in a variable at different points in time. It shows time (measured in units such as years or months) on the horizontal axis and the frequencies (percentages or rates) of another variable on the vertical axis.

Figure 2.11 shows a time series from 1900 to 2050 of the percentage of the total U.S. population that is 65 years or older (the figures for 2000 through 2050 are projections made by the Social Security Administration, as reported by the Census Bureau). This time series enables us to see clearly the dramatic increase in the elderly population. The number of elderly increased from a little less than 5% in 1900 to about 12.4% in 2000. The rate is expected to increase to 20% of the total population. This dramatic increase in the elderly population, especially beginning in 2010, is associated with the "graying" of the baby boom generation. This group, which was 0 to 9 years old in 1955, turned 55 to 64 years old in 2010.

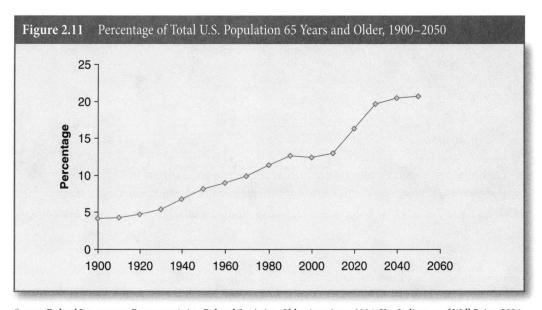

Figure 2.11　Percentage of Total U.S. Population 65 Years and Older, 1900–2050

Source: Federal Interagency Forum on Aging Related Statistics, *Older Americans 2004: Key Indicators of Well Being*, 2004.

The implications of these demographic changes are enormous. To cite just a few, there will be more pressure on the health care system and on private and public pension systems. In addition, because the voting patterns of the elderly differ from those of younger people, the "graying" of America will have major political effects.

Often, we are interested in comparing changes over time for two or more groups. Let's examine Figure 2.12, which charts the trends in the percentage of divorced elderly from 1960 to 2050 for men and women. This time-series graph shows that the percentages of divorced elderly men and elderly women were about the same until 2000. For both groups, the percentages increased from less than 2% in 1960 to about 5% in 1990.[13] According to projections, however, there will be significant increases in the percentages of men and especially women who are divorced: from 5% of all the elderly in 1990 to 8.4% of all elderly men and 13.6% of all elderly women by 2050. This sharp upturn and the gender divergence are clearly emphasized in Figure 2.12.

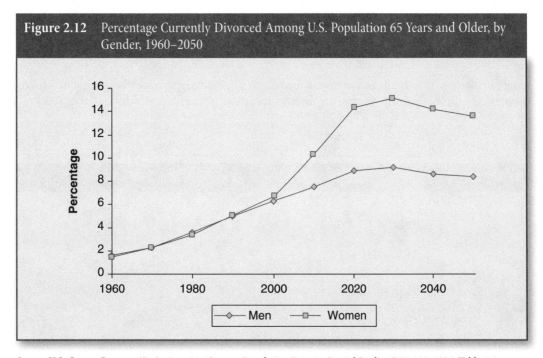

Figure 2.12   Percentage Currently Divorced Among U.S. Population 65 Years and Older, by Gender, 1960–2050

Source: U.S. Census Bureau, *65+ in America, Current Population Reports, Special Studies*, P23–190, 1996, Table 6-1.

✓ *Learning*
Check

How does the time-series chart differ from a line graph? The difference is that line graphs display frequency distributions of a single variable, whereas time-series charts display two variables. In addition, time is always one of the variables displayed in a time-series chart.

▣   A Cautionary Note: Distortions in Graphs

In this chapter, we have seen that statistical graphs can give us a quick sense of the main patterns in the data. However, graphs can not only quickly inform us they can also quickly deceive us. Because we are often more interested in general impressions than in detailed analyses of the numbers, we are more vulnerable to being swayed by distorted graphs. Edward Tufte, in his 1983 book *The Visual Display of Quantitative Information*, not only demonstrates the advantages of working with graphs but also offers a detailed discussion of some of the pitfalls in the application and interpretation of graphics.

Probably the most common distortions in graphical representations occur when the distance along the vertical or horizontal axis is altered either by not using 0 as the baseline (as demonstrated in Figures 2.13a and 2.13b) or in relation to the other axis. Axes may be stretched or shrunk to create any desired result to exaggerate or disguise a pattern in the data. In Figures 2.13a and 2.13b, 2009 international data on female representation in national parliaments are presented. Without altering the data in any way, notice how the difference between the countries is exaggerated by using 30 as a baseline (as in Figure 2.13b).

Remember: Always interpret a graph in the context of the numerical information the graph represents.

Figure 2.13   Female Representation in National Parliaments, 2009; (a) Using 0 as the Baseline and (b) Using 30 as the Baseline

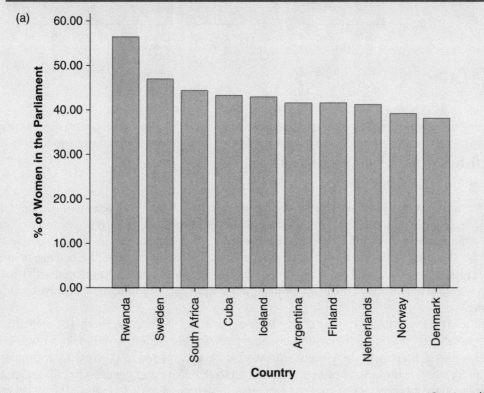

(Continued)

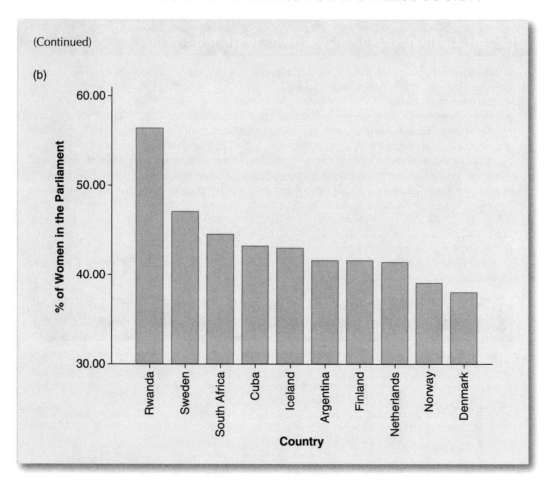

(Continued)

(b)

Statistics in Practice: The Graphic Presentation of Education

We now illustrate some additional ways in which graphics can be used to highlight diversity visually. In particular, we show how graphs can help us (a) explore the differences and similarities among the many social groups coexisting within American society and (b) emphasize the rapidly changing composition of the U.S. population. Indeed, because of the heterogeneity of American society, the most basic question to ask when you look at data is "compared with what?" This question not only is at the heart of quantitative thinking[14] but underlies inclusive thinking as well.

Three types of graphs—the bar chart, the line graph, and the time-series chart—are particularly suitable for making comparisons among groups. In this section, we will take a closer look at educational attainment. Let's begin with the bar chart displayed in Figure 2.14. It compares the percentage of those with college degrees by race/ethnicity and gender. Overall, the group with the highest percentage of college graduates is Asian and Pacific Islanders. The group with the lowest is Hispanic. A higher percentage of white and Asian and Pacific Islander men had a college degree in comparison with white and Asian and Pacific Islander women. In contrast, a higher percentage of black and Hispanic females had college degrees compared with black and Hispanic men.

The line graph provides another way of looking at differences based on gender, race/ethnicity, or other attributes such as class, age, or sexual orientation. Figure 2.15 compares the educational attainment of three age cohorts, reflecting the development of mass education in the United States during the past 50 years.

The data illustrate that the percentage of Americans who completed 5 to 8 years of education has declined by age cohort, from 5% among Americans 55 years and older to 2.6% for those 25 to 34 years old. The corresponding trend illustrated in Figure 2.15 is the increase in the percentage of Americans who have completed 13 to 15 years or 16 years or more. The percentage is highest at 34% among those 25 to 34 years of age and lowest (27.6%) for Americans 55 years or older.

Finally, Figure 2.16 is a time-series chart showing changes over time in college graduation rates among whites, blacks, Asian and Pacific Islanders, and Hispanics.

To conclude, the three examples of graphs in this section as well as other examples throughout this chapter have illustrated how graphical techniques can portray the complexities of the social world by emphasizing the distinct characteristics of age, gender, and ethnic groups. By depicting similarities and differences, graphs help us better grasp the richness and complexities of the social world.

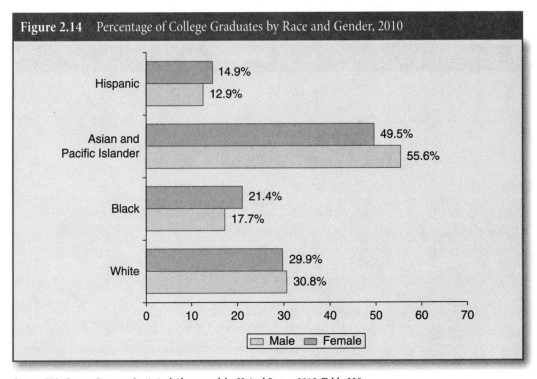

Figure 2.14   Percentage of College Graduates by Race and Gender, 2010

Source: U.S. Census Bureau, *Statistical Abstract of the United States: 2012*, Table 230.

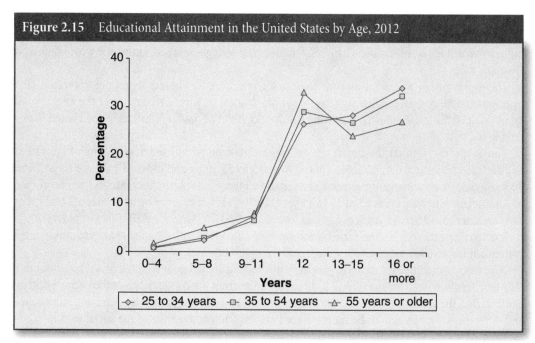

Figure 2.15 Educational Attainment in the United States by Age, 2012

Source: U.S. Census Bureau, *Educational Attainment,* CPS Historical Time Series Tables, 2012, Table A-1.

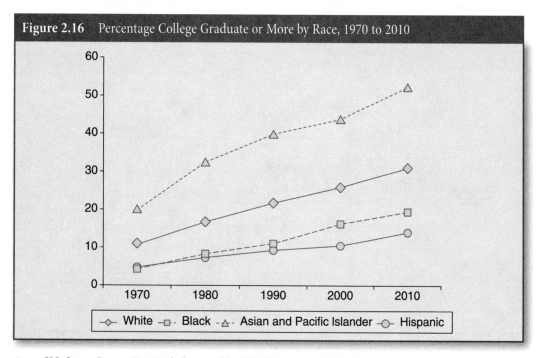

Figure 2.16 Percentage College Graduate or More by Race, 1970 to 2010

Source: U.S. Census Bureau, *Statistical Abstract of the United States: 2012,* Table 229.

Note: Persons of Hispanic origin may be any race.

MAIN POINTS

- The most basic method for organizing data is to classify the observations into a frequency distribution—a table that reports the number of observations that fall into each category of the variable being analyzed.

- Constructing a frequency distribution is usually the first step in the statistical analysis of data.

- To obtain frequency distributions for nominal and ordinal variables, count and report the number of cases that fall into each category of the variable along with the total number of cases (N).

- To construct frequency distributions for interval-ratio variables with wide ranges of values, first combine the scores into a smaller number of groups—known as class intervals—each containing a number of scores.

- Proportions and percentages are relative frequencies. To construct a proportion, divide the frequency (f) in each category by the total number of cases (N). To obtain a percentage, divide the frequency (f) in each category by the total number of cases (N) and multiply by 100.

- Percentage distributions are tables that show the percentage of observations that fall into each category of the variable. Percentage distributions are routinely added to almost any frequency table and are especially important if comparisons between groups are to be considered.

- Cumulative frequency distributions allow us to locate the relative position of a given score in a distribution. They are obtained by adding to the frequency in each category the frequencies of all the categories below it.

- Cumulative percentage distributions have wider applications than cumulative frequency distributions. A cumulative percentage distribution is constructed by adding to the percentage in each category the percentages of all the categories below it.

- One other method of expressing raw frequencies in relative terms is known as a rate. Rates are defined as the number of actual occurrences in a given time period divided by the number of possible occurrences. Rates are often multiplied by some power of 10 to eliminate decimal points and make the numbers easier to interpret.

- A pie chart shows the differences in frequencies or percentages among categories of a nominal or an ordinal variable. The categories of the variable are segments of a circle whose pieces add up to 100% of the total frequencies.

- A bar graph shows the differences in frequencies or percentages among categories of a nominal or an ordinal variable. The categories are displayed as rectangles of equal width with their heights proportional to the frequencies or percentages of the categories.

- Histograms display the differences in frequencies or percentages among categories of interval-ratio variables. The categories are displayed as contiguous bars each with width proportional to the width of the category and heights proportional to the frequencies or percentages of that category.

- A line graph shows the differences in frequencies or percentages among categories of an interval-ratio variable. Points representing the frequency of each category are placed above the midpoints of the categories (interval). Adjacent points are then joined by a straight line.

- A time-series chart displays changes in a variable at different points in time. It displays two variables: (1) time, which is labeled across the horizontal axis, and (2) another variable of interest whose values (e.g., frequencies, percentages, or rates) are labeled along the vertical axis.

KEY TERMS

bar graph	frequency distribution	pie chart
cumulative frequency	histogram	proportion
distribution	line graph	rate
cumulative percentage	percentage	time-series chart
distribution	percentage distribution	

⑤SAGE edge™

Sharpen your skills with SAGE edge at **edge.sagepub.com/ssdsess2e**. **SAGE edge for students** provides a personalized approach to help you accomplish your coursework goals in an easy-to-use learning environment.

CHAPTER EXERCISES

1. Suppose you have surveyed 30 people and asked them whether they are white (W) or nonwhite (N), and how many traumas (serious accidents, rapes, or crimes) they have experienced in the past year. You also asked them to tell you whether they perceive themselves as being in the upper, middle, working, or lower class. Your survey resulted in the raw data presented in the table below:

Race	Class	Trauma	Race	Class	Trauma
W	L	1	W	W	0
W	M	0	W	M	2
W	M	1	W	W	1
N	M	1	W	W	1
N	L	2	N	W	0
W	W	0	N	M	2
N	W	0	W	M	1
W	M	0	W	M	0
W	M	1	N	W	1
N	W	1	W	W	0
N	W	2	W	W	0
N	M	0	N	M	0

Race	Class	Trauma	Race	Class	Trauma
N	L	0	N	W	0
W	U	0	N	W	1
W	W	1	W	W	0

Source: Data based on GSS files for 1987 to 1991.

Notes: Race: W, white; N, nonwhite; Class: L, lower class; M, middle class; U, upper class; W, working class.

   a. What level of measurement is the variable race? Class?
   b. Construct raw frequency tables for race and for class.
   c. What proportion of the 30 individuals is nonwhite? What percentage is white?
   d. What proportion of the 30 individuals identified themselves as middle class?

2. Using the data and your raw frequency tables from Exercise 1, construct a frequency distribution for class.
   a. Which is the smallest perceived class?
   b. Which two classes include the largest percentages of people?

3. Using the data from Exercise 1, construct a frequency distribution for trauma.
   a. What level of measurement is used for the trauma variable?
   b. Are people more likely to have experienced no traumas or only one trauma in the past year?
   c. What proportion has experienced one or more traumas in the past year?

4. Using the data from Exercise 1, construct bar graphs showing percentage distributions for race, class, and trauma.

5. A question on whether immigrants are good for America was included in the GSS 2010. Results are provided in the table below, noting the percentages who agree or strongly agree by political party (not all responses are reported here, so totals will not add up to 100%). Do these data support the statement that people's views on immigration are related to their political party affiliations? Why or why not?

	Strong Democrat	Independent	Strong Republican
	%	%	%
Agree strongly	9.1	7.7	2.1
Agree	47.7	34.6	47.9

6. How many hours per week do you spend on e-mail? In 2010, the GSS included a question on number of hours spent on e-mail. Data are presented here for a sample of 99 men and women.

Exercises

E-mail Hours per Week	Frequency
0	19
1	20
2	13
3	5
4	2
5	6
6	5
7	2
8	3
9	1
10 or more	23

a. Compute the cumulative frequency and cumulative percentage distribution for the data.
b. What proportion of the sample spent 3 hours or less per week on e-mail?
c. What proportion of the sample spent 6 or more hours per week on e-mail?
d. Construct a graph or chart that best displays these data. Explain why the graph you selected is appropriate.

7. The tables below present the frequency distributions for education by gender and race based on the GSS 2010. Use them to answer the following questions.

	Race	
Education	White (f)	Black (f)
Less than high school	72	26
High school graduate	272	59
Junior college	46	10
Bachelor	118	16
Graduate	77	7

	Gender	
Education	Male (f)	Female (f)
Less than high school	46	67
High school graduate	151	214
Junior college	24	37
Bachelor	65	81
Graduate	43	49

a. Construct tables based on percentages and cumulative percentages of educational attainment for race and gender.
b. What percentage of men have continued their education beyond high school? What is the comparable percentage for women?
c. What percentage of whites have completed high school or less? What is the comparable percentage for blacks?
d. Are the cumulative percentages more similar for men and women or for the racial and ethnic groups? (In other words, where is there more inequality?) Explain.

8. Policy analysts have noted that the number of those without health insurance is increasing in the United States. Access to health insurance has been identified as an important social issue. Data from the National Center for Health Statistics (2013) are presented below (see Table 2.16), measuring the percentage of persons with no health insurance for at least part of 2011 by selected characteristics. Note: Characteristic totals will not equal 100%.
a. What can be said about who did not have health insurance in 2011? How does the percentage of those without health insurance vary by each demographic characteristic?
b. For each variable, what would be the best way to graphically present the data?

Table 2.16   The Uninsured Population Below 65 Years of Age by Selected Characteristics, 2011

Characteristic	Percentage
Age (years)	
Below 18	7.0
18–44	25.4
45–64	15.4
Percentage of the poverty level	
Below 100%	28.4
100%–199%	30.0
200%–399%	16.5
400% or more	5.2
Race origin	
White only	16.7
Black only	20.6
Asian only	16.5
American Indian or Alaska Native only	34.2
Two or more races	16.0

(Continued)

Exercises

(Continued)

Table 2.16 The Uninsured Population Below 65 Years of Age by Selected Characteristics, 2011

Characteristic	Percentage
Geographic region	
Northeast	11.8
Midwest	13.4
South	20.4
West	20.0

Source: National Center for Health Statistics, *Health, United States, 2012: With Special Feature on Emergency Care* (Hyattsville, MD: National Center for Health Statistics, 2013).

9. The time-series chart shown in Figure 2.17 displays trends for presidential election voting rates by race and Hispanic origin from 1996 to 2012. Analysts noted how for the first time, in the 2012 presidential election, black voting rates exceeded the rates for non-Hispanic whites. Overall votes cast were higher in 2012 than 2008 (131,948,000 vs. 131,144,000—data not reported in the figure), an increase attributed to minority voters. Describe the variation in voting rates for the four racial and Hispanic-origin groups.

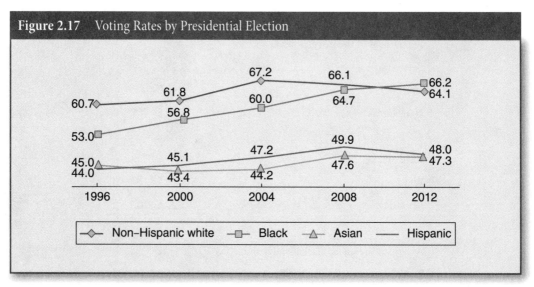

Figure 2.17 Voting Rates by Presidential Election

Source: Thom File, *The Diversifying Electorate—Voting Rates by Race and Hispanic Origin in 2012 (and other recent elections),* Current Population Survey (P20-568), 2013, Figure 1.

10. The 2010 GSS data on educational level can be further broken down by race as follows:
    a. Construct two histograms for education, using percentages for whites ($N = 584$) and for blacks ($N = 118$).
    b. Describe the differences in educational attainment by race.

Years of Education	Whites	Blacks
0	2	0
1	0	0
2	1	1
3	0	0
4	2	1
5	2	0
6	7	1
7	4	2
8	11	3
9	15	2
10	21	11
11	21	10
12	168	27
13	42	8
14	71	23
15	21	7
16	106	11
17	21	3
18	32	3
19	14	3
20	23	2

11. In this exercise, we examine the rate of sexual violence against females reported by Michael Planty and his colleagues (2013). The table includes the rate of victimization per 1,000 females age 12 or older for three time periods.
    a. The rates are highest for which age-group over the three time periods?
    b. How would you characterize the relationship between victim age and rate of victimization? Explain the reason for your answer.

Age (Years)	1994–1998	1999–2004	2005–2010
12–17	11.3	7.6	4.1
18–34	7.0	5.3	3.7
35–64	2.0	1.8	1.5
65 or older	0.1*	0.2*	0.2*

*Interpret with caution; estimate based on 10 or fewer sample cases, or coefficient of variation is greater than 50%.

Source: Michael Planty, Lynn Langton, Christopher Krebs, Marcus Berzofsky, and Hope Smiley-McDonald, "Female Victims of Sexual Violence, 1994–2010" (2013). Retrieved from http://bjs.gov/content/pub/pdf/fvsv9410.pdf.

12. In 2011 Gallup reported that for the first time a majority of Americans believed same-sex marriage should be legally recognized. In its report, data for different demographic groups and changes in their levels of support were reported for 2010 and 2011. Review each demographic variable and summarize the percentage change between years.

	% Should Be Legal, 2010	% Should Be Legal, 2011
Sex and age (years)		
Men, 18–49	48	61
Men, 50+	32	35
Women, 18–49	58	65
Women, 50+	37	45
Age (years)		
18–34	54	70
35–54	50	53
55+	33	39
Political affiliation		
Democrats	56	69
Independents	49	59
Republicans	28	28

Exercises

	% Should Be Legal, 2010	% Should Be Legal, 2011
Political views		
Liberals	70	78
Moderates	56	65
Conservatives	25	28

Source: Frank Newport, "For First Time, Majority of Americans Favor Legal Gay Marriage," May 20, 2011. Retrieved from http://www.gallup.com/poll/147662/First-Time-Majority-Americans-Favor-Legal-Gay-Marriage.aspx. Copyright © 2011 Gallup, Inc. All rights reserved. The content is used with permission; however, Gallup retains all rights of republication.

13. We present pie charts based on two GSS 2010 variables: GUNLAW (whether the respondent favors or opposes stricter gun laws) by DEGREE (respondent's highest degree earned). In addition to creating individual pie charts for each degree category, we've also summarized the GUNLAW percentages for each DEGREE category.

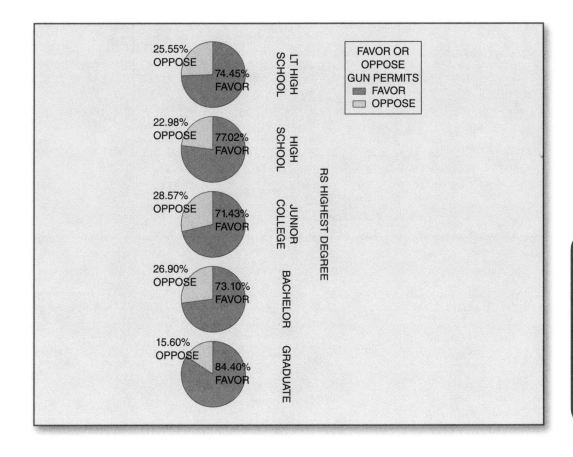

Review the pie charts. What can you conclude about the relationship between position on stricter gun laws and educational attainment?

14. In 2010, GSS respondents were asked to rate their agreement with the following statement: "I worry that if I run out of money or health insurance I will get second class health care" (PAYHLTH). We present a bar graph comparing men's and women's responses (in percentages).
    a. What is the level of measurement for PAYHLTH?
    b. Explain the differences between men's and women's responses. Who is more worried about receiving second-class health care?

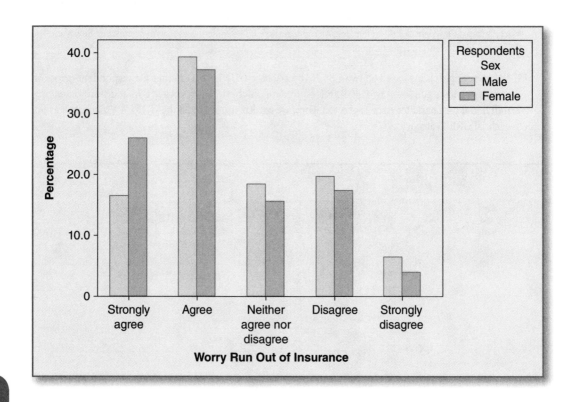

Chapter 3

Measures of Central Tendency

In Chapter 2, we learned that frequency distributions and graphical techniques are useful tools for describing data. Another way of describing a distribution is by selecting a single number that describes or summarizes the distribution more concisely. Such numbers describe what is typical about the distribution, for example, the average income among Latinos who are college graduates or the most common party identification among the rural poor. These are called **measures of central tendency**. In this chapter, we will learn about three measures of central tendency: the mode, the median, and the mean. The choice of an appropriate measure of central tendency for representing a distribution depends on three factors: (1) the way the variables are measured (their levels of measurement), (2) the shape of the distribution, and (3) the purpose of the research.

Measures of central tendency   Categories or scores that describe what is average or typical of the distribution.

▣ THE MODE

The **mode** is the category or score with the largest frequency or percentage in the distribution.

Mode The category or score with the highest frequency (or percentage) in the distribution.

We can use the mode to determine, for example, the most common foreign language spoken in the United States today. What is the most common foreign language spoken in the United States today? To answer this question, look at Table 3.1, which lists the 10 most commonly spoken foreign languages in the United States and the number of people who speak each language. The table shows that Spanish is the most common; more than 35 million people speak Spanish. In this example, we refer to "Spanish" as the mode—the category with the largest frequency in the distribution.

The mode is used to describe nominal variables, those variables that allow classification based on only a qualitative and not a quantitative property. By describing the most commonly occurring category of a nominal variable (such as Spanish in our example), the mode thus reflects the most important element of the distribution of a variable measured at the nominal level. The mode is the only measure of central tendency that can be used with nominal-level variables.

Table 3.1 Ten Most Common Foreign Languages
Spoken in the United States, 2009

Language	Number of Speakers
Spanish	35,468,501
Chinese	2,600,150
Tagalog	1,513,734
French	1,305,503
Vietnamese	1,251,468
German	1,109,216
Korean	1,039,021
Russian	881,723
Arabic	845,396
Italian	753,992

Source: U.S. Census Bureau, *Statistical Abstract of the United States: 2012,*
    Table 53.

◫ THE MEDIAN

The **median** is a measure of central tendency that can be calculated for variables that are at least at an ordinal level of measurement. The median represents the exact middle of a distribution; it is the score that divides the distribution into two equal parts so that half the cases are above it and half below it. For example, according to the U.S. Bureau of Labor Statistics, the median weekly earnings of full-time wage and salary workers in 2012 were $768.[1] This means that half the workers in the United States earned more than $768 a week and half earned less than $768. Since many variables used in social research are ordinal, the median is an important measure of central tendency.

Median   The score that divides the distribution into two equal parts so that half the cases are above it and half below it.

The median is a suitable measure for those variables whose categories or scores can be arranged in order of magnitude from the lowest to the highest. Therefore, the median can be used with ordinal or interval-ratio variables, for which scores can be at least rank ordered, but cannot be calculated for variables measured at the nominal level.

Finding the Median in Sorted Data

It is very easy to find the median. In most cases, it can be done by a simple inspection of the sorted data. The location of the median score differs somewhat, depending on whether the number of observations is odd or even. Let's consider an example with an odd number of cases.

Suppose we are looking at the responses of five people to the question "Thinking about the economy, how would you rate economic conditions in this country today?" Following are the responses of these five hypothetical persons:

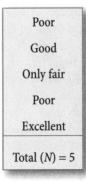

Poor
Good
Only fair
Poor
Excellent
Total (*N*) = 5

First arrange the responses in order from the lowest to the highest (or the highest to the lowest):

Poor
Poor
Only fair
Good
Excellent
Total (*N*) = 5

The median is the response associated with the middle case. Find the middle case when *N* is odd by adding 1 to *N* and dividing by 2: $(N + 1)/2$. Since *N* is 5, you calculate $(5 + 1)/2 = 3$. The middle case is thus the third case, and the median is "only fair," the response associated with the third case. Notice that the median divides the distribution exactly in half so that there are two respondents who are more satisfied and two respondents who are less satisfied.

✓ *Learning*
Check

When working with an even number of cases, the middle case will not be a whole number. If the data are ordinal, simply report both of the category names that constitute the middle cases. If the data are interval-ratio, average the two middle cases together and report that as the median.

Finding the Median in Frequency Distributions

Often our data are arranged in frequency distributions. Take, for instance, the frequency distribution displayed in Table 3.2. It shows the political views of General Social Survey (GSS) respondents in 2010.

We begin by specifying *N*, the total number of respondents. In this particular example, $N = 1,457$. We then use the formula $(N + 1)/2$, or $(1,457 + 1)/2 = 729$. The median is the value of the category associated with the 729th case. The cumulative frequency (Cf) of the 729th case falls in the category "moderate"; thus, the median is "moderate." This may seem odd; however, the median is always the value of the response category, not the frequency.

A second approach to locating the median in a frequency distribution is to use the cumulative percentages column, as shown in the last column of Table 3.2. In this example, the percentages are cumulated from "extremely liberal" to "extremely conservative." We could also cumulate the other way, from "extremely conservative" to "extremely liberal." To find the median, we identify the response category that contains a cumulative percentage value equal to 50%. The median is the value of the category associated with this observation.[2] Looking at Table 3.2, the percentage value equal to 50% falls within the category "moderate." The median for this distribution is therefore "moderate." If you are not sure why the middle of the distribution—the 50% point—is associated with the category "moderate," look again at the cumulative percentage column (C%). Notice that 28.3% of the observations are accumulated below the category "moderate" and that 65.0% are accumulated up to and including the category "moderate." We know, then, that the percentage value equal to 50% is located somewhere within the "moderate" category.

Table 3.2   Political Views of GSS Respondents, 2010

Political View	Frequency (f)	Cf	Percentage	C%
Extremely liberal	57	57	3.9	3.9
Liberal	168	225	11.5	15.4
Slightly liberal	187	412	12.8	28.3
Moderate	535	947	36.7	65.0
Slightly conservative	213	1,160	14.6	79.6
Conservative	239	1,399	16.4	96.0
Extremely conservative	58	1,457	4.0	100.0
Total (N)	1,457		100.0	

For a review of cumulative distributions, refer to Chapter 2.

✓ Learning
Check

▣ STATISTICS IN PRACTICE: GENDERED INCOME INEQUALITY

We can use the median to compare groups. Consider the significant changes that have taken place during the past few decades in the income levels of men and women in the United States. Income levels profoundly influence our lives both socially and economically. Higher income is associated with increased education and work experience for both men and women.

Figure 3.1 compares the median incomes for men and women in 1973 and in 2011. Because the median is a single number summarizing central tendency in the distribution, we can use it to note differences between subgroups of the population or changes over time. In this example, the increase in median income from 1973 to 2011 clearly shows a significant income gain for women. However, that said, in 2011 women still made, on average, about $10,000 less than men.

Locating Percentiles in a Frequency Distribution

The median is a special case of a more general set of measures of location called *percentiles*. A **percentile** is a score at or below which a specific percentage of the distribution falls. The *n*th percentile is a score below which *n*% of the distribution falls. For example, the 75th percentile is a score that divides the distribution so that 75% of the cases are below it. The median is the 50th percentile. It is a score that divides the distribution so that 50% of the cases fall below it. Like the median, percentiles require that data be ordinal or higher in level of measurement. Percentiles are easy to identify when the data are arranged in frequency distributions.

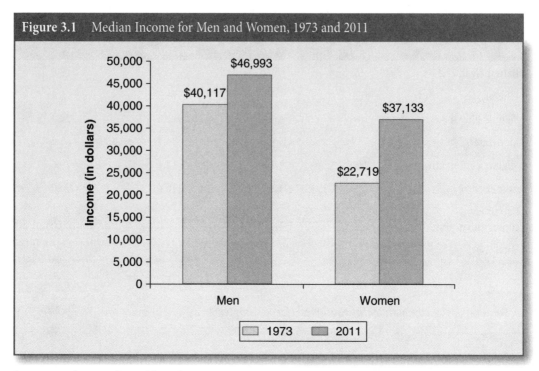

Figure 3.1 Median Income for Men and Women, 1973 and 2011

Sources: Data for 1973 obtained from the U.S. Census Bureau, Current Population Reports P60-226, *Income, Poverty and Health Insurance Coverage in the United States: 2003.* Data for 2011 obtained from the U.S. Census Bureau, American Community Survey, 2011.

Percentile A score below which a specific percentage of the distribution falls.

To help illustrate how to locate percentiles in a frequency distribution, we display in Table 3.3 the frequency distribution, the percentage distribution, and the cumulative percentage distribution of opinion about police job performance of respondents for the 2008 Monitoring the Future (MTF) survey.

The 50th percentile (the median) is "fair," meaning that 50% of the respondents view police job performance above "fair" and 50% of the respondents view police job performance as "fair" or below "fair" (as you can see from the cumulative percentage column, 50% falls somewhere in the third category, associated with the category "fair"). Similarly, the 20th percentile is "poor" because 20% of the respondents view police job performance as "poor" or below "poor."

Percentiles are widely used to evaluate relative performance on standardized achievement tests, such as the SAT or ACT. Let's suppose that your ACT score was 29. To evaluate your performance for the college admissions officer, the testing service translated your score into a percentile rank. Your percentile rank was determined by comparing your score with the scores

Table 3.3   Frequency Distribution for Police Job Performance: 2008 MTF Respondents

Police Job Performance	Frequency (f)	Percentage	C%
Very poor	125	9.9	9.9
Poor	201	15.9	25.8
Fair	452	35.7	61.5
Good	380	30.0	91.5
Very good	108	8.5	100.0
Total (N)	1,266	100.0	

of all other students who took the test at the same time. Suppose for a moment that 90% of all students received lower ACT scores than you (and 10% scored above you). Your percentile rank would have been 90. If, however, there were more students who scored better than you—let's say that 15% scored above you and 85% scored lower than you—your percentile rank would have been 85.

Another widely used measure of location is the *quartile*. The lower quartile is equal to the 25th percentile and the upper quartile is equal to the 75th percentile. (Can you locate the upper quartile in Table 3.3?) A college admissions office interested in accepting the top 25% of its applicants based on their SAT scores could calculate the upper quartile (the 75th percentile) and accept everyone whose score is equivalent to the 75th percentile or higher. (Note that they would be calculating percentiles based on the scores of their applicants, not of all students in the nation who took the SAT.)

▣ THE MEAN

The arithmetic **mean** is by far the best known and most widely used measure of central tendency. The mean is what most people call the "average." The mean is typically used to describe central tendency in interval-ratio variables such as income, age, and education. You are probably already familiar with how to calculate the mean. Simply add up all the scores and divide by the total number of scores.

Mean   A measure of central tendency that is obtained by adding up all the scores and dividing by the total number of scores. It is the arithmetic average.

Firearm statistics, for example, can be analyzed using the mean. Table 3.4 shows the 2011 gun ownership rates (per 100 population) for 15 of the most populous countries in the world. Because the variable "gun ownership rate" is an interval-ratio variable, we will select the arithmetic mean as our measure of central tendency.

Table 3.4 2011 Gun Ownership Rates per 100 People for 15 of the Most Populous Countries

Country	Gun Ownership Rate per 100 Population
China	4.9
India	4.2
United States	88.8
Indonesia	0.5
Brazil	8.0
Pakistan	11.6
Nigeria	1.5
Russia	8.9
Japan	0.6
Mexico	15.0
Vietnam	1.7
Egypt	3.5
Germany	30.3
Turkey	12.5
Iran	7.3

Source: United Nations Office on Drugs and Crime, *2011 Annual Report.*

To find the mean gun ownership rate (number of guns per 100 people) for the data presented in Table 3.4, add up the gun ownership rates for all the countries and divide the sum by the number of countries:

$$\text{Mean} = \frac{\begin{array}{c}(4.9+4.2+88.8+0.5+8.0+11.6+1.5+\\8.9+0.6+15.0+1.7+3.5+30.3+12.5+7.3)\end{array}}{15} = \frac{199}{15} = 13.3$$

The mean gun ownership rate for 15 of the most populous countries in the world is 13.3.[3]

Calculating the Mean

Another way to calculate the arithmetic mean is to use a formula. Beginning with this section, we introduce a number of formulas that will help you calculate some of the statistical concepts that we are going to present. A formula is a shorthand way to explain what operations we need to follow to obtain a certain result. So instead of saying "add all the scores together and then divide by the number of scores," we can define the mean by the following formula:

$$\bar{Y} = \frac{\Sigma Y}{N} \qquad (3.1)$$

Let's take a moment to consider these new symbols, because we continue to use them in later chapters. We use Y to represent the raw scores in the distribution of the variable of interest; $\bar{Y}$ is pronounced "Y-bar" and is the mean of the variable of interest. The symbol represented by the Greek letter Σ is pronounced "sigma," and it will be used often from now on. It is a summation sign (just like the $+$ sign) and directs us to sum whatever comes after it. Therefore, ΣY means "add up all the raw Y scores." Finally, the letter N, as you know by now, represents the number of cases (or observations) in the distribution.

Let's summarize the symbols as follows:

Y = the raw scores of the variable Y

$\bar{Y}$ = the mean of Y

ΣY = the sum of all the Y scores

N = the number of observations or cases

Now that we know what the symbols mean, let's work through another example. The following are the ages of the 10 students in a graduate research methods class:

21, 32, 23, 41, 20, 30, 36, 22, 25, 27

What is the mean age of the students?

For these data, the ages included in this group are represented by Y; $N = 10$, the number of students in the class; and ΣY is the sum of all the ages:

$$\Sigma Y = 21 + 32 + 23 + 41 + 20 + 30 + 36 + 22 + 25 + 27 = 277$$

Thus, the mean age is

$$\bar{Y} = \frac{\Sigma Y}{N} = \frac{277}{10} = 27.7$$

The mean can also be calculated when the data are arranged in a frequency distribution.

Finding the Mean in a Frequency Distribution

When data are arranged in a frequency distribution, we must give each score its proper weight by multiplying it by its frequency. We can use the following modified formula to calculate the mean:

$$\overline{Y} = \frac{\Sigma fY}{N}$$

where

Y = the raw scores of the variable Y

$\overline{Y}$ = the mean of Y

ΣfY = the sum of all the fYs

N = the number of observations or cases

✓ *Learning*
Check

The following distribution is the same as the one you used to calculate the median in an earlier Learning Check: 22, 15, 18, 33, 17, 5, 11, 28, 40, 19, 8, 20. Calculate the mean. Is it the same as the median, or is it different?

Understanding Some Important Properties of the Arithmetic Mean

The following three mathematical properties make the mean the most important measure of central tendency. It is, in fact, a concept that is basic to numerous and more complex statistical operations.

Interval-Ratio Level of Measurement

Because it requires the mathematical operations of addition and division, the mean can be calculated only for variables measured at the interval-ratio level.

Center of Gravity

Because the mean (unlike the mode and the median) incorporates all the scores in the distribution, we can think of it as the center of gravity of the distribution. That is, the mean is the point that perfectly balances all the scores in the distribution. If we subtract the mean from each score and add up all the differences, the sum will always be zero!

Sensitivity to Extremes

The examples we have used to show how to compute the mean demonstrate that, unlike with the mode or the median, every score enters into the calculation of the mean. This property makes the mean sensitive to extreme scores in the distribution. The mean is pulled in the direction of either very high or very low values. A glance at Figure 3.2 should convince you of that. Figures 3.2a and 3.2b show the incomes of 10 individuals. In Figure 3.2b, the income of one individual has shifted

Figure 3.2   The Value of the Mean Is Affected by Extreme Scores: (a) No Extreme Scores and (b) One Extreme Score

(a) No extreme scores: The mean is $3,000

Income (Y)	Frequency (f)	fY
1,000	1	1,000
2,000	2	4,000
3,000	4	12,000
4,000	2	8,000
5,000	1	5,000
	N = 10	ΣfY = 30,000

Mean = $\frac{\Sigma fY}{N} = \frac{30,000}{10}$ = $3,000

Median = $3,000

(b) One extreme score: The mean is $6,000

Income (Y)	Frequency (f)	fY
1,000	1	1,000
2,000	2	4,000
3,000	4	12,000
4,000	2	8,000
35,000	1	35,000
	N = 10	ΣfY = 60,000

Mean = $\frac{\Sigma fY}{N} = \frac{60,000}{10}$ = $6,000

Median = $3,000

from $5,000 to $35,000. Notice the effect it has on the mean; it shifts from $3,000 to $6,000! The mean is disproportionately affected by the relatively high income of $35,000 and is misleading as a measure of central tendency for this distribution. Notice that the median's value is not affected by this extreme score; it remains at $3,000. Thus, the median gives us better information on the typical income for this group. In the next section, we will see that because of the sensitivity of the mean, it is not suitable as a measure of central tendency in distributions that have a few very extreme values on one side of the distribution. (A few extreme values are no problem if they are not mostly on one side of the distribution.)

▣ THE SHAPE OF THE DISTRIBUTION: TELEVISION, EDUCATION, AND SIBLINGS

In this chapter, we have looked at the way in which the mode, median, and mean reflect central tendencies in the distribution. Distributions (this discussion is limited to distributions of interval-ratio variables) can also be described by their general shape, which can be easily represented visually. A distribution can be either symmetrical or skewed, depending on whether there are a few extreme values at one end of the distribution.

A distribution is **symmetrical** (Figure 3.3a) if the frequencies at the right and left tails of the distribution are identical, so that if it is divided into two halves, each will be the mirror image of the other. In a unimodal, symmetrical distribution, the mean, median, and mode are identical. In skewed distributions, there are a few extreme values on one side of the distribution. Distributions that have a few extremely low values are referred to as negatively skewed (Figure 3.3b). A defining feature of a negatively skewed distribution is that the value of the mean is always less than the value of the median; in other words, the mean is pulled in the direction of the lower scores. Alternatively, distributions with a few extremely high values are said to be positively skewed (Figure 3.3c). A defining feature of a positively skewed distribution is that the value of the mean is always greater than the value of the median; the mean is pulled toward the higher scores.

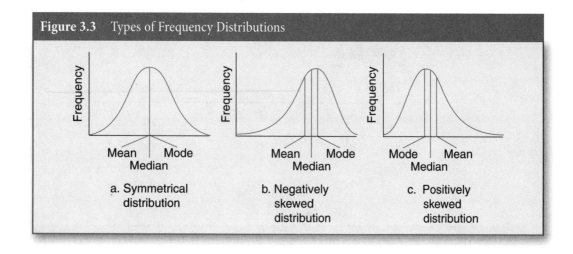

Figure 3.3   Types of Frequency Distributions

Table 3.5   Hours Spent per Day Watching Television

Hours Spent per Day Watching TV	Frequency (f)	fY	Percentage	C%
1	204	204	31.9	31.9
2	247	494	38.7	70.6
3	188	564	29.4	100.0
Total	639	$\sum fY = 1,262$	100.0	

$$\bar{Y} = \frac{\sum fY}{N} = \frac{1,262}{639} = 1.97$$

Median = 2.0
Mode = 2.0

Figure 3.4   Hours Spent per Day Watching Television

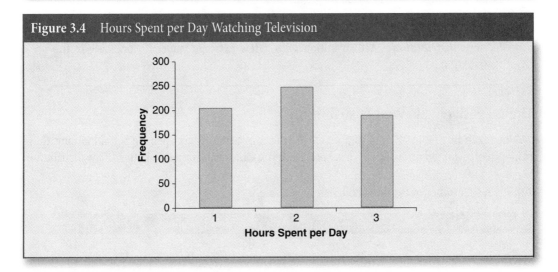

Symmetrical distribution   The frequencies at the right and left tails of the distribution are identical; each half of the distribution is the mirror image of the other.

We can illustrate the differences among the three types of distributions by examining three variables in the 2010 GSS.[4] The frequency distributions for these three variables are presented in Tables 3.5 through 3.7, and the corresponding graphs are depicted in Figures 3.4 through 3.6.

The Symmetrical Distribution

First, let's examine Table 3.5 and Figure 3.4, displaying the distribution of the number of hours per day spent watching television. Notice that the largest number (247) watch television 2 hours/day (mode = 2.0), and a fairly similar number (204 and 188, respectively) reported either

1 or 3 hours of watching television per day. As shown in Figure 3.4, the mode, the median, and the mean are almost identical, and they coincide at about the middle of the distribution. The distribution is nearly symmetrical.

The Positively Skewed Distribution

Now let's examine Table 3.6 and Figure 3.5, displaying the distribution of the number of siblings of a respondent. Notice that in this distribution, the mean, the median, and the mode have different values, with the mode having the lowest value (mode = 2.00), the median having the second lowest value (median = 3.0), and the mean having the highest value (mean = 3.07).

The distribution as depicted in Table 3.6 and Figure 3.5 is **positively skewed**. As a general rule, for **skewed distributions** the mean, the median, and the mode do not coincide. The mean, which is always pulled in the direction of extreme scores, falls closest to the tail of the distribution where a small number of extreme scores are located.

Positively skewed distribution   A distribution with a few extremely high values.

Skewed distribution   A distribution with a few extreme values on one side of the distribution.

The Negatively Skewed Distribution

Now examine Table 3.7 and Figure 3.6 for the number of years spent in school among those respondents who did not finish high school. Here you can see the opposite pattern. The distribution

Table 3.6   Number of Brothers and Sisters

Number of Siblings	Frequency (f)	fY	Percentage	C%
0	55	0	4.0	4.0
1	292	292	21.1	25.1
2	312	624	22.5	47.6
3	229	687	16.5	64.1
4	186	744	13.4	77.5
5	122	610	8.8	86.4
6	71	426	5.1	91.5
7	72	504	5.2	96.7
8	46	368	3.3	100.0
Total	1,385	4,255	100.0	

$$\bar{Y} = \frac{\Sigma fY}{N} = \frac{4,255}{1,385} = 3.07$$

Median = 3.0
Mode = 2.0

Figure 3.5 Number of Brothers and Sisters

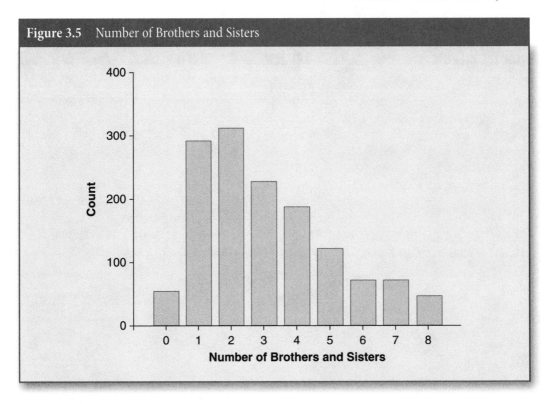

is a **negatively skewed distribution**. First, note that the largest number of years spent in school are concentrated at the high end of the scale (11 years) and that there are fewer respondents at the low end. The mean, the median, and the mode also differ in values, as they did in the previous example. However, here the mode has the highest value (mode = 11.0), the median has the second highest value (median = 10.0), and the mean has the lowest value (mean = 8.89).

Negatively skewed distribution   A distribution with a few extremely low values.

Guidelines for Identifying the Shape of a Distribution

Following are some useful guidelines for identifying the shape of a distribution.

1. In unimodal distributions, when the mode, the median, and the mean coincide or are almost identical, the distribution is symmetrical.

2. When the mean is higher than the median (or is positioned to the right of the median), the distribution is positively skewed.

3. When the mean is lower than the median (or is positioned to the left of the median), the distribution is negatively skewed.

Table 3.7 Years of School Among Respondents Without High School Degrees

Years of School	Frequency (f)	fY	Percentage	C%
0	5	0	2.0	2.0
2	2	4	0.8	2.8
3	1	3	0.4	3.2
4	6	24	2.4	5.6
5	4	20	1.6	7.2
6	24	144	9.6	16.8
7	10	70	4.0	20.8
8	26	208	10.4	31.2
9	38	342	15.2	46.4
10	66	660	26.4	72.8
11	68	748	27.2	100.0
Total	250	2,223	100.0	

$$\overline{Y} = \frac{\sum fY}{N} = \frac{2,223}{250} = 8.89$$

Median = 10.0
Mode = 11.0

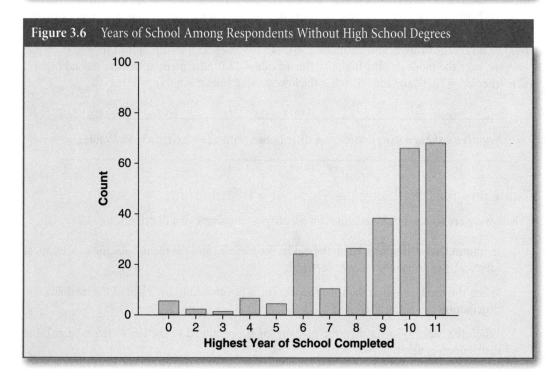

Figure 3.6 Years of School Among Respondents Without High School Degrees

▣ CONSIDERATIONS FOR CHOOSING A MEASURE OF CENTRAL TENDENCY

So far, we have considered three basic kinds of averages: the mode, the median, and the mean. Each can represent the central tendency of a distribution. But which one should we use? There is no simple answer to this question. However, in general, we tend to use only one of the three measures of central tendency, and the choice of the appropriate one involves a number of considerations. These considerations and how they affect our choice of the appropriate measure are presented in the form of a decision tree in Figure 3.7.

Level of Measurement

One of the most basic considerations in choosing a measure of central tendency is the variable's level of measurement. The valid use of any of the three measures requires that the data be measured at the level appropriate for that measure or higher. Thus, as shown in Figure 3.7, with nominal variables our choice is restricted to the mode as a measure of central tendency.

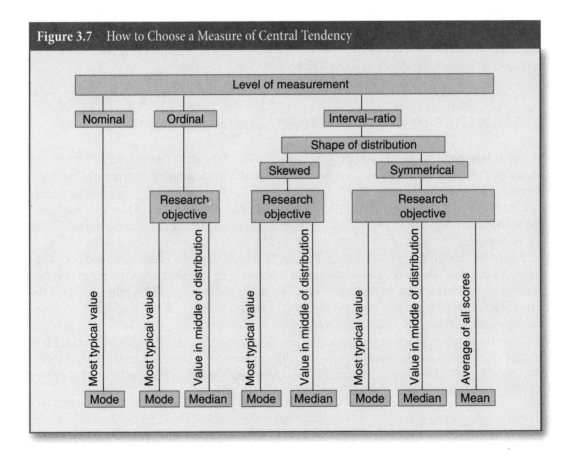

Figure 3.7 How to Choose a Measure of Central Tendency

However, with ordinal data, we have a choice: the mode or the median (or sometimes both). Our choice depends on what we want to know about the distribution. If we are interested in showing what is the most common or typical value in the distribution, then our choice is the mode. If, however, we want to show which value is located exactly in the middle of the distribution, then the median is our measure of choice.

When the data are measured on an interval-ratio level, the choice between the appropriate measures is a bit more complex and is restricted by the shape of the distribution.

Skewed Distribution

When the distribution is skewed, the mean may give misleading information on the central tendency because its value is affected by extreme scores in the distribution. The median or the mode can be chosen as the preferred measure of central tendency because neither is influenced by extreme scores. Thus, either one could be used as an "average," depending on the research objective.

Symmetrical Distribution

When the distribution we want to analyze is symmetrical, we can use any of the three averages. Again, our choice depends on the research objective and what we want to know about the distribution. In general, however, the mean is our best choice because it contains the greatest amount of information and is easier to use in more advanced statistical analyses.

▣ STATISTICS IN PRACTICE: REPRESENTING INCOME

Personal income is frequently positively skewed because there are a few people with very high incomes; therefore, the mean may not be the most appropriate measure to represent "average" income. For example, the 2011 American Community Survey—an ongoing survey of economic and income statistics—reported the 2011 mean and median annual earnings of white, black, and Latino households in the United States. In Figure 3.8, we compare the mean and median income for each group.

As shown, for all groups, the reported mean is higher than the median. This discrepancy indicates that household income in the United States is highly skewed, with the mean overrepresenting those households in the upper-income bracket and misrepresenting the income of the average household. A preferable alternative (also shown) is to use the median annual earnings of these groups.

Since the earnings of whites are the highest in comparison with all other groups, it is useful to look at each group's median earnings relative to the earnings of whites. For example, blacks were paid just 58 cents ($32,229/$55,412) and Latinos were paid 70 cents ($38,624/$55,412) for every $1 paid to whites.

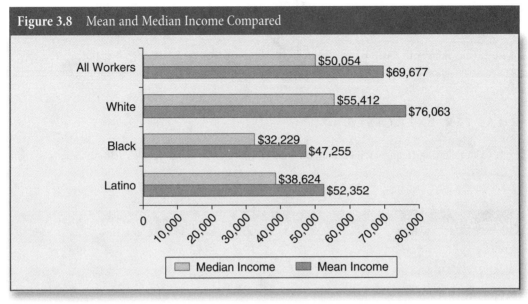

Figure 3.8 Mean and Median Income Compared

Source: U.S. Census Bureau, American Community Survey, 2011.

MAIN POINTS

• The mode, the median, and the mean are measures of central tendency—numbers that describe what is average or typical about the distribution.

• The mode is the category or score with the largest frequency (or percentage) in the distribution. It is often used to describe the most commonly occurring category of a nominal level variable.

• The median is a measure of central tendency that represents the exact middle of the distribution. It is calculated for variables measured on at least an ordinal level of measurement.

• The mean is typically used to describe central tendency in interval-ratio variables, such as income, age, or education. We obtain the mean by summing all the scores and dividing by the total (N) number of scores.

• In a symmetrical distribution, the frequencies at the right and left tails of the distribution are identical. In skewed distributions, there are either a few extremely high (positive skew) or a few extremely low (negative skew) values.

KEY TERMS

mean
measures of
   central tendency
median
mode

negatively skewed
   distribution
percentile
positively skewed
   distribution

skewed distribution
symmetrical
   distribution

⑤SAGE edge™

Sharpen your skills with SAGE edge at **edge.sagepub.com/ssdsess2e**. **SAGE edge for students** provides a personalized approach to help you accomplish your coursework goals in an easy-to-use learning environment.

CHAPTER EXERCISES

1. The following frequency distribution contains information about people's self-evaluations of their lives.

Respondent Assessment of Life	Frequency	Percentage	Cumulative Percentage
Exciting	470	48.3	48.3
Routine	444	45.6	93.9
Dull	60	6.1	100.0
Total	974	100.0	

Source: GSS, 2010.

    a. Find the mode.
    b. Find the median.
    c. Interpret the mode and the median.
    d. Why would you not want to report the mean for this variable?

2. Same-sex unions have increasingly become a heated political issue. The 2010 GSS asked respondents' opinions on homosexual relations. Four response categories ranged from "always wrong" to "not wrong at all." See the following frequency distribution:

Homosexual Relations	Frequency	Percentage	Cumulative Percentage
Always wrong	467	50.2	50.2
Almost always wrong	41	4.4	54.6
Sometimes wrong	76	8.2	62.8
Not wrong at all	346	37.2	100.0
Total	930	100.0	

    a. At what level is this variable measured? What is the mode for this variable?

    b. Calculate the median for this variable. In general, how would you characterize the public's attitude about homosexual relations?

3. The following frequency distribution contains information on the number of hours worked last week for a sample of 32 Latino adults.

Hours Worked Last Week	Frequency	Percentage	Cumulative Percentage
20	3	9.4	9.4
25	2	6.3	15.6
28	1	3.1	18.8
29	1	3.1	21.9
30	3	9.4	31.3
32	1	3.1	34.4
40	14	43.8	78.1
50	2	6.3	84.4
52	1	3.1	87.5
55	1	3.1	90.6
60	1	3.1	93.8
64	1	3.1	96.9
70	1	3.1	100.0
Total	32	100.0	

Source: GSS, 2010.

    a. What is the level of measurement, mode, and median for "hours worked last week"?

    b. Construct quartiles for weeks worked last year. What is the 25th percentile? The 50th percentile? The 75th percentile? Why don't you need to calculate the 50th percentile to answer this question?

4. Using data from the 2010 GSS, the following is the frequency distribution for respondent opinion of whether racism is no longer present in the United States:

Exercises

Racism in the Past	Frequency
Strongly agree	94
Somewhat agree	142
Somewhat disagree	160
Strongly disagree	84

    a. Calculate the median category for this variable.

    b. Also, report which category contains the 20th and 80th percentiles.

5. Does health status vary with age? The following table, taken from the 2010 GSS, depicts the health status across various age-groups (not all ages are displayed).

| Health Status | Age-Group (years) | | | |
	18–29	30–39	40–49	50–59
Excellent	56	55	41	38
Good	71	77	87	74
Fair	34	34	34	44
Poor	3	3	6	17

    a. Calculate the median and mode for each age-group.

    b. Use this information to characterize whether health status varies by age. Does the median or mode provide a better description of the data? Do the statistics support the idea that some age-groups are healthier than others?

6. The number of Americans on Medicare is increasing as expected with the aging baby boomer population. The following table shows the number of Americans on Medicare in 2005 and 2009 for eight U.S. states. *Note:* The numbers listed are in thousands.

State	2005	2009
Alabama	740	828
Delaware	125	145
Florida	3,008	3,289
Illinois	1,674	1,806
Minnesota	691	767
New Hampshire	185	217
New York	2,758	2,937
Washington	807	938

Source: U.S. Census Bureau, *The 2012 Statistical Abstract*, Table 147.

Exercises

a. Calculate the mean number of Americans on Medicare in these eight states for both 2005 and 2009. How would you characterize the difference in the number of Americans on Medicare between 2005 and 2009? Does the mean adequately represent the central tendency of the distribution of Americans on Medicare in each year for these eight states? Why or why not?

b. Recalculate the mean for each year after removing Florida, Illinois, and New York from the table. Is the mean now a better representation of central tendency for the remaining five states? Explain.

7. U.S. households have become smaller over the years. The following table from the 2010 GSS contains information on the number of people currently aged 18 years or older living in a respondent's household. Calculate the mean number of people living in a U.S. household in 2010.

Household Size	Frequency
1	381
2	526
3	227
4	200
5	96
6	42
7	19
8	5
9	2
10	2
Total	1,500

8. In Exercise 6, you calculated the mean number of Americans on Medicare. We now want to test whether the distribution of Americans on Medicare is symmetrical or skewed.
   a. Calculate the median and mode for each year, using all eight states. Based on these results and the means, how would you characterize the distribution of Americans on Medicare for each year?
   b. Does the mean or median best represent the central tendency of each distribution? Why?
   c. If you found the distributions to be skewed, what might be the statistical cause?

9. In Exercise 7, you examined U.S. household size in 2010. Using these data, construct a histogram to represent the distribution of household size.
   a. From the appearance of the histogram, would you say the distribution is positively or negatively skewed? Why?
   b. Now calculate the median for the distribution and compare this value with the value of the mean from Exercise 7. Do these numbers provide further evidence to support your decision about how the distribution is skewed? Why do you think the distribution of household size is asymmetrical?

10. Exercise 3 used GSS data on the number of hours worked per week for a sample of 32 Latino adults.
    a. Calculate the mean number of hours worked per week.
    b. Compare the value of the mean with those you have already calculated for the median. Without constructing a histogram, describe whether and how the distribution of weeks worked per year is skewed.

Exercises

11. You listen to a debate between two politicians discussing the economic health of the United States. One politician says that the average income of all workers in the United States is $72,235; the other says that American workers make, on average, only $52,029, so Americans are not as well off as the first politician claims. Is it possible for both these politicians to be correct? If so, explain how.

12. Do male murder rates vary with country population? Investigate this question using the following data for selected countries grouped by population size, the top 10 countries and the bottom 10. Calculate the mean and median for each group of countries. Where is the murder rate for men the highest? Do the mean and median have the same pattern for the two groups?

2008–2010 Male Murder Rate per 100,000			
Top 10 by Population	*Murder Rate*	*Bottom 10 by Population*	*Murder Rate*
China	2.2	Vietnam	2.6
India	3.9	Egypt	2.2
United States	6.6	Germany	0.9
Indonesia	13.9	Turkey	8.6
Brazil	54.7	Iran	2.3
Pakistan	4.3	Democratic Republic of the Congo	35.8
Nigeria	18.2	France	1.9
Russia	29.1	United Kingdom	1.7
Japan	0.4	Italy	1.6
Mexico	23.0	South Korea	2.2

Source: United Nations Office on Drugs and Crime, *2011 Annual Report.*

13. Many policymakers are interested in different measures of participation in the labor force. Use the information from the 2010 GSS to explore how labor force participation varies by sex (not all categories shown; some collapsed).

Participation in Labor Force	Men	Women
Working	390	441
Not working/unemployed	66	42
Keeping house	18	163
Total	474	646

Calculate the appropriate measures of central tendency for men and women. How would you describe their labor force participation rates? Are they similar or different and why? Use the appropriate measures of central tendency that you calculated to support your answer.

14. Infant mortality rates are a key indicator of the quality of health care a country can provide for its citizens. The following table shows infant mortality rates for selected countries based on estimates reported in the Central Intelligence Agency's *World Factbook 2010*.

Country	Infant Mortality Rate
Afghanistan	121.63
Canada	4.85
Colombia	15.92
Finland	3.40
Luxembourg	4.39
Panama	11.32
Rwanda	62.51
Syria	15.12
Turkey	23.07
United States	6.00
Zimbabwe	28.23

a. Calculate the mean and median infant mortalities reported for the 11 countries presented above.
b. Is this distribution skewed? If yes, how do you know?
c. Provide two possible explanations that might explain why some countries have very low infant mortality rates while others have very high infant mortality rates.

15. Provided below is SPSS output from the 2010 GSS for whether white and black respondents agree with the statement that racism is in the past.

RACISM IS IN THE PAST

WHAT IS RS RACE 1ST MENTION			Frequency	Percent	Valid Percent	Cumulative Percent
WHITE	Valid	Strongly agree	74	6.5	20.3	20.3
		Somewhat Agree	119	10.4	32.6	52.9
		Somewhat Disagree	119	10.4	32.6	85.5
		Strongly disagree	53	4.6	14.5	100.0
		Total	365	31.9	100.0	
	Missing	IAP	770	67.3		
		DON' T KNOW	4	.3		
		NA	5	.4		
		Total	779	68.1		
	Total		1144	100.0		
BLACK OR AFRICAN AMERICAN	Valid	Strongly agree	11	5.1	14.5	14.5
		Somewhat Agree	9	4.2	11.8	26.3
		Somewhat Disagree	30	14.0	39.5	65.8
		Strongly disagree	26	12.1	34.2	100.0
		Total	76	35.3	100.0	
	Missing	IAP	138	64.2		
		DON' T KNOW	1	.5		
		Total	139	64.7		
	Total		215	100.0		

    a. What is the level of measurement of this variable?

    b. Find all appropriate measures of central tendency for both whites and blacks. (You can ignore the "missing" responses.)

    c. Do whites and blacks differ in whether they consider racism to be in the past? Support your answer by citing appropriate measures of central tendency.

16. Provided below is SPSS output from the 2010 GSS measuring respondents' ideal number of children.

IDEAL NUMBER OF CHILDREN

		Frequency	Percent	Valid Percent	Cumulative Percent
Valid	0	8	.5	.9	.9
	1	29	1.9	3.4	4.3
	2	477	31.8	55.1	59.4
	3	238	15.9	27.5	86.9
	4	83	5.5	9.6	96.5
	5	18	1.2	2.1	98.6
	6	9	.6	1.0	99.7
	7	3	.2	.3	100.0
	Total	865	57.7	100.0	
Missing	IAP	481	32.1		
	DK,NA	37	2.5		
	System	117	7.8		
	Total	635	42.3		
Total		1500	100.0		

    a. What category contains the 90th percentile?

    b. What is the median?

    c. What is the mean? Ignore the "missing" cases in your calculation.

    d. Given your findings, is the distribution symmetrical, positively skewed, or negatively skewed? What do you think accounts for this?

Chapter 4

Measures of Variability

Chapter Learning Objectives

❖ Understanding the importance of measuring variability
❖ Learning how to calculate and interpret the range, interquartile range, variance, and standard deviation
❖ Understanding the criteria for choosing a measure of variation

I n the previous chapter, we looked at measures of central tendency: the mean, the median, and the mode. With these measures, we can use a single number to describe what is average for or typical of a distribution. Although measures of central tendency can be very helpful, they tell only part of the story. In fact, when used alone, they may mislead rather than inform. Another way of summarizing a distribution of data is by selecting a single number that describes how much variation and diversity there is in the distribution. Numbers that describe diversity or variation are called measures of variability. Researchers often use measures of central tendency along with **measures of variability** to describe their data.

Measures of variability   Numbers that describe diversity or variability in the distribution of a variable.

In this chapter, we discuss four measures of variability: the range, the interquartile range, the standard deviation, and the variance. Before we discuss these measures, let's explore why they are important.

◙ THE IMPORTANCE OF MEASURING VARIABILITY

The importance of looking at variation and diversity can be illustrated by thinking about the differences in the experiences of U.S. women. Are women united by their similarities or divided by their differences? The answer is *both*. To address the similarities without dealing with differences is "to misunderstand and distort that which separates as well as that which binds women together."[1] Even when we focus on one particular group of women, it is important to look at the differences as well as the commonalities. Take, for example, Asian American women. As a group, they share a number of characteristics.

> Their participation in the workforce is higher than that of women in any other ethnic group. Many . . . live life supporting others, often allowing their lives to be subsumed by the needs of the extended family. . . . However, there are many circumstances when these shared experiences are not sufficient to accurately describe the condition of a particular Asian-American woman. Among Asian-American women there are those who were born in the United States . . . and . . . those who recently arrived in the United States. Asian-American women are diverse in their heritage or country of origin: China, Japan, the Philippines, Korea . . . and . . . India. . . . Although the majority of Asian-American women are working class—contrary to the stereotype of the "ever successful" Asians—there are poor, "middle-class," and even affluent Asian-American women.[2]

As this example illustrates, one basis of stereotyping is treating a group as if it were totally represented by its central value, ignoring the diversity within the group. Sociologists often contribute to this type of stereotyping when their empirical generalizations, based on a statistical difference between averages, are interpreted in an overly simplistic way. All this argues for the importance of using measures of variability as well as central tendency whenever we want to characterize or compare groups. Whereas the similarities and commonalities in the experiences of Asian American women are depicted by a measure of central tendency, the diversity of their experiences can be described only by using measures of variation.

The concept of variability has implications not only for describing the diversity of social groups such as Asian American women but also for issues that are important in your everyday life. One of the most important issues facing the academic community is how to reconstruct the curriculum to make it more responsive to the needs of students. Let's consider the issue of statistics instruction at the college level.

Let's suppose that a university committee is examining the issue of how to better respond to the needs of students. In its attempt to evaluate statistics courses offered in different departments, the committee compares the grading policies in two courses. The first, offered in the sociology department, is taught by Professor Brown; the second, offered through the school of social work, is taught by Professor Yamato. The committee finds that over the years, the average grade for Professor Brown's class has been C+. The average grade in Professor Yamato's class is also C+. We could easily be misled by these statistics into thinking that the grading policies of both instructors are about the same. However, we need to look more closely into how the grades are distributed in each of the classes. The differences in the distribution of grades are illustrated in Figure 4.1, which displays the frequency polygon for the two classes.

Figure 4.1 Distribution of Grades for Professors Brown's and Yamato's Statistics Classes

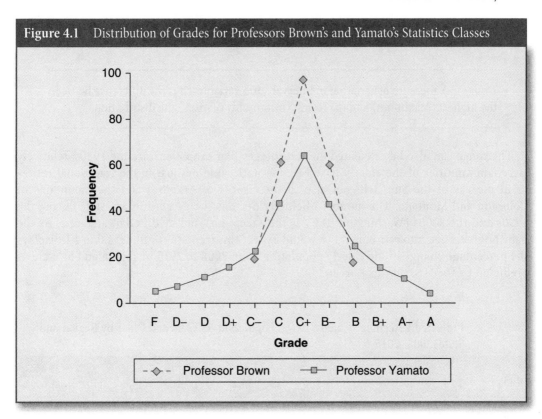

Compare the shapes of these two distributions. Notice that while both distributions have the same mean, they are shaped very differently. The grades in Professor Yamato's class are more spread out, ranging from A to F, whereas the grades for Professor Brown's class are clustered around the mean and range only from B to C. Although the means for both distributions are identical, the grades in Professor Yamato's class vary considerably more than the grades given by Professor Brown. The comparison between the two classes is more complex than we first thought it would be.

As this example demonstrates, information on how scores are spread from the center of a distribution is as important as information about the central tendency in a distribution. This type of information is obtained by measures of variability.

▣ THE RANGE

The simplest and most straightforward measure of variation is the **range**, which measures variation in interval-ratio variables. It is the difference between the highest (maximum) and the lowest (minimum) scores in the distribution:

$$\text{Range} = \text{Highest score} - \text{Lowest score}$$

In the 2010 General Social Survey (GSS), the oldest person included in the study was 89 years old and the youngest was 18. Thus, the range was 89 − 18 = 71 years.

Range   A measure of variation in interval-ratio variables. It is the difference between the highest (maximum) and the lowest (minimum) scores in the distribution.

The range can also be calculated on percentages. For example, since the 1980s, relatively large communities of the elderly have become noticeable not just in the traditional retirement meccas of the Sun Belt[3] but also in the Ozarks of Arkansas and the mountains of Colorado and Montana. The number of elderly persons increased in every state during the 1990s and the 2000s (Washington, D.C., is the exception), but by different amounts. As the baby boomers age into retirement, we would expect this trend to continue. Table 4.1 displays the percentage change in the elderly population from 2008 to 2015 by region and by state as predicted by the U.S. Census Bureau.[4]

Table 4.1   Projected Percentage Change in the Population 65 Years and Older by Region and State, 2008–2015

Region, Division, and State	Percentage Change	Region, Division, and State	Percentage Change
Northeast	16.0	Missouri	14.5
		Nebraska	12.4
Connecticut	20.9	North Dakota	12.6
Delaware	22.3	Ohio	12.4
Maine	25.6	South Dakota	9.4
Massachusetts	17.7	Wisconsin	17.6
New Hampshire	28.4		
New Jersey	20.4		
New York	12.8	South	22.8
Pennsylvania	12.5	Alabama	15.1
Rhode Island	18.2	Arkansas	14.7
Vermont	31.4	Florida	29.7
		Georgia	21.1
Midwest	14.0	Kentucky	12.5
		Louisiana	23.0
Indiana	11.3	Maryland	23.1
Illinois	12.9	Mississippi	16.4
Iowa	11.2	North Carolina	20.7
Kansas	14.2	Oklahoma	12.8
Michigan	15.5	South Carolina	22.1
Minnesota	19.2		

Region, Division, and State	Percentage Change	Region, Division, and State	Percentage Change
South (cont.)		California	27.1
		Colorado	22.7
Tennessee	18.2	Hawaii	18.8
Texas	25.9	Idaho	20.2
Virginia	26.8	Montana	27.0
Washington, D.C.	−14.1	Nevada	42.1
West Virginia	15.4	New Mexico	31.9
		Oregon	17.1
West	27.0	Utah	13.8
		Washington	23.2
Alaska	50.0	Wyoming	36.9
Arizona	36.8		

Source: U.S. Census Bureau, *Statistical Abstract of the United States: 2010*, Tables 16 and 18.

To find the ranges in a distribution, simply pick out the highest and the lowest scores in the distribution and subtract. Alaska has the highest percentage change, with 50%, and Washington, D.C., has the lowest change, with −14.1%. The range is 64.1 percentage points, or 50% to −14.1%.

Although the range is simple and quick to calculate, it is a rather crude measure because it is based on only the lowest and the highest scores. These two scores might be extreme and rather atypical, which might make the range a misleading indicator of the variation in the distribution. For instance, note that among the 50 states and Washington, D.C., listed in Table 4.1, no state has a percentage decrease as large that of Washington, D.C., and only Nevada has a percentage increase nearly as high as Alaska's. The range of 64.1 percentage points does not give us information about the variation in states between Washington, D.C., and Alaska.

✓ **Learning Check**

Why can't we use the range to describe diversity in nominal variables? The range can be used to describe diversity in ordinal variables (e.g., we can say that responses to a question ranged from "somewhat satisfied" to "very dissatisfied"), but it has no quantitative meaning. Why not?

▣ THE INTERQUARTILE RANGE: INCREASES IN ELDERLY POPULATIONS

To remedy the limitation of the range, we can employ an alternative—the *interquartile range*. The **interquartile range (IQR)**, a measure of variation for interval-ratio variables, is the width of the middle 50% of the distribution. It is defined as the difference between the lower and upper quartiles (Q_1 and Q_3).

$$IQR = Q_3 - Q_1$$

Recall that the first quartile (Q_1) is the 25th percentile, the point at which 25% of the cases fall below it and 75% above it. The third quartile (Q_3) is the 75th percentile, the point at which 75% of the cases fall below it and 25% above it. The IQR, therefore, defines variation for the middle 50% of the cases.

Interquartile range (IQR)   The width of the middle 50% of the distribution. It is defined as the difference between the lower and upper quartiles (Q_1 and Q_3).

Like the range, the IQR is based on only two scores. However, because it is based on intermediate scores, rather than on the extreme scores in the distribution, it avoids some of the instability associated with the range.

These are the steps for calculating the IQR:

1. To find Q_1 and Q_3, order the scores in the distribution from the highest to the lowest score, or vice versa. Table 4.2 presents the data of Table 4.1 arranged in order from Alaska (50.0%) to Washington, D.C. (−14.1%).

Table 4.2   Projected Percentage Change in the Population 65 Years and Older, 2008–2015, by State, Ordered From the Highest to the Lowest

State	Percentage Change	State	Percentage Change	State	Percentage Change
Alaska	50.0	Delaware	22.3	Alabama	15.1
Nevada	42.1	South Carolina	22.1	Arkansas	14.7
Wyoming	36.9	Georgia	21.1	Missouri	14.5
Arizona	36.8	Connecticut	20.9	Kansas	14.2
New Mexico	31.9	North Carolina	20.7	Utah	13.8
Vermont	31.4	New Jersey	20.4	Illinois	12.9
Florida	29.7	Idaho	20.2	New York	12.8
New Hampshire	28.4	Minnesota	19.2	Oklahoma	12.8
California	27.1	Hawaii	18.8	North Dakota	12.6
Montana	27.0	Rhode Island	18.2	Kentucky	12.5
Virginia	26.8	Tennessee	18.2	Pennsylvania	12.5
Texas	25.9	Massachusetts	17.7	Nebraska	12.4

State	Percentage Change	State	Percentage Change	State	Percentage Change
Maine	25.6	Wisconsin	17.6	Ohio	12.4
Washington	23.2	Oregon	17.1	Indiana	11.3
Maryland	23.1	Mississippi	16.4	Iowa	11.2
Louisiana	23.0	Michigan	15.5	South Dakota	9.4
Colorado	22.7	West Virginia	15.4	Washington, D.C.	−14.1

Source: U.S. Census Bureau, *Statistical Abstract of the United States: 2010*, Tables 16 and 18.

2. Next, we need to identify the first quartile, Q_1, or the 25th percentile. We have to identify the percentage increase in the elderly population associated with the state that divides the distribution so that 25% of the states are below it. To find Q_1, we multiply N by 0.25:

$$(N)(0.25) = (51)(0.25) = 12.75$$

The first quartile falls between the 12th and the 13th states. Counting from the bottom, the 12th state is Illinois, and the percentage increase associated with it is 12.9. The 13th state is Utah, with a percentage increase of 13.8. To find the first quartile, we take the average of 12.9 and 13.8. Therefore, $(12.9 + 13.8)/2 = 13.35$ is the first quartile (Q_1).

3. To find Q_3, we have to identify the state that divides the distribution in such a way that 75% of the states are below it. We multiply N this time by 0.75:

$$(N)(0.75) = (51)(0.75) = 38.25$$

The third quartile falls between the 38th and the 39th states. Counting from the bottom, the 38th state is Washington, and the percentage increase associated with it is 23.2. The 39th state is Maine, with a percentage increase of 25.6. To find the third quartile, we take the average of 23.2 and 25.6. Therefore, $(23.2 + 25.6)/2 = 24.4$ is the third quartile (Q_3).

4. We are now ready to find the IQR:

$$IQR = Q_3 - Q_1 = 24.4 - 13.35 = 11.05$$

The IQR of percentage increase in the elderly population is 11.05 percentage points.

Figure 4.2 The Range Versus the Interquartile Range: Number of Children Among Two Groups of Women

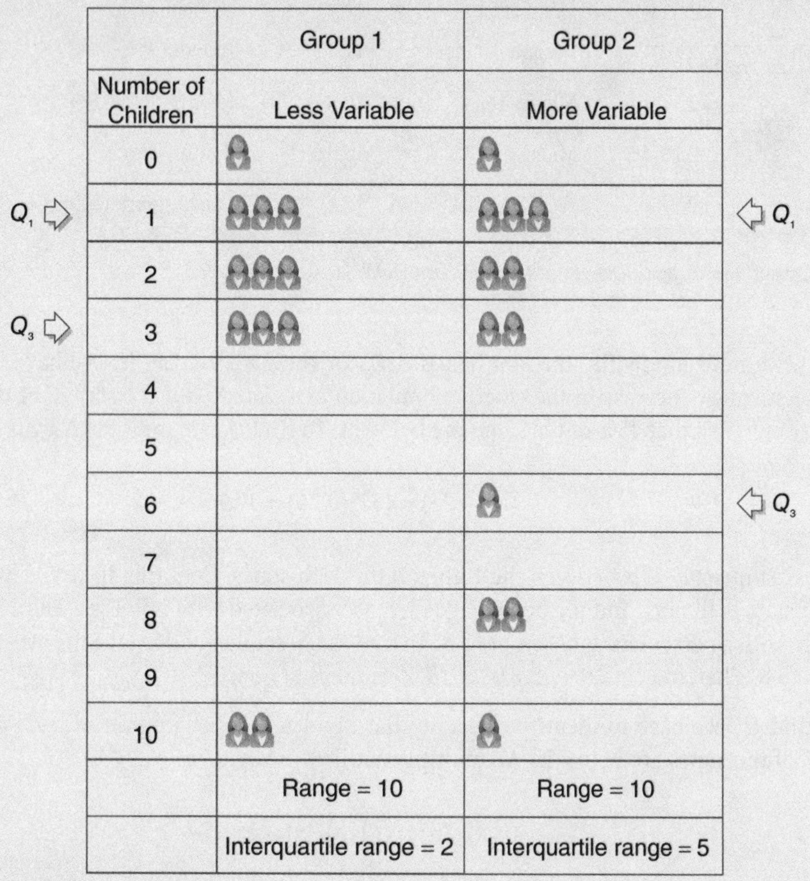

Number of Children	Group 1 Less Variable	Group 2 More Variable
0		
1		
2		
3		
4		
5		
6		
7		
8		
9		
10		
	Range = 10	Range = 10
	Interquartile range = 2	Interquartile range = 5

Notice that the IQR gives us better information than the range. The range gave us a 64.1-point spread, from 50% to −14.1%, but the IQR tells us that half the states are clustered between 24.4 and 13.35—a much narrower spread. The extreme scores represented by Alaska and Washington, D.C., have no effect on the IQR because they fall at the extreme ends of the distribution. This difference between the range and the interquartile range is also illustrated in Figure 4.2, in which the extreme values for 10 children affect the value of the range but not of the IQR.

✓ *Learning Check*

Why is the IQR better than the range as a measure of variability, especially when there are extreme scores in the distribution? To answer this question, you may want to examine Figure 4.2.

▣ THE VARIANCE AND THE STANDARD DEVIATION: CHANGES IN THE ELDERLY POPULATION

As of 2010, the elderly population in the United States was 13 times as large as in 1900, and it is projected to continue to increase.[5] The pace and direction of these demographic changes will create compelling social, economic, and ethical choices for individuals, families, and governments.

Table 4.3 presents the projected percentage change in the elderly population for all regions of the United States.

Table 4.3 Projected Percentage Change in the Elderly Population by Region, 2008–2015

Region	Percentage
Northeast	16.0
South	22.8
Midwest	14.0
West	27.0
Mean ($\bar{Y}$)	19.95

Source: U.S. Census Bureau, *Statistical Abstract of the United States: 2010*, Tables 16 and 18.

Table 4.3 shows that between 2008 and 2015, the size of the elderly population in the United States is projected to increase by an average of 19.95%. But this average increase does not inform us about the regional variation in the elderly population. For example, will the northeastern states show a smaller-than-average increase because of the out-migration of the elderly population to the warmer climate of the Sun Belt states? Is the projected increase higher in the South because of the immigration of the elderly?

Although it is important to know the average projected percentage increase for the nation as a whole, you may also want to know whether regional increases might differ from the national average. If the regional projected increases are close to the national average, the figures will cluster around the mean, but if the regional increases deviate much from the national average, they will be widely dispersed around the mean.

Table 4.3 suggests that there is considerable regional variation. The percentage change ranges from 27.0% in the West to 14.0% in the Midwest, so the range is 13.0% (27.0% − 14.0% = 13.0%). Moreover, most of the regions are projected to deviate considerably from the national average of 19.95%. How large are these deviations on average? We want a measure that will give us information about the overall variation among all regions in the United States and, unlike the range or the IQR, will not be based on only two scores.

Such a measure will reflect how much, on average, each score in the distribution deviates from some central point, such as the mean. We use the mean as the reference point rather than other kinds of averages (the mode or the median) because the mean is based on all the scores in the distribution. Therefore, it is more useful as a basis from which to calculate average deviation. The sensitivity of the mean to extreme values carries over the calculation of the average deviation, which is based on the mean. Another reason for using the mean as a reference point is that more advanced measures of variation require the use of algebraic properties that can be assumed only by using the arithmetic mean.

The *variance* and the *standard deviation* are two closely related measures of variation that increase or decrease based on how closely the scores cluster around the mean. The **variance** is the average of the squared deviations from the center (mean) of the distribution, and the **standard deviation** is the square root of the variance. Both measure variability in interval-ratio variables.

Variance   A measure of variation for interval-ratio variables; it is the average of the squared deviations from the mean.

Standard deviation   A measure of variation for interval-ratio variables; it is equal to the square root of the variance.

Calculating the Deviation From the Mean

Consider again the distribution of the percentage change in the elderly population for the four regions of the United States. Because we want to calculate the average difference of all the regions from the national average (the mean), it makes sense to first look at the difference between each region and the mean. This difference, called a deviation from the mean, is symbolized as $(Y - \overline{Y})$. The sum of these deviations can be symbolized as $\Sigma(Y - \overline{Y})$.

The calculations of these deviations for each region are displayed in Table 4.4 and Figure 4.3. We have also summed these deviations. Note that each region has either a positive or a negative deviation score. The deviation is positive when the percentage change in the elderly home population is above the mean. It is negative when the percentage change is below the mean. Thus, for example, the Northeast's deviation score of -3.95 means that its percentage change in the elderly population was 3.95 percentage points below the mean.

You may wonder if we could calculate the average of these deviations by simply adding up the deviations and dividing them. Unfortunately we cannot, because the sum of the deviations of scores from the mean is always zero, or algebraically $\Sigma(Y - \overline{Y})$. In other words, if we were to subtract the mean from each score and then add up all the deviations as we did in Table 4.4, the sum would be zero, which in turn would cause the average deviation (i.e., average difference) to compute to zero. This is always true because the mean is the center of gravity of the distribution.

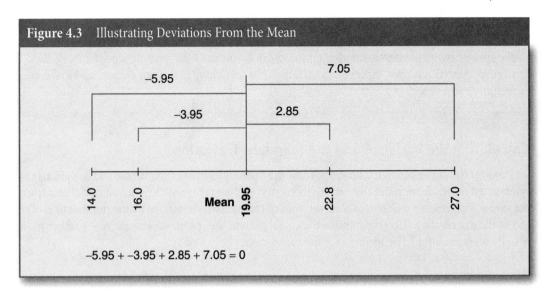

Figure 4.3   Illustrating Deviations From the Mean

$$-5.95 + -3.95 + 2.85 + 7.05 = 0$$

Table 4.4   Projected Percentage Change in the Elderly Population, 2008–2015, by Region, Deviation From the Mean, and Deviation From the Mean Squared

Region	*Percentage*	$(Y - \bar{Y})$	$(Y - \bar{Y})^2$
Northeast	16.0	16.0 – 19.95 = –3.95	15.60
South	22.8	22.8 – 19.95 = 2.85	8.12
Midwest	14.0	14.0 – 19.95 = –5.95	35.40
West	27.0	27.0 – 19.95 = 7.05	49.70
	$\Sigma(Y) = 79.8$	$\Sigma(Y-\bar{Y}) = 0$	$\Sigma(Y-\bar{Y})^2 = 108.82$

$$\text{Mean} = \bar{Y} = \frac{\Sigma Y}{N} = \frac{79.8}{4} = 19.95$$

Mathematically, we can overcome this problem by squaring the deviations—that is, multiplying each deviation by itself to get rid of the negative sign.

The final column of Table 4.4 squares the actual deviations from the mean and adds together the squares. The sum of the squared deviations is symbolized as $\Sigma(Y-\bar{Y})^2$. Note that by squaring the deviations, we end up with a sum representing the deviation from the mean, which is positive. (Note that this sum will equal zero if all the cases have the same value as the mean.) In our example, this sum is $\Sigma(Y-\bar{Y})^2 = 108.82$.

Examine Table 4.4 again and note the disproportionate contribution of the western region to the sum of the squared deviations from the mean (it actually accounts for about 45% of the sum of squares). Can you explain why? (Hint: It has something to do with the sensitivity of the mean to extreme values.)

Calculating the Variance and the Standard Deviation

The average of the squared deviations from the mean is known as the *variance*. The variance is symbolized as S_Y^2. Remember that we are interested in the *average* of the squared deviations from the mean. Therefore, we need to divide the sum of the squared deviations by the number of scores (*N*) in the distribution. However, unlike the calculation of the mean, we will use $N-1$ rather than N in the denominator.[6] The formula for the variance can be stated as

$$S_Y^2 = \frac{\Sigma\left(Y - \overline{Y}\right)^2}{N-1} \tag{4.1}$$

where

S_Y^2 = the variance

$(Y - \overline{Y})$ = the deviation from the mean

$\Sigma(Y - \overline{Y})^2$ = the sum of the squared deviations from the mean

N = the number of scores

Note that the formula incorporates all the symbols we defined earlier. This formula means that the variance is equal to the average of the squared deviations from the mean.

Follow these steps to calculate the variance:

1. Calculate the mean, $\overline{Y} = \Sigma(Y)/N$.

2. Subtract the mean from each score to find the deviation, $Y - \overline{Y}$.

3. Square each deviation, $(Y - \overline{Y})^2$.

4. Sum the squared deviations, $\Sigma(Y - \overline{Y})^2$.

5. Divide the sum by $N-1$, $\Sigma(Y - \overline{Y})^2/(N-1)$.

6. The answer is the variance.

To assure yourself that you understand how to calculate the variance, go back to Table 4.4 and follow this step-by-step procedure for calculating the variance. Now plug the required quantities into Formula 4.1. Your result should look like this:

$$S_Y^2 = \frac{\Sigma(Y-\bar{Y})^2}{(N-1)} = \frac{108.82}{3} = 36.27$$

One problem with the variance is that it is based on squared deviations and therefore is no longer expressed in the original units of measurement. For instance, it is difficult to interpret the variance of 36.27, which represents the distribution of the percentage change in the elderly population, because this figure is expressed in squared percentages. Thus, we often take the square root of the variance and interpret it instead. This gives us the *standard deviation*, S_Y.

The standard deviation, symbolized as S_Y, is the square root of the variance, or

$$S_Y = \sqrt{S_Y^2}$$

The standard deviation for our example is

$$S_Y = \sqrt{S_Y^2} = \sqrt{36.27} = 6.02$$

The formula for the standard deviation uses the same symbols as the formula for the variance:

$$S_Y = \sqrt{\frac{\Sigma(Y-\bar{Y})^2}{(N-1)}} \tag{4.2}$$

As we interpret the formula, we can say that the standard deviation is equal to the square root of the average of the squared deviations from the mean.

The advantage of the standard deviation is that unlike the variance, it is measured in the same units as the original data. For instance, the standard deviation for our example is 6.02. Because the original data were expressed in percentages, this number is expressed as a percentage as well. In other words, you could say, "The standard deviation is 6.02%." But what does this mean? The actual number tells us very little by itself, but it allows us to evaluate the dispersion of the scores around the mean.

In a distribution where all the scores are identical, the standard deviation is zero. Zero is the lowest possible value for the standard deviation; in an identical distribution, all the points would be the same, with the same mean, mode, and median. There is no variation or dispersion in the scores.

The more the standard deviation departs from zero, the more variation there is in the distribution. There is no upper limit to the value of the standard deviation. In our example, we can conclude that a standard deviation of 6.02% means that the projected percentage change in the elderly population for the four regions of the United States is widely dispersed around the mean of 19.95%.

The standard deviation can be considered a standard against which we can evaluate the positioning of scores relative to the mean and to other scores in the distribution. As we will see in more detail in Chapter 5, in most distributions, unless they are highly skewed, about 34% of all scores fall between the mean and 1 standard deviation above the mean. Another 34% of scores fall between the mean and 1 standard deviation below it. Thus, we would expect the majority of scores (68%) to fall within 1 standard deviation of the mean.

For example, let's consider the gross domestic products (GDPs) of 70 countries. The SPSS output is shown below.

Descriptive Statistics

	N	Minimum	Maximum	Mean	Std. Deviation
Gross Domestic Product per capita	70	400	80700	19247.14	17628.277
Valid N (listwise)	70				

The columns tell us that there were 70 countries in our sample and that the minimum GDP per capita was $400 and the maximum was $80,700. This is quite a gap between the poorest and richest countries in our sample. The mean and standard deviation are listed in the final two columns.

The mean GDP per capita is $19,247.14, with a standard deviation of $17,628.28. We can expect about 68% of these countries to have GDP per capita values within a range of $1,618.86 ($19,247.14 − $17,628.28) to $36,875.42 ($19,274.14 + $17,628.28). Hence, based on the mean and the standard deviation, we have a pretty good indication of what would be considered a typical GDP per capita value for the majority of countries in our sample. For example, we would consider a country with a GDP per capita value of $80,700 to be extremely wealthy in comparison with other countries.

Another way to interpret the standard deviation is to compare it with another distribution. For instance, Table 4.5 displays the means and standard deviations of employee age for two samples drawn from a *Fortune* 100 corporation. Samples are divided into female clerical and female technical. Note that the mean ages for both samples are about the same—approximately 39 years. However, the standard deviations suggest that the distribution of age is dissimilar between the two groups. Figure 4.4 loosely illustrates this dissimilarity in the two distributions.

The relatively low standard deviation for female technical indicates that this group is relatively homogeneous in age. That is to say, most of the women's ages, while not identical, are fairly similar. The average deviation from the mean age of 39.87 is 3.75 years. In contrast, the standard deviation for female clerical employees is about twice the standard deviation for female technical. This suggests a wider dispersion or greater heterogeneity in the ages of clerical workers. We can say that the average deviation from the mean age of 39.46 is 7.80 years for clerical workers. The larger standard deviation indicates a wider dispersion of points below or above the mean. On average, clerical employees are further in age from their mean of 39.46.[7]

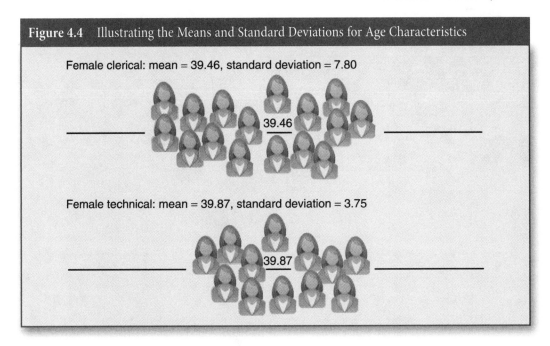

Figure 4.4   Illustrating the Means and Standard Deviations for Age Characteristics

Female clerical: mean = 39.46, standard deviation = 7.80

39.46

Female technical: mean = 39.87, standard deviation = 3.75

39.87

Table 4.5   Age Characteristics of Female Clerical and Technical Employees

Characteristic	Female Clerical (N = 22)	Female Technical (N = 39)
Mean age	39.46	39.87
Standard deviation	7.80	3.75

Source: Adapted from Marjorie Armstrong-Stassen, "The Effect of Gender and Organizational Level on How Survivors Appraise and Cope With Organizational Downsizing," *Journal of Applied Behavioral Science,* 34, no. 2 (June 1998): 125–142. Reprinted with permission.

▣ CONSIDERATIONS FOR CHOOSING A MEASURE OF VARIATION

So far, we have considered four measures of variation: (1) the range, (2) the IQR, (3) the variance, and (4) the standard deviation. Each measure can represent the degree of variability in a distribution. But which one should we use? There is no simple answer to this question. However, in general, we tend to use only one measure of variation, and the choice of the appropriate one involves a number of considerations. These considerations and how they affect our choice of the appropriate measure are presented in the form of a decision tree in Figure 4.5.

Figure 4.5　How to Choose a Measure of Variation

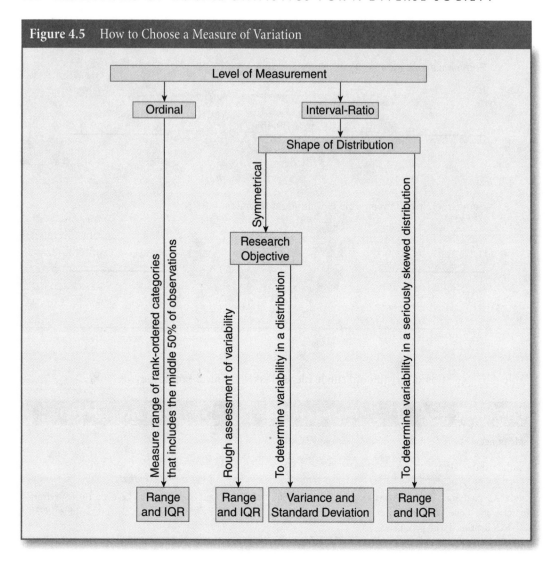

As in choosing a measure of central tendency, one of the most basic considerations in choosing a measure of variability is the variable's level of measurement. Valid use of any of the measures requires that the data be measured at the level appropriate for that measure or higher, as shown in Figure 4.5.

Ordinal level: You can use the IQR. However, the IQR relies on distance between two scores to express variation, information that cannot be obtained from ordinal-measured scores. The compromise is to use the IQR (reporting Q_1 and Q_3) alongside the median, interpreting the IQR as the range of rank-ordered values that includes the middle 50% of the observations.[8]

Interval-ratio level: For interval-ratio variables, you can choose the variance (or standard deviation), the range, or the IQR. Because the range, and to a lesser extent the IQR, is based on only two scores in the distribution (and therefore tends to be sensitive if either of the two points is extreme), the variance and/or standard deviation is usually preferred. However, if a distribution is extremely skewed so that the mean is no longer representative of the central tendency in the distribution, the range and the IQR can be used. The range and the IQR will also be useful when you are reading tables or quickly scanning data to get a rough idea of the extent of dispersion in the distribution.

▣ READING THE RESEARCH LITERATURE: DIFFERENCES IN COLLEGE ASPIRATIONS AND EXPECTATIONS AMONG LATINO ADOLESCENTS

In Chapter 2, we discussed how frequency distributions are presented in the professional literature. We noted that most statistical tables presented in the social science literature are considerably more complex than those we describe in this book. The same can be said about measures of central tendency and variation. Most research articles use measures of central tendency and variation in ways that go beyond describing the central tendency and variation of a single variable. In this section, we refer to both the mean and the standard deviation because in most research reports the standard deviation is given along with the mean.

Table 4.6 displays data taken from a research article published in *Social Problems*.[9] This table illustrates a common research application of the mean and standard deviation. The authors of this article examine how ethnicity plays into one's college aspirations and expectations. Their major focus is to explore "potential differences in college aspirations and expectations across the three largest Latino groups and the potential sources of such differences."[10] We focus only on their data for Cubans and Mexicans to present a simplified example of the mean and standard deviation.

Table 4.6 Ethnicity and College Aspirations and Expectations

	Cubans		Mexicans	
	Mean	*Standard Deviation*	*Mean*	*Standard Deviation*
How much respondent wants to go to college	4.50	1.80	4.20	1.30
How likely respondent will go to college	4.30	2.00	3.70	1.40

Source: Adapted from Stephanie A. Bohon, Monica Kirkpatrick Johnson, and Bridget K. Gorman, "College Aspirations and Expectations Among Latino Adolescents in the United States," *Social Problems* 53, no. 2 (2006): 207–225. Published by the University of California Press.

Note: The authors examine variation among other ethnicities in their paper. However, to simplify this example, we focus on Cubans and Mexicans.

Understanding the relationship between ethnicity and college aspirations and expectations is, nonetheless, critical in that the U.S. Latino population is growing faster than any other minority group, yet Latinos remain the least educated of all other people of color.

Data for this study come from the National Longitudinal Study of Adolescent Health survey. This survey is a representative sample of American adolescents in Grades 7 to 12. Complex sampling strategies are employed to ensure a representative sample. Thus, factors such as variation in geographic location, type of school, racial makeup, and so on are accounted for during data collection.

Respondents were asked a variety of questions, but the authors focused specifically on questions about college aspirations and expectations. Their measure of college aspirations is based on a scale of 1 to 5 derived from the question "How much do you want to go to college?" An answer of 1 indicated a low desire to go to college, while an answer of 5 indicated a high desire to go to college. Their measure of college expectations was also based on the same scale ranging from 1 (*low*) to 5 (*high*). However, it was derived from the question "How likely is it that you will go to college?"

The first thing we should look at is the means. Are they similar or different? For both the expectations and aspirations measure, we can see that Mexicans have slightly lower aspirations (4.20) and expectations (3.70) than Cubans (4.50 and 4.30, respectively). Furthermore, the standard deviations indicate that there is more variability in each of these measures for Cubans than for Mexicans.

The researchers of this study described the data displayed in Table 4.6 as follows:

They show strong aspirations for and expectations of college attendance across each of the five groups. Important differences across ethnic groups exist, however. As anticipated, Mexicans have weaker than average . . . and Cubans have stronger than average aspirations and expectations.[11]

Why might this be? The authors conclude their discussion of the data presented in Table 4.6 by arguing as follows:

Differential aspirations and expectations may be explained by the considerable differences in family and household characteristics, parental hopes for their child's educational success, and academic skills and disengagement.[12]

MAIN POINTS

- Measures of variability are numbers that describe how much variation or diversity there is in a distribution.

- The range measures variation in interval-ratio variables and is the difference between the highest (maximum) and the lowest (minimum) scores in the distribution. To find the range, subtract the lowest from the highest score in a distribution. For an ordinal variable, just report the lowest and the highest values without subtracting.

- The interquartile range (IQR) measures the width of the middle 50% of the distribution. It is defined as the difference between the lower and upper quartiles (Q_1 and Q_3). For an ordinal variable, just report Q_1 and Q_3 without subtracting.

- The variance and the standard deviation are two closely related measures of variation for interval-ratio variables that increase or decrease based on how closely the scores cluster around the mean. The variance is the average of the squared deviations from the center (mean) of the distribution; the standard deviation is the square root of the variance.

KEY TERMS

interquartile range
(IQR)

measures of variability
range

standard deviation
variance

$SAGE edge™

Sharpen your skills with SAGE edge at **edge.sagepub.com/ssdsess2e**. **SAGE edge for students** provides a personalized approach to help you accomplish your coursework goals in an easy-to-use learning environment.

CHAPTER EXERCISES

1. Public corruption continues to be a concern. Let's examine data from the U.S. Department of Justice to explore the variability in public corruption in 1990 and 2009. All the numbers below are of those convicted of public corruption.

Number of Public Corruption Convictions by Year

1990		2009	
Govt. Level	*No. of Convictions*	*Govt. Level*	*No. of Convictions*
Federal	583	Federal	426
State	79	State	102
Local	225	Local	257

Source: U.S. Census Bureau, *Statistical Abstract of the United States: 2012*, Table 338.

a. What is the range of convictions in 1990? In 2009? Which is greater?
b. What is the mean number of convictions in 1990 and 2009?
c. Calculate the standard deviation for 1990 and 2009.
d. Which year appears to have more variability in number of convictions as measured by the standard deviation? Are the results consistent with what you found using the range?

2. The output below depicts data for projected elderly population change in midwestern and western states between 2008 and 2015 from Table 4.1.

Descriptives

Region			Statistic
Population_Change	Midwest	Mean	13.600
		Std. Deviation	2.7831
		Minimum	9.4
		Maximum	19.2
		Range	9.8
		Interquartile Range	3.7
	West	Mean	28.277
		Std. Deviation	10.6948
		Minimum	13.8
		Maximum	50.0
		Range	36.2
		Interquartile Range	17.3

a. Compare the range for the western states with that of the Midwest. Which region had a greater range?

b. Examine the IQR for each region. Which is greater?

c. Compare the standard deviation for western states with that of the Midwest. Which area has a greater standard deviation?

d. Use the statistics to characterize the variability in population increase of the elderly in the two regions. Does one region have more variability than another? If yes, why do you think that is?

3. Occupational prestige is a statistic developed by sociologists to measure the status of one's occupation. Occupational prestige is also a component of what sociologists call socioeconomic status, a composite measure of one's status in society. On average, people with more education tend to have higher occupational prestige than people with less education. We investigate this using the 2010 GSS variable PRESTG80 and the Explore procedure to generate the selected SPSS output shown in Figure 4.6.

Figure 4.6 Descriptive Statistics for Occupational Prestige Score by Highest Degree Earned

PRESTG80			Statistic
RS OCCUPAIONAL PRESIGE SCORE (1980)	High School Diploma	Mean	40.59
		Median	40.00
		Std. Deviation	11.419
		Minimum	17
		Maximum	75
		Range	58
		Interquartile Range	17
	Bachelor's Degree	Mean	50.95
		Median	51.00
		Std. Deviation	12.930
		Minimum	23
		Maximum	75
		Range	52
		Interquartile Range	23

Exercises

a. Note that SPSS supplies the IQR, the median, and the minimum and maximum values of each group. Looking at the values of the mean and median, do you think the distribution of prestige is skewed for respondents with a high school diploma? For respondents with a bachelor's degree? Why or why not?

b. Explain why you think there is more variability of prestige for either group, or why the variability of prestige is similar for the two groups.

4. The Census Bureau collects information about divorce rates. The following table summarizes the divorce rate for 10 U.S. states in 2007. Use the table to answer the questions that follow.

State	Divorce Rate per 1,000 Population
Alaska	4.3
Florida	4.7
Idaho	4.9
Maine	4.5
Maryland	3.1
Nevada	6.5
New Jersey	3.0
Texas	3.3
Vermont	3.8
Wisconsin	2.9

Source: U.S. Census Bureau, *Statistical Abstract of the United States: 2010*, Table 126.

a. Calculate and interpret the range and the IQR. Which is a better measure of variability? Why?

b. Calculate and interpret the mean and standard deviation.

c. Identify two possible explanations for the variation in divorce rates across the 10 states.

5. The respondents of the 2007 Health Information National Trends Survey reported their psychological distress on a scale between 0 and 24. In the table below, you will see separate data on two groups of respondents' distress scores: those who have ever been diagnosed as having cancer, and those who have not.

	Psychological Distress Score	
	Diagnosed	*Not Diagnosed*
$\bar{Y}$	3.9	4.87
ΣY	729	5,849
$\Sigma(Y-\bar{Y})^2$	3,059.14	25,180.20
N	187	1,200

  a. Calculate the variance and standard deviation from these statistics for both groups.
  b. What can you say about the variability in the distress scores for those respondents who have been diagnosed as having cancer and those who have not? Why might there be a difference? Why might there be more variability for one group than for the other?
  c. Was it necessary in this problem to provide you with the mean value to calculate the variance and standard deviation?

6. You are interested in studying the variability of crimes committed (including violent and property crimes) and police expenditures in the eastern and midwestern United States. The Census Bureau collected the following statistics on these two variables for 21 states in the East and Midwest in 2008.

The SPSS output showing the mean and the standard deviation for both variables is presented below.

Descriptive Statistics

	N	Minimum	Maximum	Mean	Std. Deviation
Number of Crimes per 100,000 Population	21	2181	4188	3038.90	583.004
Police Protection Expenditures (in millions of dollars)	21	120	8164	1703.95	1895.214
Valid N (listwise)	21				

  a. What are the means? The standard deviations?
  b. Compare the mean with the standard deviation for each variable. Does there appear to be more variability in the number of crimes or in police expenditures per capita in these states? Which states contribute more to this greater variability?
  c. Suggest why one variable has more variability than the other. In other words, what social forces would cause one variable to have a relatively large standard deviation?

7. Average life expectancy for women in 2010–2011 is reported for 10 countries. Calculate the appropriate measures of central tendency and variability for both European countries and non-European countries. Is there more variability in life expectancy for European countries or non-European countries? If so, what might explain these differences?

Country	Life Expectancy at Birth[a]
European countries	
France	84.8
Germany	82.6
Netherlands	82.7
Spain	84.7
Turkey	76.3

Country	Life Expectancy at Birth[a]
Non-European countries	
Japan	86.4
Australia	84.0
Mexico	79.2
Iceland	83.5
Israel	83.4

Source: GLOBAL13SSDS.

[a] Data for each country collected from either 2010 or 2011.

8. You have been asked to prepare a brief statement about the labor force participation rate for men and women (percentage of population age 15–64 years). Using statistics from the following table obtained from the GLOBAL13SSDS data set, write a paragraph or two about labor force participation rates throughout the world. In your answer, be sure to identify at least two explanations for your findings.

Labor Force Participation Rates	Participating Countries	Mean	Standard Deviation
Men	70	79.44	6.20
Women	70	57.02	18.54

9. The following table summarizes the racial differences in education and the ideal number of children for Chinese Americans and Filipino Americans. Based on the means and standard deviations (in parentheses), what conclusions can be drawn about differences in the ideal number of children?

	Chinese Americans	Filipino Americans
Education (years)	15.55 (3.643)	13.42 (3.704)
Ideal number of children	2.88 (2.167)	4.00 (2.098)

Source: GSS, 2010.

10. Provided below is SPSS output from the 2010 GSS measuring variability in respondents' ages and respondents' ages when their first children were born.

Exercises

Statistics

		AGE OF RESPONDENT	R'S AGE WHEN 1ST CHILD BORN
N	Valid	1483	1106
	Missing	17	394
Mean		49.21	23.72
Median		49.00	22.00
Mode		29	20
Std. Deviation		17.557	5.933
Variance		308.248	35.199
Minimum		18	13
Maximum		89	53
Percentiles	25	35.00	20.00
	50	49.00	22.00
	75	62.00	27.00

a. What is the range for each variable? What is the interquartile range for each variable? What is the standard deviation for each variable?

b. Citing some measures of variability you found in (a), which of the two variables has more variability? From a social perspective, why would we expect one to have more variability than the other?

Chapter 5

The Normal Distribution

Chapter Learning Objectives

❖ Recognizing the importance and the use of the normal distribution in statistics
❖ Describing the properties of the normal distribution
❖ Transforming a raw score into a standard (Z) score and vice versa
❖ Using the standard normal table
❖ Transforming a Z score into a proportion (or percentage) and vice versa

W hile in the preceding chapters we have described empirical distributions that are based on real data, in this chapter we will describe a theoretical distribution known as the *normal curve* or the **normal distribution**. A *theoretical distribution* is similar to an empirical distribution in that it can be organized into frequency distributions, displayed using graphs, and described by its central tendency and variation using measures such as the mean and the standard deviation. However, unlike an empirical distribution, a theoretical distribution is based on theory rather than on real data. The value of the theoretical normal distribution lies in the fact that many empirical distributions that we study seem to approximate it. We can often learn a lot about the characteristics of these empirical distributions based on our knowledge of the theoretical normal distribution.

Normal distribution   A bell-shaped and symmetrical theoretical distribution with the mean, the median, and the mode all coinciding at its peak and with the frequencies gradually decreasing at both ends of the curve.

▣ PROPERTIES OF THE NORMAL DISTRIBUTION

The normal curve (Figure 5.1) looks like a bell-shaped frequency polygon. Because of this property, it is sometimes called the *bell-shaped curve*. One of the most striking characteristics of the normal distribution is its perfect symmetry. Notice that if you fold Figure 5.1 exactly in the middle, you have two equal halves, each the mirror image of the other. This means that precisely half the observations fall on each side of the middle of the distribution. In addition, the midpoint of the normal curve is the point having the maximum frequency. This is also the point at which three measures coincide: the mode (the point of the highest frequency), the median (the point that divides the distribution into two equal halves), and the mean (the average of all the scores). Notice also that most of the observations are clustered around the middle, with the frequencies gradually decreasing at both ends of the distribution.

Empirical Distributions Approximating the Normal Distribution

The normal curve is a theoretical ideal, and real-life distributions never match this model perfectly. However, researchers study many variables (e.g., standardized tests such as the SAT, ACT, or GRE; height; athletic ability; and numerous social and political attitudes) that closely resemble this theoretical model. When we say that a variable is "normally distributed," we mean that the graphic display will reveal an approximately bell-shaped and symmetrical distribution closely resembling the idealized model shown in Figure 5.1. This property makes it possible for us to describe many empirical distributions based on our knowledge of the normal curve.

An Example: Final Grades in Statistics

For example, let's examine the frequencies and the bar chart presented in Table 5.1. These data are the final scores of 1,200 students of social statistics. We overlaid a normal curve on the distribution shown in Table 5.1. Notice how closely our empirical distribution of statistics scores approximates the normal curve!

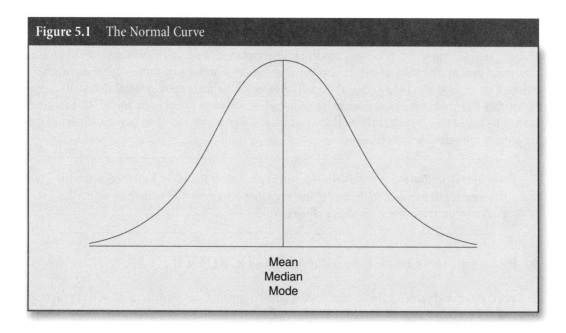

Figure 5.1 The Normal Curve

Mean
Median
Mode

Table 5.1 Final Grades in Social Statistics of 1,200 Students (1983–1993): A Near Normal Distribution

Midpoint Score	Frequency Bar Chart	f	Cf	%	C%
40	*	4	4	0.33	0.33
50	************	78	82	6.50	6.83
60	**************************	275	357	22.92	29.75
70	*******************************	483	840	40.25	70.00
80	**************************	274	1,114	22.83	92.83
90	***********	81	1,195	6.75	99.58
100	*	5	1,200	0.42	100.00

| 10 | 50 | 100 | 200 | 300 | 400 | 500 |

Mean $(\overline{Y}) = 70.07$ Median $= 70.00$ Mode $= 70.00$
Standard deviation $(S_r) = 10.27$

Note that 70 is the most frequent score obtained by the students, and therefore, it is the mode of the distribution. Because about half the students are either above (49.99%) or below (50.01%) this score (based on raw frequencies), both the mean (70.07) and the median (70) are approximately 70. Also shown in Table 5.1 is the gradual decrease in the number of students who scored either above or below 70. Very few students scored higher than 90 or lower than 50.

When we use the term *normal curve*, we are not referring to identical distributions. The shape of a normal distribution varies, depending on the mean and standard deviation of the particular distribution. (The symbol μ_Y is the population notation for the mean, while σ_Y stands for the population standard deviation. We will discuss these symbols in more detail in the next chapter.) For example, in Figure 5.2, we present two normally shaped distributions with identical means ($\mu_Y = 12$) but with different standard deviations ($\sigma_{Y1} = 3$, $\sigma_{Y2} = 5$). Note that the distribution with the larger standard deviation appears relatively wider and flatter.

Areas Under the Normal Curve

Regardless of the precise shape of the distribution, in all normal or nearly normal curves we find a constant proportion of the area under the curve lying between the mean and any given distance

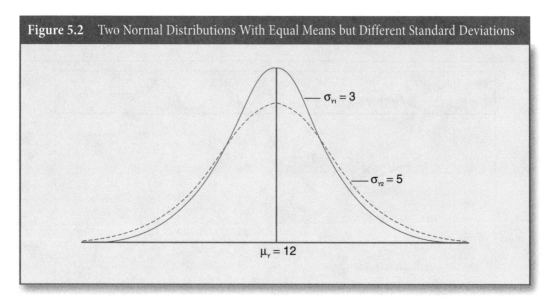

Figure 5.2   Two Normal Distributions With Equal Means but Different Standard Deviations

from the mean when measured in standard deviation units. The area under the normal curve may be conceptualized as a proportion or percentage of the number of observations in the sample. Thus, the entire area under the curve is equal to 1.00 or 100% (1.00 × 100) of the observations. Because the normal curve is perfectly symmetrical, exactly 0.50 or 50% of the observations lie above or to the right of the center, which is the mean of the distribution, and 50% lie below or to the left of the mean.

In Figure 5.3, note the percentage of cases that will be included between the mean and 1, 2, and 3 standard deviations above and below the mean. The mean of the distribution divides it exactly into half; 34.13% is included between the mean and 1 standard deviation to the right of the mean, and the same percentage is included between the mean and 1 standard deviation to the left of the mean. The plus signs indicate standard deviations above the mean; the minus signs denote standard deviations below the mean. Thus, between the mean and ±1 standard deviation, 68.26% of all the observations in the distribution occur; between the mean and ±2 standard deviations, 95.46% of all observations in the distribution occur; and between the mean and ±3 standard deviations, 99.72% of the observations occur.

✓ *Learning*
Check

Review and confirm the properties of the normal curve. What is the area underneath the curve equal to? What percentage of the distribution is within 1 standard deviation? Within 2 and 3 standard deviations? Verify the percentage of cases by summing the percentages in Figure 5.3.

Interpreting the Standard Deviation

The fixed relationship between the distance from the mean and the areas under the curve represents a property of the normal curve that has highly practical applications. As long as a distribution is

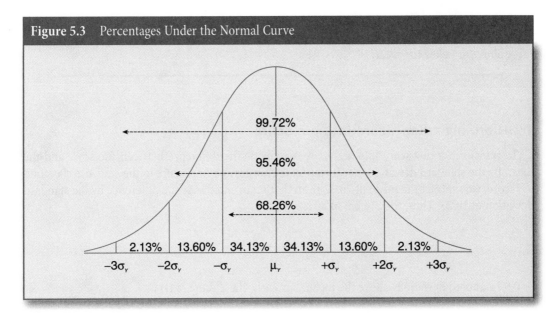

Figure 5.3 Percentages Under the Normal Curve

normal and we know the mean and the standard deviation, we can determine the relative frequency (proportion or percentage) of cases that fall between any score and the mean.

This property provides an important interpretation for the standard deviation of empirical distributions that are approximately normal. For such distributions, when we know the mean and the standard deviation, we can determine the percentage of scores that are within any distance, measured in standard deviation units, from that distribution's mean. For example, we know that scores on college entrance tests such as the SAT and ACT are normally distributed. For instance, the SAT math and verbal sections each have a set mean of 500 and a standard deviation of 100 (http://professionals.collegeboard.com/testing/sat-reasoning/scores/sat-data-tables). This means that approximately 68% of the students who take the test obtain scores between 400 (1 standard deviation below the mean) and 600 (1 standard deviation above the mean) on each section. We can also anticipate that approximately 95% of the students who take the test will score between 300 (2 standard deviations below the mean) and 700 (2 standard deviations above the mean).

Not every empirical distribution is normal. We've learned that the distributions of some common variables, such as income, are skewed and therefore not normal. The fixed relationship between the distance from the mean and the areas under the curve applies only to distributions that are normal or approximately normal.

▣ STANDARD (Z) SCORES

We can express the difference between any score in a distribution and the mean in terms of *standard scores*, also known as *Z scores*. A **standard (Z) score** is the number of standard deviations that a given raw score (or the observed score) is above or below the mean. A raw score can be transformed into a Z score to find how many standard deviations it is above or below the mean.

Standard (Z) score   The number of standard deviations that a given raw score is above or below the mean.

Transforming a Raw Score Into a Z Score

To transform a raw score into a *Z* score, we divide the difference between the score and the mean by the standard deviation. For instance, to transform a final score in the statistics class into a *Z* score, we subtract the mean of 70.07 from that score and divide the difference by the standard deviation of 10.27. Thus, the *Z* score of 80 is

$$\frac{80 - 70.07}{10.27} = 0.97$$

or 0.97 standard deviations above the mean. Similarly, the *Z* score of 60 is

$$\frac{60 - 70.07}{10.27} = -0.98$$

or 0.98 standard deviations below the mean; the negative sign indicates that this score is below the mean.

This calculation, in which the difference between a raw score and the mean is divided by the standard deviation, gives us a method of standardization known as *transforming a raw score into a Z score* (also known as a standard score). The *Z* score formula is

$$Z = \frac{Y - \bar{Y}}{S_Y} \tag{5.1}$$

A *Z* score allows us to represent a raw score in terms of its relationship to the mean and to the standard deviation of the distribution. It represents how far a given raw score is from the mean in standard deviation units. A positive *Z* indicates that a score is larger than the mean, and a negative *Z* indicates that it is smaller than the mean. The larger the *Z* score, the larger the difference between the score and the mean (Table 5.2).

Transforming a Z Score Into a Raw Score

For some normal curve applications, we need to reverse the process, transforming a *Z* score into a raw score instead of transforming a raw score into a *Z* score. A *Z* score can be converted to a raw score to find the score associated with a particular distance from the mean when this distance is expressed in standard deviation units. For example, suppose we are interested in finding out the

Table 5.2   Final Social Science Statistics Scores Converted to
Z Scores

Final Score	Z Score
40	$Z = \dfrac{40 - 70.07}{10.27} = \dfrac{-30.07}{10.27} = -2.93$
60	$Z = \dfrac{60 - 70.07}{10.27} = \dfrac{-10.07}{10.27} = -0.98$
80	$Z = \dfrac{80 - 70.07}{10.27} = \dfrac{9.93}{10.27} = 0.97$
100	$Z = \dfrac{100 - 70.07}{10.27} = \dfrac{29.93}{10.27} = 2.91$
$\overline{Y} = 70.07$	$S_Y = 10.27$

final score in the statistics class that lies 1 standard deviation above the mean. To solve this problem, we begin with the Z score formula:

$$Z = \frac{Y - \overline{Y}}{S_Y}$$

Note that for this problem, we have the following values for Z ($Z = 1$), the mean, and the standard deviation $S_Y = 10.27$, but we need to determine the value of Y:

$$1 = \frac{Y - 70.07}{10.27}$$

Through simple algebra, we solve for Y:

$$Y = 70.07 + 1(10.27) = 70.07 + 10.27 = 80.34$$

The score of 80.34 lies 1 standard deviation (or 1 Z score) above the mean of 70.07.

The general formula for transforming a Z score into a raw score is

$$Y = \bar{Y} + Z(S_Y) \qquad (5.2)$$

Thus, to transform a Z score into a raw score, multiply the Z score by the standard deviation and add this product to the mean.

Now, what statistics score lies 1.5 standard deviations below the mean? Because the score lies below the mean, the Z score is negative. Thus,

$$Y = 70.07 + (-1.5)(10.27) = 70.07 - 15.41 = 54.66$$

The score of 54.66 lies 1.5 standard deviations below the mean of 70.07.

✓ *Learning*
Check

Transform the Z scores in Table 5.2 back into raw scores. Your answers should agree with the raw scores listed in the table.

▣ THE STANDARD NORMAL DISTRIBUTION

When a normal distribution is represented in standard scores (Z scores), we call it the **standard normal distribution**. Standard scores, or Z scores, are the numbers that tell us the distance between an actual score and the mean in terms of standard deviation units. The standard normal distribution has a mean of 0.0 and a standard deviation of 1.0.

Standard normal distribution   A normal distribution represented in standard (Z) scores, with mean $= 0$ and standard deviation $= 1$.

Figure 5.4 shows a standard normal distribution with areas under the curve associated with 1, 2, and 3 standard scores above and below the mean. To help you understand the relationship between raw scores of a distribution and standard Z scores, we also show the raw scores in the statistics class that correspond to these standard scores. For example, notice that the mean for the statistics score distribution is 70.07 and the corresponding Z score—the mean of the standard normal distribution—is 0. The score of 80.34 is 1 standard deviation above the mean ($70.07 + 10.27 = 80.34$); therefore, its corresponding Z score is +1. Similarly, the score of 59.80 is 1 standard deviation below the mean ($70.07 - 10.27 = 59.80$), and its Z-score equivalent is −1.

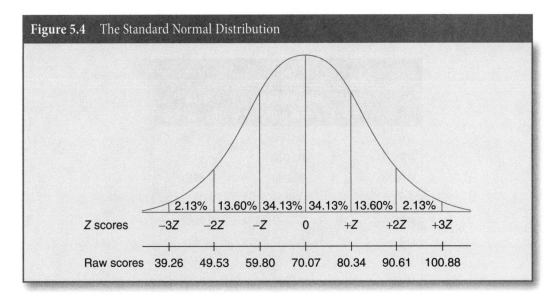

Figure 5.4 The Standard Normal Distribution

◉ THE STANDARD NORMAL TABLE

We can use Z scores to determine the proportion of cases that are included between the mean and any Z score in a normal distribution. The areas or proportions under the standard normal curve, corresponding to any Z score or its fraction, are organized into a special table called the **standard normal table.** The table is presented in Appendix A. In this section, we discuss how to use this table.

Standard normal table A table showing the area (as a proportion, which can be translated into a percentage) under the standard normal curve corresponding to any Z score or its fraction.

The Structure of the Standard Normal Table

Table 5.3 reproduces a small part of the standard normal table. Note that the table consists of three columns.

Column A lists positive Z scores. Because the normal curve is symmetrical, the proportions that correspond to positive Z scores are identical to the proportions corresponding to negative Z scores.

Column B shows the area included between the mean and the Z score listed in Column A. Note that when Z is positive, the area is located on the right side of the mean (see Figure 5.5a), whereas for a negative Z score, the same area is located left of the mean (Figure 5.5b).

Table 5.3   Excerpt of the Standard Normal Table

(A)	(B)	(C)
Z	Area Between Mean and Z	Area Beyond Z
0.00	0.0000	0.5000
0.01	0.0040	0.4960
0.02	0.0080	0.4920
0.03	0.0120	0.4880
0.04	0.0160	0.4840
0.05	0.0199	0.4801
0.06	0.0239	0.4761
0.07	0.0279	0.4721
0.08	0.0319	0.4681
0.09	0.0359	0.4641
0.10	0.0398	0.4602

Column C shows the proportion of the area that is beyond the Z score listed in Column A. Areas corresponding to positive Z scores are on the right side of the curve (see Figure 5.5a). Areas corresponding to negative Z scores are identical except that they are on the left side of the curve (Figure 5.5b).

Transforming Z Scores Into Proportions (or Percentages)

We illustrate how to use Appendix A with some simple examples using our data on students' final statistics scores (see Table 5.1). The examples in this section are applications that require the transformation of Z scores into proportions (or percentages).

Finding the Area Between the Mean and a Specified Positive Z Score

Use the standard normal table to find the area between the mean and a specified positive Z score. To find the percentage of students whose scores range between the mean (70.07) and 85, follow these steps:

1. Convert 85 to a Z score:

$$Z = \frac{85 - 70.07}{10.27} = 1.45$$

Figure 5.5 Areas Between Mean and Z (B) and Beyond Z (C)

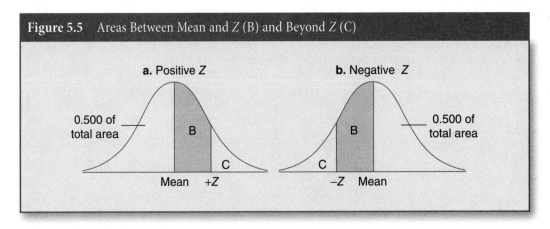

2. Look up 1.45 in Column A (in Appendix A) and find the corresponding area in Column B, 0.4265. We can translate this proportion into a percentage (0.4265 × 100 = 42.65%) of the area under the curve included between the mean and a Z score of 1.45 (Figure 5.6).

3. Thus, 42.65% of the students scored between 70.07 and 85.

To find the actual number of students who scored between 70.07 and 85, multiply the proportion 0.4265 by the total number of students. Thus, approximately 512 students (0.4265 × 1,200 = 512) obtained scores between 70.07 and 85.

Figure 5.6 Finding the Area Between the Mean and a Specified Positive Z Score

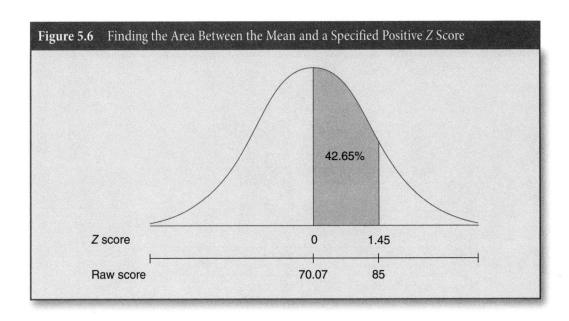

Finding the Area Between the Mean and a Specified Negative Z Score

What is the percentage of students whose scores ranged between 65 and 70.07? We can use the standard normal table and the following steps to find out.

1. Convert 65 to a *Z* score:

$$Z = \frac{65 - 70.07}{10.27} = -0.49$$

2. Because the proportions that correspond to positive *Z* scores are identical to the proportions corresponding to negative *Z* scores, we ignore the negative sign of *Z* and look up 0.49 in Column A. The area corresponding to a *Z* score of 0.49 is 0.1879. This indicates that 0.1879 of the area under the curve is included between the mean and a *Z* score of −0.49 (Figure 5.7). We convert this proportion to 18.79% (0.1879 × 100 = 18.79%).

3. Thus, approximately 225 (0.1879 × 1,200 = 225) students obtained scores between 65 and 70.07.

Finding the Area Above a Positive Z Score or Below a Negative Z Score

We can compare students who have done very well or very poorly to get a better idea of how they compare with other students in the class.

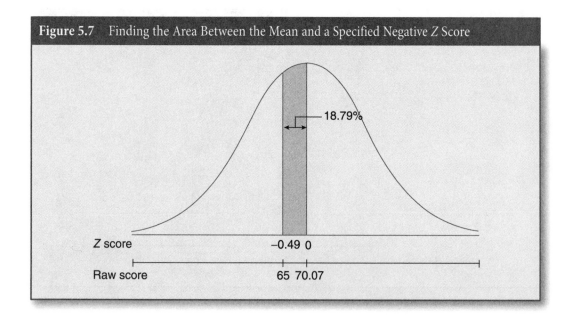

Figure 5.7 Finding the Area Between the Mean and a Specified Negative *Z* Score

To identify students who did very well, we selected all students who scored above 85. To find how many students scored above 85, first convert 85 to a Z score:

$$Z = \frac{85 - 70.07}{10.27} = 1.45$$

Thus, the Z score corresponding to a final score of 85 in statistics is equal to 1.45.

The area beyond a Z score of 1.45 includes all students who scored above 85. This area is shown in Figure 5.8. To find the proportion of students whose scores fall in this area, refer to the entry in Column C that corresponds to a Z score of 1.45, 0.0735. This means that 7.35% (0.0735 × 100 = 7.35%) of the students scored above 85. To find the actual number of students in this group, multiply the proportion 0.0735 by the total number of students. Thus, there were 1,200 × 0.0735, or about 88 students, who scored above 85 over the 10-year period.

A similar procedure can be applied to identify the number of students who did not do well in the class. The cutoff point for poor performance in this class was the score of 50. To determine how many students did poorly, we first converted 50 to a Z score:

$$Z = \frac{50 - 70.07}{10.27} = -1.95$$

The Z score corresponding to a final score of 50 is equal to −1.95. The area beyond a Z score of −1.95 includes all students who scored below 50. This area is also shown in Figure 5.8. Locate the proportion of students in this area in Column C in the entry corresponding to a Z score of 1.95. (Remember that the proportions corresponding to positive or negative Z scores are identical.)

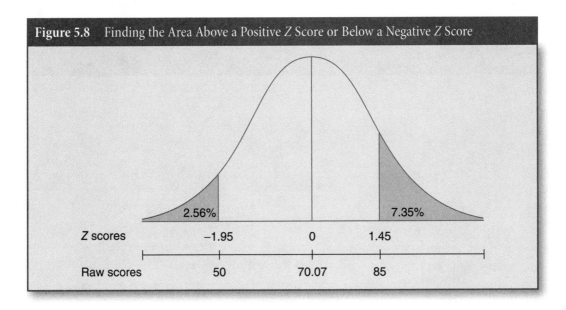

Figure 5.8   Finding the Area Above a Positive Z Score or Below a Negative Z Score

This proportion is equal to 0.0256. Thus, 2.56% (0.0256 × 100 = 2.56%) of the group, or about 31 (0.0256 × 1,200) students, performed poorly in statistics.

Transforming Proportions (or Percentages) Into Z Scores

The examples in this section are applications that require transforming proportions (or percentages) into Z scores.

Finding a Z Score Bounding an Area Above It

Assuming that a grade of A is assigned to the top 10% of the students, what would it take to get an A in the class? To answer this question, we need to identify the cutoff point for the top 10% of the class. This problem involves two steps:

1. Find the Z score that bounds the top 10% or 0.1000 (0.1000 × 100 = 10%) of all the students who took statistics (Figure 5.9).

   Refer to the areas under the normal curve shown in Appendix A. First, look for an entry of 0.1000 (or the value closest to it) in Column C. The entry closest to 0.1000 is 0.1003. Then, locate the Z score in Column A that corresponds to this proportion. The Z score associated with the proportion 0.1003 is 1.28.

2. Find the final score associated with a Z score of 1.28.

   This step involves transforming the Z score into a raw score. We learned earlier in this chapter (Formula 5.2) that to transform a Z score into a raw score we multiply the score by the standard deviation and add that product to the mean. Thus,

$$Y = 70.07 + 1.28(10.27) = 70.07 + 13.15 = 83.22$$

The cutoff point for the top 10% of the class is a score of 83.22.

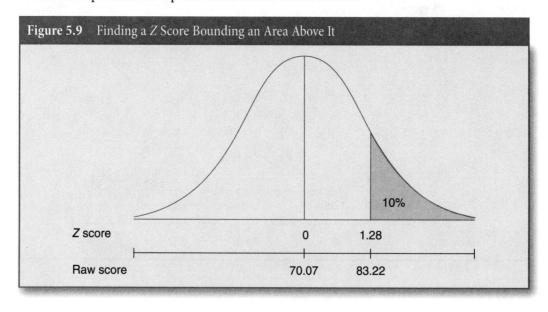

Figure 5.9   Finding a Z Score Bounding an Area Above It

Finding a Z Score Bounding an Area Below It

Now, let's assume that a grade of F was assigned to the bottom 5% of the class. What would be the cutoff point for a failing score in statistics? Again, this problem involves two steps:

1. Find the Z score that bounds the lowest 5% or 0.0500 of all the students who took the class (Figure 5.10).

   Refer to the areas under the normal curve, and look for an entry of 0.0500 (or the value closest to it) in Column C. The entry closest to 0.0500 is 0.0495. Then, locate the Z score in Column A that corresponds to this proportion, 1.65. Because the area we are looking for is on the left side of the curve—that is, below the mean—the Z score is negative. Thus, the Z score associated with the lowest 0.0500 (or 0.0495) is −1.65.

2. To find the final score associated with a Z score of −1.65, convert the Z score to a raw score:

$$Y = 70.07 + (-1.65)(10.27) = 70.07 - 16.95 = 53.12$$

The cutoff for a failing score in statistics is 53.12.

Can you find the number of students who got a score of at least 90 in the statistics course? How many students got a score below 60?

✓ *Learning Check*

Figure 5.10   Finding a *Z* Score Bounding an Area Below It

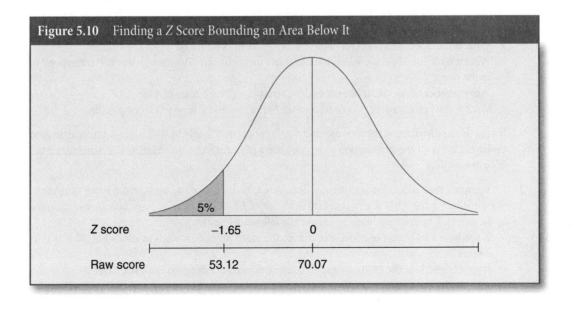

MAIN POINTS

- The normal distribution is central to the theory of inferential statistics. It also provides a model for many empirical distributions that approximate normality.

- In all normal or nearly normal curves, we find a constant proportion of the area under the curve lying between the mean and any given distance from the mean when measured in standard deviation units.

- The standard normal distribution is a normal distribution represented in standard scores, or Z scores, with mean $= 0$ and standard deviation $= 1$. Z scores express the number of standard deviations that a given score is above or below the mean. The proportions corresponding to any Z score or its fraction are organized into a special table called the standard normal table.

KEY TERMS

normal distribution
standard normal distribution

standard normal table
standard (Z) score

$SAGE edge™

Sharpen your skills with SAGE edge at **edge.sagepub.com/ssdsess2e**. **SAGE edge for students** provides a personalized approach to help you accomplish your coursework goals in an easy-to-use learning environment.

CHAPTER EXERCISES

1. We discovered that 1,013 General Social Survey (GSS) respondents in 2010 watched television for an average of 3.01 hr/day, with a standard deviation of 2.65 hr/day. Answer the following questions, assuming the distribution of the number of television hours is normal.
   a. What is the Z score for a person who watches more than 8 hr/day?
   b. What proportion of people watch television less than 5 hr/day? How many does this correspond to in the sample?
   c. What number of television hours per day corresponds to a Z score of $+1$?
   d. What is the percentage of people who watch between 1 and 6 hr of television per day?

2. If a particular distribution that you are studying is not normal, it may be difficult to determine the area under the curve of the distribution or to translate a raw score into a Z value. Is this statement true? Why or why not?

3. Let's assume that education is normally distributed. Using GSS data, we find the mean number of years of education is 13.47 with a standard deviation of 3.1. A total of 1,496 respondents were included in the survey. Use these numbers to answer the following questions.
   a. If you have 13.47 years of education, that is, the mean number of years of education, what is your Z score?
   b. If your friend is in the 60th percentile, how many years of education does she have?

4. The 2010 GSS provides the following statistics for the average years of education for lower-, working-, middle-, and upper-class respondents and their associated standard deviations.

	Mean	Standard Deviation	N
Lower class	11.61	2.67	123
Working class	12.80	2.85	697
Middle class	14.45	3.08	626
Upper class	15.45	2.98	38

a. Assuming that years of education is normally distributed in the population, what proportion of working-class respondents have 12 to 16 years of education? What proportion of upper-class respondents have 12 to 16 years of education?

b. What is the probability that a working-class respondent, drawn at random from the population, will have more than 16 years of education? What is the equivalent probability for a middle-class respondent drawn at random?

c. What is the probability that a lower- or upper-class respondent will have less than 12 years of education?

d. Find the upper and lower limits, centered on the mean, that will include 50% of all working-class respondents.

e. If years of education is actually positively skewed in the population, how would that change your other answers?

5. The following table displays unemployment information for each region of the United States in March 2013. The unemployment numbers listed in the table are in thousands. For example, the Midwest was home to 2,525,600 unemployed residents when data were collected.

Unemployment (in thousands) in the United States: 2013			
Midwest	2,525.6	West	3,004.6
Northeast	2,240.2	South	4,106.9

Source: U.S. Bureau of Labor Statistics, *News Release*, April 19, 2013, USDL-13-0672, Table 1.

a. What are the mean and standard deviation unemployment numbers for all regions?

b. Using information from (a), how many regions fall more than 1 standard deviation above the mean? How does this number compare with the number expected from the theoretical normal curve distribution?

c. Create a histogram representing the unemployment figures for all regions. Does the distribution appear to be normal? Explain your answer.

6. Information on the occupational prestige scores for blacks and whites is presented in the following table.

Exercises

	Mean	Standard Deviation	N
Whites	45.03	13.93	1,100
Blacks	40.83	13.07	195

Source: GSS, 2010.

a. What percentage of whites should have occupational prestige scores above 60? How many whites in the sample should have occupational prestige scores above 60?

b. What percentage of blacks should have occupational prestige scores above 60? How many blacks should have occupational prestige scores above 60?

c. What proportion of whites have prestige scores between 30 and 70? How many whites have prestige scores between 30 and 70?

d. How many blacks in the sample should have occupational prestige scores between 30 and 60?

7. SAT scores are normed so that, in any year, the mean of the verbal or math test should be 500 and the standard deviation 100. Assuming this is true (it is only approximately true, both because of variation from year to year and because scores have decreased since the SAT tests were first developed), answer the following questions.

a. What percentage of students score above 625 on the math SAT in any given year?

b. What percentage of students score between 400 and 600 on the verbal SAT?

8. The Hate Crime Statistics Act of 1990 requires the attorney general to collect national data about crimes that manifest evidence of prejudice based on race, religion, sexual orientation, or ethnicity, including the crimes of murder and non-negligent manslaughter, forcible rape, aggravated assault, simple assault, intimidation, arson, and destruction, damage, or vandalism of property. The hate crime data collected in 2007 reveal, based on a randomly selected sample of 300 incidents, that the mean number of victims of a particular type of hate crime was 1.28, with a standard deviation of 0.82. Assuming that the number of victims of was normally distributed, answer the following questions.

a. What proportion of crime incidents had more than 2 victims?

b. What is the probability that there was more than 1 victim in an incident?

c. What proportion of crime incidents had fewer than 4 victims?

9. The number of hours people work each week varies widely for many reasons. Using the 2010 GSS, you find that the mean number of hours worked last week was 40.62, with a standard deviation of 15.26, based on a sample size of 838.

a. Assume that hours worked is approximately normally distributed in the sample. What is the probability that someone in the sample will work 60 hours or more in a week? How many people in the sample of 894 should have worked 60 hours or more?

b. What is the probability that someone will work 30 hours or fewer in a week (i.e., work part-time)? How many people does this represent in the sample?

10. The National Collegiate Athletic Association (NCAA) has a public access database on each Division I sports team in the United States, which contains data on team-level academic progress rates (APRs), eligibility rates, and retention rates. The mean APR of 359 men's basketball teams for the 2010–2011 academic year was 950.35 (based on a 1,000-point scale), with a standard deviation of 30.58. Assuming that the distribution of APRs for the teams is approximately normal,

a. Would a team be at the upper quartile (the top 25%) of the APR distribution with an APR score of 975?

b. What APR score should a team have to be more successful than 75% of all the teams?

c. What is the Z value for this score?

11. The following table, based on the same NCAA data, gives the means and standard deviations of eligibility and retention rates (based on a 1,000-point scale) for the 2010–2011 academic year, along with the scores for two men's basketball teams, A and B. Assume that test scores are normally distributed.

	Mean	Standard Deviation	Team A	Team B
Eligibility	974.3	40.4	917	962
Retention	970.2	38.8	913	962

a. On which criterion (eligibility or retention) did Team A do better, relative to the other teams? Calculate appropriate statistics to answer this question.

b. On which criterion (eligibility or retention) did Team B do better, relative to the other teams? Calculate appropriate statistics to answer this question.

c. What proportion of the teams have retention rates below the retention rate of Team B?

12. What is the value of the mean score for any standard normal distribution? What is the value of the standard deviation for any standard normal distribution? Explain why this is true for any standard normal distribution.

13. You are asked to do a study of shelters for abused and battered women to determine the necessary capacity in your city to provide housing for most of these women. After recording data for a whole year, you find that the mean number of women in shelters each night is 250, with a standard deviation of 75. Fortunately, the distribution of the number of women in the shelters each night is normal, so you can answer the following questions posed by the city council.

a. If the city's shelters have a capacity of 350, will that be enough places for abused women on 95% of all nights? If not, what number of shelter openings will be needed?

b. The current capacity is only 220 openings, because some shelters have been closed. What is the percentage of nights that the number of abused women seeking shelter will exceed current capacity?

14. Provided below is SPSS output from the 2010 GSS. Respondents were polled on their occupational prestige score, which researchers use as a measure of socioeconomic status.

Statistics

RS OCCUPATIONAL PRESTIGE SCOI

N	Valid	1420
	Missing	80
Mean		44.19
Median		43.00
Mode		51
Std. Deviation		13.867
Variance		192.296
Range		69
Percentiles	25	32.00
	50	43.00
	75	51.00

Assume that occupational prestige scores are normally distributed. If sociologists have a score of 60.75, what proportion of people have an occupational prestige score higher than sociologists?

15. Provided below is SPSS output from the 2007 Health Information National Trends Survey data set. Respondents were asked what their total pretax household income was for the past year. Provided are the descriptive statistics for the variable.

Statistics

What is your total pretax HH income ir

N	Valid	507
	Missing	993
Mean		65628.7968
Std. Deviation		59373.91596
Variance		3525261896

a. Assuming the data are normally distributed, what household income would someone have to earn to fare better than 99% of people?

b. A classmate points out that our earlier assumption—that the data are normally distributed—is faulty. She says that, making this assumption, we would expect a large percentage of the sample to earn less than $0 annually, which is impossible. What percentage of the sample would earn less than $0 annually, were it normally distributed? Does this undermine our assumption?

Chapter 6

Sampling and Sampling Distributions

Chapter Learning Objectives

❖ Understanding the aims of sampling and basic principles of probability

❖ Understanding and applying the concept of the sampling distribution

❖ Understanding the nature of the central limit theorem

Until now, we have ignored the question of who or what should be observed when we collect data or whether the conclusions based on our observations can be generalized to a larger group of observations. The truth is that we are rarely able to study or observe everyone or everything we are interested in. Although we have learned about various methods to analyze observations, remember that these observations represent only a tiny fraction of all the possible observations we might have chosen. Consider the following examples.

Example 1: The Muslim Student Association on your campus is interested in conducting a study of experiences with campus diversity. You only have enough funds to survey 300 students on your campus. Given that your campus is home to more than 20,000 students, what should you do?

Example 2: Local environmental activists want to assess recycling practices on your campus to develop a proposal to reduce unnecessary waste. Since the university serves more than 30,000 students, faculty, and staff, how should the activists proceed?

Example 3: The student union on your campus is trying to find out how it can better address the needs of commuter students and has commissioned you to conduct a needs assessment survey. You have been given enough money to survey about 500 students. Given that there are nearly 15,000 commuters on your campus, is this an impossible task?

What do these problems have in common? In all situations, the major problem is that there is too much information and not enough resources to collect and analyze all of it.

▣ AIMS OF SAMPLING[1]

Researchers in the social sciences almost never have enough time or money to collect information about the entire group that interests them. Known as the **population**, this group includes all the cases (individuals, objects, or groups) in which the researcher is interested. For example, in our first illustration, there are more than 20,000 students; the population in the second illustration consists of all 30,000 faculty, staff, and students; and in the third illustration, the population is all 15,000 commuter students.

Population   A group that includes all the cases (individuals, objects, or groups) in which the researcher is interested.

Fortunately, we can learn a lot about a population if we carefully select a subset of it. This subset is called a **sample**. Through the process of *sampling*—selecting a subset of observations from the population of interest—we attempt to generalize the characteristics of the larger group (population) based on what we learn from the smaller group (the sample). This is the basis of *inferential statistics*—making predictions or inferences about a population from observations based on a sample.

Sample   A subset of cases selected from a population.

The term **parameter**, associated with the population, refers to measures used to describe the distribution of the population we are interested in. For instance, the average commuting time for *all* of the 15,000 students on your campus is a population parameter because it refers to a population characteristic. In previous chapters, we have learned the many ways of describing a distribution, such as a proportion, a mean, or a standard deviation. When used to describe the population distribution, these measures are referred to as parameters. Thus, a population mean, a population proportion, and a population standard deviation are all parameters.

Parameter   A measure (e.g., mean or standard deviation) used to describe the population distribution.

We use the term **statistic** when referring to a corresponding characteristic calculated for the sample. For example, the average commuting time for a *sample* of commuter students is a sample statistic. Similarly, a sample mean, a sample proportion, and a sample standard deviation are all statistics.

Statistic   A measure (e.g., mean or standard deviation) used to describe the sample distribution.

In this chapter as well as in Chapters 7 and 8, we discuss some of the principles involved in generalizing results from samples to the population. In our discussion, we will use different notations when referring to sample statistics and population parameters. Table 6.1 presents the sample notation and the corresponding population notation.

The distinctions between a sample and a population and between a parameter and a statistic are illustrated in Figure 6.1. We've included for illustration the population parameter of 0.60—the proportion of white respondents in the population. However, since we almost never have enough resources to collect information about the population, it is rare that we

Table 6.1   Sample and Population Notations

Measure	Sample Notation	Population Notation
Mean	$\overline{Y}$	μ_Y
Proportion	p	π
Standard deviation	S_Y	σ_Y
Variance	S_Y^2	σ_Y^2

Figure 6.1   The Proportion of White Respondents in a Population and in a Sample

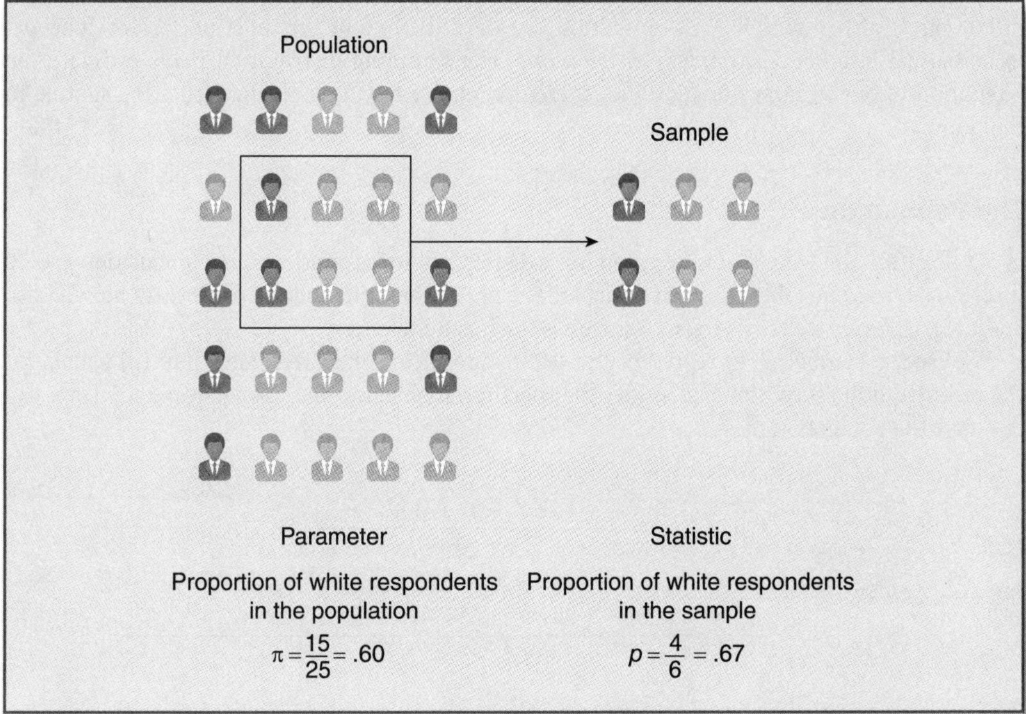

Population

Sample

Parameter
Proportion of white respondents
in the population
$$\pi = \frac{15}{25} = .60$$

Statistic
Proportion of white respondents
in the sample
$$p = \frac{4}{6} = .67$$

know the value of a parameter. The goal of most research is to find the population parameter. Researchers usually select a sample from the population to obtain an estimate of the population parameter. Thus, the major objective of sampling theory and statistical inference is to provide estimates of unknown parameters from sample statistics that can be easily obtained and calculated.

✓ Learning
 Check

> *It is important that you understand what the terms* population, sample, parameter, *and* statistic *mean. Use your own words so that the meaning makes sense to you. If you cannot clearly define these terms, review the preceding material. You will see these sample and population notations over and over again. If you memorize them, you will find it much easier to understand the formulas used in inferential statistics.*

▣ THE CONCEPT OF THE SAMPLING DISTRIBUTION

We began this chapter with a few examples illustrating why researchers in the social sciences almost never collect information on the entire population that interests them. Instead, they usually select a sample from that population and use the principles of statistical inference to estimate the characteristics, or parameters, of that population based on the characteristics, or statistics, of the sample. In this section, we describe one of the most important concepts in statistical inference—*sampling distribution*. The sampling distribution helps estimate the likelihood of our sample statistics and, therefore, enables us to generalize from the sample to the population.

The Population

To illustrate the concept of the sampling distribution, let's consider as our population the 20 individuals listed in Table 6.2.[2] Our variable, *Y*, is the income (in dollars) of these 20 individuals, and the parameter we are trying to estimate is the mean income.

We use the symbol μ_Y to represent the population mean; the Greek letter mu (μ) stands for the mean, and the subscript *Y* identifies the specific variable, income. Using Formula 3.1, we can calculate the population mean:

$$\mu_Y = \frac{\Sigma Y}{Y} = \frac{Y_1 + Y_2 + Y_3 + Y_4 + Y_5 + \ldots + Y_{20}}{20}$$

$$= \frac{11,350 + 7,859 + 41,654 + 13,445 + 17,458 + \ldots + 25,671}{20}$$

$$= 22,766$$

Table 6.2   The Population: Personal Income (in dollars) for
20 Individuals (hypothetical data)

Individual	Income (Y)
Case 1	11,350 (Y_1)
Case 2	7,859 (Y_2)
Case 3	41,654 (Y_3)
Case 4	13,445 (Y_4)
Case 5	17,458 (Y_5)
Case 6	8,451 (Y_6)
Case 7	15,436 (Y_7)
Case 8	18,342 (Y_8)
Case 9	19,354 (Y_9)
Case 10	22,545 (Y_{10})
Case 11	25,345 (Y_{11})
Case 12	68,100 (Y_{12})
Case 13	9,368 (Y_{13})
Case 14	47,567 (Y_{14})
Case 15	18,923 (Y_{15})
Case 16	16,456 (Y_{16})
Case 17	27,654 (Y_{17})
Case 18	16,452 (Y_{18})
Case 19	23,890 (Y_{19})
Case 20	25,671 (Y_{20})
Mean (μ_Y) = 22,766	Standard deviation (σ_Y) = 14,687

Using Formula 4.2, we can also calculate the standard deviation for this population distribution. We use the Greek symbol sigma (σ) to represent the population's standard deviation and the subscript Y to stand for our variable, income:

$$\sigma_Y = 14{,}687$$

Of course, most of the time, we do not have access to the population. So instead, we draw one sample, compute the mean—the statistic—for that sample, and use it to estimate the population mean—the parameter.

The Sample

Let's pretend that μ_Y is unknown and that we estimate its value by drawing a random sample of three individuals ($N = 3$) from the population of 20 individuals and calculate the mean income for that sample. The incomes included in that sample are as follows:

Case 8	18,342
Case 16	16,456
Case 17	27,654

Now let's calculate the mean for that sample:

$$\overline{Y} = \frac{18,342 + 16,456 + 27,654}{3} = 20,817$$

Note that our sample mean, $\overline{Y}$ = $20,817, differs from the actual population parameter, $22,766. This discrepancy is due to sampling error. **Sampling error** is the discrepancy between a sample estimate of a population parameter and the real population parameter. By comparing the sample statistic with the population parameter, we can determine the sampling error. The sampling error for our example is 1,949 (22,766 − 20,817 = 1,949).

Sampling error   The discrepancy between a sample estimate of a population parameter and the real population parameter.

Now let's select another random sample of three individuals. This time, the incomes included are as follows:

Case 15	18,923
Case 5	17,458
Case 17	27,654

The mean for this sample is

$$\overline{Y} = \frac{18,923 + 17,458 + 27,654}{3} = 21,345$$

The sampling error for this sample is 1,421 (22,766 − 21,345 = 1,421), somewhat less than the error for the first sample we selected.

The Dilemma

Unfortunately, we rarely have information about the actual population parameter. If we did, we would not need to conduct a study! Moreover, few, if any, sample estimates correspond exactly to the actual population parameter. This, then, is our dilemma: If sample estimates vary and if most estimates result in some sort of sampling error, how much confidence can we place in the estimate? On what basis can we infer from the sample to the population?

The Sampling Distribution

The answer to this dilemma is to use a device known as the sampling distribution. The **sampling distribution** is a theoretical probability distribution of all possible sample values for the statistic in which we are interested. If we were to draw all possible random samples of the same size from our population of interest, compute the statistic for each sample, and plot the frequency distribution for that statistic, we would obtain an approximation of the sampling distribution.

Sampling distribution A theoretical probability distribution of all possible sample values for the statistics in which we are interested.

▣ THE SAMPLING DISTRIBUTION OF THE MEAN

Sampling distributions are theoretical distributions, which means that they are never really observed. However, to help grasp the concept of the sampling distribution, let's illustrate how one could be generated from a limited number of samples.

An Illustration

For our illustration, we use one of the most common sampling distributions—the sampling distribution of the mean. The **sampling distribution of the mean** is a theoretical distribution of sample means that would be obtained by drawing from the population all possible samples of the same size.

Sampling distribution of the mean A theoretical probability distribution of sample means that would be obtained by drawing from the population all possible samples of the same size.

Let's go back to our example in which our population is made up of 20 individuals and their incomes. From that population (Table 6.2), we now randomly draw 50 possible samples of size 3, computing the mean income for each sample and replacing it before drawing another.

In our first sample of size 3, we draw three incomes: $8,451, $41,654, and $18,923. The mean income for this sample is

$$\overline{Y} = \frac{8,451 + 41,654 + 18,923}{3} = 23,009$$

We repeat this process 49 more times, each time computing the sample mean and restoring the sample to the original list. Table 6.3 lists the means of the first five and the 50th samples of $N = 3$ that were drawn from the population of 20 individuals. (Note that $\Sigma \overline{Y}$ refers to the sum of all the means computed for each of the samples and M refers to the total number of samples that were drawn.)

A histogram for all 50 samples is displayed in Figure 6.2. This distribution is an example of a sampling distribution of the mean. Note that in its structure, the sampling distribution resembles a frequency distribution of raw scores, except that here each score is a sample mean, and the corresponding frequencies are the number of samples with that particular mean value.

Remember that the distribution depicted in Figure 6.2 is an empirical distribution, whereas the sampling distribution is a theoretical distribution.

The Mean of the Sampling Distribution

Like the sample and population distributions, the sampling distribution can be described in terms of its mean and standard deviation. We use the symbol $\mu_{\overline{Y}}$ to represent the mean of the sampling distribution. The subscript indicates that the variable of this distribution is the mean. To obtain the mean of the sampling distribution, add all the individual sample means ($\Sigma \overline{Y} = 1,237,482$) and divide by the number of samples ($M = 50$). Thus, the mean of the sampling distribution of the mean is actually the mean of means:

$$\mu_{\overline{Y}} = \frac{\Sigma \overline{Y}}{M} = \frac{1,237,482}{50} = 24,750$$

The Standard Error of the Mean

The standard deviation of the sampling distribution is also called the **standard error of the mean**. The standard error of the mean, $\sigma_{\overline{Y}}$, describes how much dispersion there is in the sampling distribution, or how much variability there is in the value of the mean from sample to sample:

$$\sigma_{\overline{Y}} = \frac{\sigma_Y}{\sqrt{N}}$$

Table 6.3   Mean Income of 50 Samples of Size 3

Sample	Mean ($\overline{Y}$)
First	23,009
Second	19,079
Third	18,873
Fourth	26,885
Fifth	21,847
⋮	⋮
Fiftieth	26,645
Total (M) = 50	$\Sigma \overline{Y}$ = 1,237,482

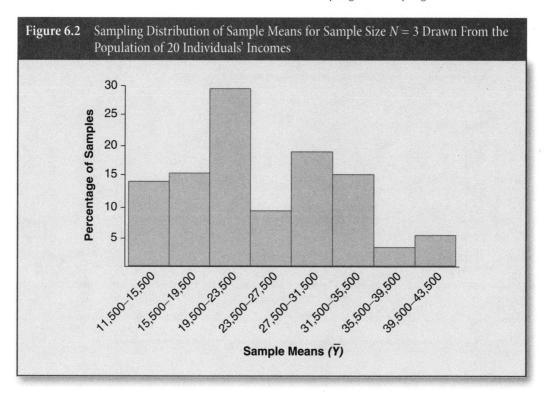

Figure 6.2 Sampling Distribution of Sample Means for Sample Size $N = 3$ Drawn From the Population of 20 Individuals' Incomes

This formula tells us that the standard error of the mean is equal to the standard deviation of the population σ_Y divided by the square root of the sample size (N). For our example, because the population standard deviation is 14,687 and our sample size is 3, the standard error of the mean is

$$\sigma_{\bar{Y}} = \frac{14,687}{\sqrt{3}} = 8,480$$

Standard error of the mean The standard deviation of the sampling distribution of the mean. It describes how much dispersion there is in the sampling distribution of the mean.

▣ THE CENTRAL LIMIT THEOREM

In Figures 6.3a and 6.3b, we compare the histograms for the population and sampling distributions of Tables 6.2 and 6.3. Figure 6.3a shows the population distribution of 20 incomes, with a mean $\mu_Y = 22{,}766$ and a standard deviation $\sigma_Y = 14{,}687$. Figure 6.3b shows the sampling

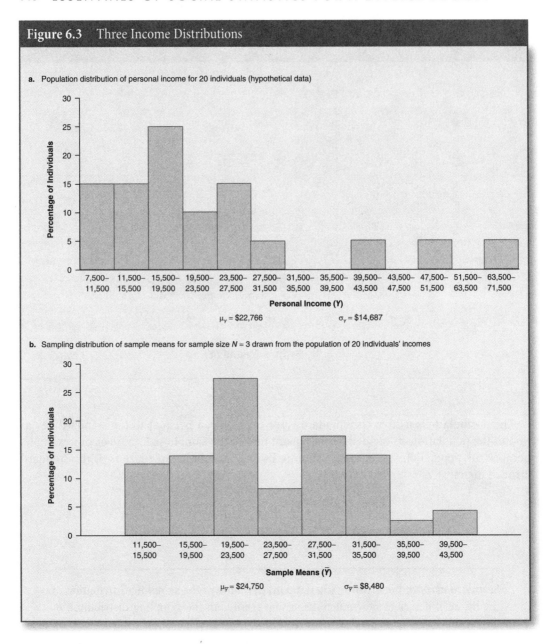

Figure 6.3   Three Income Distributions

a. Population distribution of personal income for 20 individuals (hypothetical data)

$\mu_Y = \$22,766$ $\sigma_Y = \$14,687$

b. Sampling distribution of sample means for sample size $N = 3$ drawn from the population of 20 individuals' incomes

$\mu_{\bar{Y}} = \$24,750$ $\sigma_{\bar{Y}} = \$8,480$

distribution of the means from 50 samples of $N = 3$ with a mean $\mu_{\bar{Y}} = 24,750$ and a standard deviation (the standard error of the mean) $\sigma_{\bar{Y}} = 8,480$. These two figures illustrate some of the basic properties of sampling distributions in general and the sampling distribution of the mean in particular.

First, as can be seen from Figures 6.3a and 6.3b, the shapes of the two distributions differ considerably. Whereas the population distribution is skewed to the right, the sampling distribution of the mean is less skewed—that is, closer to symmetry and a normal distribution.

Second, whereas only a few of the sample means coincide exactly with the population mean, $22,766, the sampling distribution centers on this value. The mean of the sampling distribution is a pretty good approximation of the population mean.

In the discussions that follow, we make frequent references to the mean and standard deviation of the three distributions. To distinguish among the different distributions, we use certain conventional symbols to refer to the means and standard deviations of the sample, the population, and the sampling distribution. Note that we use Greek letters to refer to both the sampling and the population distributions.

The population: We began with the *population distribution* of 20 individuals. This distribution actually exists. It is an empirical distribution that is usually unknown to us. We are interested in estimating the mean income for this population.

The sample: We drew a sample from that population. The *sample distribution* is an empirical distribution that is known to us and is used to help us estimate the mean of the population. We selected 50 samples of $N = 3$ and calculated the mean income. We usually use the sample mean as an estimate of the population mean μ_Y.

The sampling distribution of the mean: For illustration, we generated an approximation of the sampling distribution of the mean, consisting of 50 samples of $N = 3$. *The sampling distribution of the mean* does not really exist. It is a theoretical distribution.

	Mean	*Standard Deviation*
Sample distribution	$\overline{Y}$	S_Y
Population distribution	μ_Y	σ_Y
Sampling distribution of $\overline{Y}$	$\mu_{\overline{Y}}$	$\sigma_{\overline{Y}}$

Third, the variability of the sampling distribution is considerably smaller than the variability of the population distribution. Note that the standard deviation for the sampling distribution ($\sigma_{\overline{Y}} = 8,480$) is almost half that for the population ($\sigma_Y = 14,687$).

These properties of the sampling distribution of the mean are summarized more systematically in one of the most important statistical principles underlying statistical inference. It is called the **central limit theorem**, and it states that if all possible random samples of size N are drawn from a population with a mean μ_Y and a standard deviation σ_Y, then as N becomes larger, the sampling distribution of sample means becomes approximately normal, with mean $\mu_{\overline{Y}}$ equal to the population mean and a standard deviation equal to

$$\sigma_{\overline{Y}} = \frac{\sigma_Y}{\sqrt{N}}$$

Central limit theorem   If all possible random samples of size N are drawn from a population with a mean μ_Y and a standard deviation σ_Y, then as N becomes larger, the sampling distribution of sample means becomes approximately normal, with mean $\mu_{\bar{Y}}$ equal to the population mean and standard deviation equal to

$$\sigma_{\bar{Y}} = \frac{\sigma_Y}{\sqrt{N}}$$

The significance of the central limit theorem is that it tells us that with a *sufficient sample size,* the sampling distribution of the mean will be normal regardless of the shape of the population distribution. Therefore, even when the population distribution is skewed, we can still assume that the sampling distribution of the mean is normal, given random samples of large enough size. Furthermore, the central limit theorem also assures us that (1) as the sample size gets larger, the mean of the sampling distribution becomes equal to the population mean, and (2) as the sample size gets larger, the standard error of the mean (the standard deviation of the sampling distribution of the mean) decreases in size. The standard error of the mean tells how much variability in the sample estimates there is from sample to sample. The smaller the standard error of the mean, the closer (on average) the sample means will be to the population mean. Thus, the larger the sample, the more closely the sample statistic clusters around the population parameter.

✓ *Learning Check*

Make sure you understand the difference between the number of samples that can be drawn from a population and the sample size. Whereas the number of samples is infinite in theory, the sample size is under the control of the investigator.

The Size of the Sample

Although there is no hard-and-fast rule, a general rule of thumb is that when $N = 50$ or more, the sampling distribution of the mean will be approximately normal regardless of the shape of the distribution. However, we can assume that the sampling distribution will be normal even with samples as small as 30 if we know that the population distribution approximates normality.

✓ *Learning Check*

What is a normal population distribution? If you can't answer this question, go back to Chapter 5. You must understand the concept of a normal distribution before you can understand the techniques involved in inferential statistics.

The Significance of the Sampling Distribution and the Central Limit Theorem

To estimate the mean income of a population of 20 individuals, we drew a sample of three cases and calculated the mean income for that sample. Our sample mean, $\bar{Y} = 20{,}817$, differs from the actual population parameter, $\mu_Y = 22{,}766$. When we selected different samples, we found each time that the sample mean differed from the population mean. These discrepancies are due to sampling errors. Had we taken a number of additional samples, we probably would have found that the mean was different each time because every sample differs slightly. Few, if any, sample means would correspond exactly to the actual population mean. Usually we have only one sample statistic as our best estimate of the population parameter.

If sample estimates vary and if most result in some sort of sampling error, how much confidence can we place in the estimate? On what basis can we infer from the sample to the population?

The solution lies in the sampling distribution and its properties. Because the sampling distribution is a theoretical distribution that includes all possible sample outcomes, we can compare our sample outcome with it and estimate the likelihood of its occurrence.

Our knowledge is based on what the central limit theorem tells us about the properties of the sampling distribution of the mean. If our sample size is large enough (at least 50 cases), most sample means will be quite close to the true population mean. It is highly unlikely that our sample mean would deviate much from the actual population mean.

In Chapter 5, we saw that in all normal curves, a constant proportion of the area under the curve lies between the mean and any given distance from the mean when measured in standard deviation units, or Z scores. We can find this proportion in the standard normal table (Appendix A).

Knowing that the sampling distribution of the means is approximately normal, with a mean $\mu_{\bar{Y}}$ and a standard deviation $\sigma_Y/\sqrt{N}$ (the standard error of the mean), we can use Appendix A to determine the probability that a sample mean will fall within a certain distance—measured in standard deviation units, or Z scores—of $\mu_{\bar{Y}}$ or μ_Y. For example, we can expect approximately 68% (or we can say the probability is approximately 0.68) of all sample means to fall within ±1 standard error ($\sigma_{\bar{Y}} = \sigma_Y/\sqrt{N}$, or the standard deviation of the sampling distribution of the mean) of $\mu_{\bar{Y}}$ or μ_Y. This information helps us evaluate the accuracy of our sample estimates.

✓ *Learning Check*

Suppose a population distribution has a mean $\mu_Y = 150$ and a standard deviation $\sigma_Y = 30$, and you draw a simple random sample of N = 100 cases. What is the probability that the mean is between 147 and 153? What is the probability that the sample mean exceeds 153? Would you be surprised to find a mean score of 159? Why? (Hint: To answer these questions, you need to apply what you learned in Chapter 5 about Z scores and areas under the normal curve [Appendix A].) Remember, to translate a raw score into a Z score we used this formula:

$$Z = \frac{Y - \bar{Y}}{S_Y}$$

(Continued)

(Continued)

However, because here we are dealing with a sampling distribution, replace Y with the sample mean $\overline{Y}$, $\overline{Y}$ with the sampling distribution's mean $\mu_{\overline{Y}}$, and s_Y with the standard error of the mean:

$$Z = \frac{\overline{Y} - \mu_{\overline{Y}}}{\sigma_Y / \sqrt{N}}$$

▣ STATISTICS IN PRACTICE: THE CENTRAL LIMIT THEOREM

There are numerous applications of the central limit theorem. As varied as these applications may be, what they have in common is that the information and data they use is based on relatively small random samples taken from considerably larger and often varied populations. For example, a 2012 survey on higher education conducted by *Time Magazine*/Carnegie Corporation reported that 80% of all U.S. adults believe that "at many colleges, the education students receive is not worth what they pay for it." This observation is based on a national random sample of only 1,000 adults and 540 senior administrators at public and private colleges and universities.[3] Similarly, a public opinion poll conducted in 2012 reported that 88% of Americans favor a federal law requiring background checks on all potential gun buyers. The researchers based their conclusion on a sample of 965 adults selected from a population of about 72,000.[4]

Election polls to predict presidential election results provide another example of the benefits of the sampling distribution and the central limit theorem. In November 2012, Barack Obama was reelected president of the United States with 51% of the votes. Mitt Romney, the Republican candidate, received 47% of the vote. Weeks before the election took place, numerous polls called the election within two or three percentage points of the actual result. These predictions were based on samples not larger than about 2,000 registered voters.

What is astounding about these polls (or other empirical studies based on random samples) is not only their accuracy, but that their observations about very large populations (such as all the eligible voters in the United States) are often based on a single sample. Moreover, the size of the sample can be limited to a few hundred respondents or even fewer, regardless of the size of the population! Thus, whether we study a population of 1,000, 10,000, or 100,000, we wouldn't necessarily have to increase the size of the sample.

MAIN POINTS

- Through the process of sampling, researchers attempt to generalize the characteristics of a large group (the population) from a subset (sample) selected from that group. The term *parameter*, associated with the population, refers to the information we are interested in finding out. *Statistic* refers to a corresponding calculated sample statistic.

- A probability sample design allows us to estimate the extent to which the findings based on one sample are likely to differ from what we would find by studying the entire population.

- A simple random sample is chosen in such a way as to ensure that every member of the population and every combination of N members have an equal chance of being chosen.

- The sampling distribution is a theoretical probability distribution of all possible sample values for the statistic in which we are interested. The sampling distribution of the mean is a frequency distribution of all possible

sample means of the same size that can be drawn from the population of interest.

- According to the central limit theorem, if all possible random samples of size N are drawn from a population with a mean μ_Y and a standard deviation σ_Y, then as N becomes larger, the sampling distribution of sample means becomes approximately normal, with mean σ_Y and standard deviation $\sigma_Y/\sqrt{N}$.

- The central limit theorem tells us that with sufficient sample size, the sampling distribution of the mean will be normal regardless of the shape of the population distribution. Therefore, even when the population distribution is skewed, we can still assume that the sampling distribution of the mean is normal, given a large enough randomly selected sample size.

KEY TERMS

central limit theorem
parameter
population
sample

sampling
  distribution
sampling distribution
  of the mean

sampling error
standard error of the
  mean
statistic

ⓈSAGE edge™

Sharpen your skills with SAGE edge at **edge.sagepub.com/ssdsess2e**. **SAGE edge for students** provides a personalized approach to help you accomplish your coursework goals in an easy-to-use learning environment.

CHAPTER EXERCISES

1. Can the standard error of a variable ever be larger than, or even equal in size to, the standard deviation for the same variable? Justify your answer by means of both a formula and a discussion of the relationship between these two concepts.

2. When taking a random sample from a very large population, how does the standard error of the mean change when
   a. the sample size is increased from 100 to 1,600?
   b. the sample size is decreased from 300 to 150?
   c. the sample size is multiplied by 4?

3. The following table shows the number of active military personnel in 2009, by region (including the District of Columbia).

Pacific	229,634	Mountain	89,816	West South Central	177,336
West North Central	64,564	East North Central	26,384	East South Central	68,440
South Atlantic	376,034	Middle Atlantic	41,441	New England	8,579

Sources: U.S. Census Bureau, *Statistical Abstract of the United States: 2012*, Table 508 (data) and U.S. Census Bureau, *Census Regions and Divisions of the United States* (regions).

   a. Calculate the mean and standard deviation for the population.
   b. Now take 10 samples of size 3 from the population. Use simple random sampling. Calculate the mean for each sample.
   c. Once you have calculated the mean for each sample, calculate the mean of means (i.e., add up your 10 sample means and divide by 10). How does this mean compare with the mean for all states?
   d. How does the value of the standard deviation that you calculated in Exercise 3a compare with the value of the standard error (i.e., the standard deviation of the sampling distribution)?
   e. Construct two histograms, one for the distribution of values in the population and the other for the various sample means taken from Exercise 3b. Describe and explain any differences you observe between the two distributions.
   f. It is important that you have a clear sense of the population that we are working with in this exercise. What is the population?

4. The mean family income in Wisconsin, according to the 2011 data, is about $77,531,[5] with a standard deviation (for the population) of approximately $59,750.
   a. Imagine that you are taking a subsample of 200 state residents. What is the probability that your sample mean is between $71,000 and $77,531?
   b. For this same sample size, what is the probability that the sample mean exceeds $85,000?

5. A small population of $N = 10$ has values of 4, 7, 2, 11, 5, 3, 4, 6, 10, and 1.
   a. Calculate the mean and standard deviation for the population.
   b. Take 10 simple random samples of size 3, and calculate the mean for each.
   c. Calculate the mean and standard deviation of all these sample means. How closely does the mean of all sample means match the population mean? How is the standard deviation of the means related to the standard deviation for the population?

6. The following data summarize the 2009 external debt (in millions of dollars) for seven countries.

Country	Debt (in millions of dollars)
Brazil	276,900
Chile	71,600
Colombia	52,200

Country	Debt (in millions of dollars)
India	237,700
Mexico	192,000
South Africa	42,100
Turkey	251,400

Source: U.S. Census Bureau, *Statistical Abstract of the United States: 2012*, Table 1404.

a. Assume that $\sigma_Y = 93{,}500$. Calculate the standard error and interpret. (*Hint:* Consider the formula for the standard error. Since you are provided with the population standard deviation, calculating the standard error requires only minor calculations.)

b. Write a report wherein you discuss the following: the standard error compared with the standard deviation of the population, the shape of the sampling distribution, and suggestions for reducing the standard error.

7. Provided below is SPSS output from the 2010 General Social Survey (GSS). Respondents were polled on how many children they have.

Statistics

NUMBER OF CHILDREN

N	Valid	1496
	Missing	4
Mean		1.97
Median		2.00
Mode		0
Std. Deviation		1.734
Variance		3.007
Range		8
Percentiles	25	.00
	50	2.00
	75	3.00

a. If we were to have a sampling distribution of all possible subsamples of size 100 we could draw from this "population," what would the standard deviation of that sampling distribution (the standard error) be?

b. What are our odds (expressed as a proportion) of drawing a subsample of 100 respondents that has a mean of exactly 2.00 children, or even higher?

8. Provided below is SPSS output from the 2010 GSS. Respondents were polled on their years of education.

Statistics

HIGHEST YEAR OF SCHOOL COMPLI

N	Valid	1496
	Missing	4
Mean		13.47
Median		13.00
Mode		12
Std. Deviation		3.115
Variance		9.705
Range		20
Percentiles	25	12.00
	50	13.00
	75	16.00

If we were to take a subsample of 50 respondents from this "population," what would be the chances of obtaining a sample mean of a high school education or lower? Provide both the Z score and the probability (in the form of a percentage).

Chapter 7

Estimation

Chapter Learning Objectives

❖ Understanding the concept of estimation and the reasons for it
❖ Estimating confidence intervals for means
❖ Understanding the concept of risk and how to reduce it
❖ Estimating confidence intervals for proportions

I n this chapter, we discuss the procedures involved in estimating population means and proportions. These procedures are based on the principles of sampling and statistical inference
discussed in Chapter 6. Knowledge about the sampling distribution allows us to estimate
population means and proportions from sample outcomes and to assess the accuracy of these
estimates.

Every other year, the National Opinion Research Center conducts the General Social Survey (GSS)
on a representative sample of about 1,500 respondents. The GSS, from which many of the examples
in this book are selected, is designed to provide social science researchers with a readily accessible
database of socially relevant attitudes, behaviors, and attributes of a cross section of the U.S. adult
population. For example, in analyzing the responses to the 2010 GSS, researchers found that the average respondent's education was about 13.47 years. This average probably differs from the average of
the population from which the GSS sample was drawn. However, we can establish that in most cases
the sample mean (in this case, 13.47 years) is fairly close to the actual true average in the population.

▣ ESTIMATION DEFINED

The actual average level of education in the United States is a population parameter. The average level
of education in the United States as calculated from the GSS is a sample estimate of that population
parameter. Sample estimates are used to calculate population parameters; the mean number of years
of education of 13.47 calculated from the GSS sample can be used to estimate the mean education of
all adults in the United States.

These are all illustrations of estimation. **Estimation** is a process whereby we select a random sample from a population and use a sample statistic to estimate a population parameter. We can use sample proportions as estimates of population proportions, sample means as estimates of population means, or sample variances as estimates of population variances.

Estimation   A process whereby we select a random sample from a population and use a sample statistic to estimate a population parameter.

Reasons for Estimation

The goal of most research is to find the population parameter. Yet we hardly ever have enough resources to collect information about the entire population. We rarely know the value of the population parameter. On the other hand, we can learn a lot about a population by randomly selecting a sample from that population and obtaining an estimate of the population parameter. The major objective of sampling theory and statistical inference is to provide estimates of unknown population parameters from sample statistics. Figure 7.1 demonstrates the relationship between sample statistics and population parameters.

Point and Interval Estimation

Estimates of population characteristics can be divided into two types: point estimates and interval estimates. **Point estimates** are sample statistics used to estimate the exact value of a population parameter. When the Gallup organization reports that 59% of Americans support gay and lesbian relations, they are using a point estimate. Similarly, if we reported the average level of education of the population of adult Americans to be exactly 13.47 years, we would be using a point estimate.

Point estimate   A sample statistic used to estimate the exact value of a population parameter.

The problem with point estimates is that sample statistics vary, usually resulting in some sort of sampling error. When we use a sample statistic to estimate the exact value of a population parameter, we never know how accurate it is.

One method of increasing accuracy is to use an interval estimate rather than a point estimate. In interval estimation, we identify a range of values within which the population parameter may fall. This range of values is called a **confidence interval (CI)**. Instead of using a single value, 13.47 years, as an estimate of the mean education of adult Americans, we could say that the population mean is somewhere between 12 and 14 years.

Confidence interval (CI)   A range of values defined by the confidence level within which the population parameter is estimated to fall. Sometimes confidence intervals are referred to as margin of error.

When we use confidence intervals to estimate population parameters, such as mean educational levels, we can also evaluate the accuracy of this estimate by assessing the likelihood that any given interval will contain the mean. This likelihood, expressed as a percentage or a probability, is called a **confidence level**. Confidence intervals are defined in terms of confidence levels. Thus, by selecting a 95% confidence level, we are saying that there is a .95 probability—or

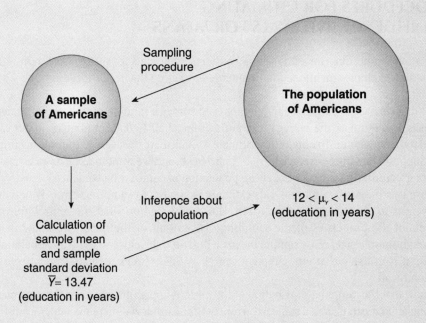

Figure 7.1   Estimation as a Type of Inference

The goal of inferential statistics is to say something meaningful about the population based entirely on information from a sample of that population. A confidence interval attempts to do just that: By knowing a sample mean, sample size, and sample standard deviation, we are able to say something about the population from which that sample was drawn.

Sampling procedure

A sample of Americans

The population of Americans

Inference about population

$12 < \mu_y < 14$
(education in years)

Calculation of sample mean and sample standard deviation
$\overline{Y} = 13.47$
(education in years)

We know exactly what our sample mean is. Combining this information with the sample standard deviation and sample size gives us a range within which we can confidently say that the population mean falls.

95 chances out of 100—that a specified interval will contain the population mean. Confidence intervals can be constructed for any level of confidence, but the most common ones are the 90%, 95%, and 99% levels. You should also know that confidence intervals are sometimes referred to in terms of **margin of error**. In short, margin of error is simply the radius of a confidence interval. If we select a 95% confidence level, we would have a 5% chance of our interval being incorrect.

Confidence level   The likelihood, expressed as a percentage or a probability, that a specified interval will contain the population parameter.

Margin of error   The radius of a confidence interval.

✓ *Learning*
Check   *What is the difference between a point estimate and a confidence interval?*

▣ PROCEDURES FOR ESTIMATING CONFIDENCE INTERVALS FOR MEANS

To illustrate the procedure for establishing confidence intervals for means, we'll reintroduce one of the research examples mentioned in Chapter 6—assessing the needs of commuting students on our campus.

Recall that we have been given enough money to survey a random sample of 500 students. One of our tasks is to estimate the average commuting time of all 15,000 commuters on our campus—the population parameter. To obtain this estimate, we calculate the average commuting time for the sample. Suppose the sample average is $\bar{Y} = 7.5$ hr/week, and we want to use it as an estimate of the true average commuting time for the entire population of commuting students.

Because it is based on a sample, this estimate is subject to sampling error. We do not know how close it is to the true population mean. However, based on what the central limit theorem tells us about the properties of the sampling distribution of the mean, we know that with a large enough sample size, most sample means will tend to be close to the true population mean. Therefore, it is unlikely that our sample mean, $\bar{Y} = 7.5$ hr/week, deviates much from the true population mean.

We know that the sampling distribution of the mean is approximately normal with a mean equal to the population mean μ_Y and a standard error $\sigma_{\bar{Y}}$ (standard deviation of the sampling distribution) as follows:

$$\sigma_{\bar{Y}} = \frac{\sigma_Y}{\sqrt{N}} \tag{7.1}$$

This information allows us to use the normal distribution to determine the probability that a sample mean will fall within a certain distance—measured in standard deviation (standard error) units or Z scores—of μ_Y or $\mu_{\bar{Y}}$. We can make the following assumptions:

- A total of 68% of all random sample means will fall within ±1 standard error of the true population mean.
- A total of 95% of all random sample means will fall within ±1.96 standard errors of the true population mean.
- A total of 99% of all random sample means will fall within ±2.58 standard errors of the true population mean.

On the basis of these assumptions and the value of the standard error, we can establish a range of values—a confidence interval—that is likely to contain the actual population mean. We can also evaluate the accuracy of this estimate by assessing the likelihood that this range of values will actually contain the population mean.

The general formula for constructing a confidence interval (CI) for any level is

$$\mathrm{CI} = \bar{Y} \pm Z(\sigma_{\bar{Y}}) \tag{7.2}$$

Note that to calculate a confidence interval, we take the sample mean and add to or subtract from it the product of a Z value and the standard error.

The Z score we choose depends on the desired confidence level. For example, to obtain a 95% confidence interval, we would choose a Z of 1.96 because we know (from Appendix A) that 95% of the area under the curve lies between ±1.96. Similarly, for a 99% confidence level, we would choose a Z score of 2.58. The relationship between the confidence level and Z is illustrated in Figure 7.2 for the 95% and 99% confidence levels.

✓ *Learning Check*

To understand the relationship between the confidence level and Z, review the material in Chapter 5. What would be the appropriate Z value for a 98% confidence interval?

Determining the Confidence Interval

To determine the confidence interval for a mean, follow these steps:

1. Calculate the standard error of the mean.

2. Decide on the level of confidence, and find the corresponding Z value.

3. Calculate the confidence interval.

4. Interpret the results.

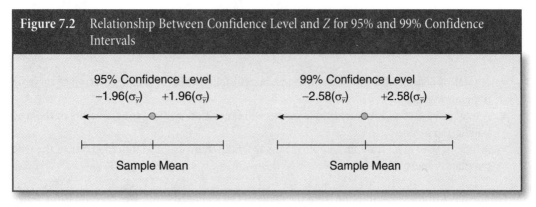

Figure 7.2   Relationship Between Confidence Level and Z for 95% and 99% Confidence Intervals

Source: David Freedman, Robert Pisani, Roger Purves, and Ani Akhikari, *Statistics*, 2nd ed. (New York: Norton, 1991).

Let's return to the problem of estimating the mean commuting time of the population of students on our campus. How would you find the 95% confidence interval?

Calculating the Standard Error of the Mean

Let's suppose that the standard deviation for our population of commuters is $\sigma_Y = 1.5$. We calculate the standard error for the sampling distribution of the mean:

$$\sigma_{\bar{Y}} = \frac{\sigma_Y}{\sqrt{N}} = \frac{1.5}{\sqrt{500}} = 0.07$$

Deciding on the Level of Confidence and Finding the Corresponding Z Value

We decide on a 95% confidence level. The Z value corresponding to a 95% confidence level is 1.96.

Calculating the Confidence Interval

The confidence interval is calculated by adding and subtracting from the observed sample mean the product of the standard error and Z:

$$95\% \text{ CI} = 7.5 \pm 1.96(0.07)$$
$$= 7.5 \pm 0.14$$
$$= 7.36 \text{ to } 7.64$$

The 95% CI for the mean commuting time is illustrated in Figure 7.3.

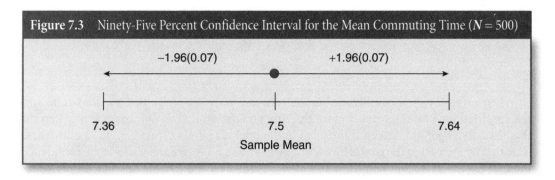

Figure 7.3 Ninety-Five Percent Confidence Interval for the Mean Commuting Time (*N* = 500)

Interpreting the Results

We can be 95% confident that the actual mean commuting time—the true population mean—is not less than 7.36 hours and not greater than 7.64 hours. In other words, if we collected a large number of samples (*N* = 500) from the population of commuting students, 95 times out of 100, the true population mean would be included within our computed interval. With a 95% confidence level, there is a 5% risk that we are wrong. Five times out of 100, the true population mean will not be included in the specified interval.

Remember that we can never be sure whether the population mean is actually contained within the confidence interval. Once the sample is selected and the confidence interval defined, the confidence interval either does or does not contain the population mean—but we will never be sure.

✓ *Learning Check*

What is the 90% confidence interval for the mean commuting time? (Hint: First, find the Z value associated with a 90% confidence level.)

To further illustrate the concept of confidence intervals, let's suppose that we draw 10 different samples (*N* = 500) from the population of commuting students. For each sample mean, we construct a 95% confidence interval. Figure 7.4 displays these confidence intervals. Each horizontal line represents a 95% confidence interval constructed around a sample mean (marked with a circle).

The vertical line represents the population mean. Note that the horizontal lines that intersect the vertical line are the intervals that contain the true population mean. Only 1 out of the 10 confidence intervals does not intersect the vertical line, meaning it does not contain the population mean. Drawing all possible samples, 95% of the intervals would include the true population mean, and 5% would not.

Reducing Risk

One way to reduce the risk of being incorrect is by increasing the level of confidence. For instance, we can increase our confidence level from 95% to 99%. The 99% confidence interval for our commuting example is as follows:

$$99\% \ CI = 7.5 \pm 2.58(0.07)$$

$$= 7.5 \pm 0.18$$

$$= 7.32 \ to \ 7.68$$

When using the 99% confidence interval, there is only a 1% risk that we are wrong and the specified interval does not contain the true population mean. We can be almost certain that the true population mean is included in the interval ranging from 7.32 to 7.68 hr/week. Note that by increasing the confidence level, we have also increased the width of the confidence interval from 0.28 (7.36–7.64) to 0.36 hours (7.32–7.68), thereby making our estimate less precise.

You can see in Table 7.1 and Figure 7.5 that there is a trade-off between achieving greater confidence in an estimate and the precision of that estimate. Although using a higher level of confidence (e.g., 99%) increases our confidence that the true population mean is included in our confidence interval, the estimate becomes less precise as the width of the interval increases. Although we are only 95% confident that the interval ranging between 7.36 and 7.64 hours includes the true population mean, it is a more precise estimate than the 99% interval ranging from 7.32 to 7.68 hours. The relationship between the confidence level and the precision of the confidence interval is illustrated in Figure 7.5. Table 7.1 lists three commonly used confidence levels along with their corresponding Z values.

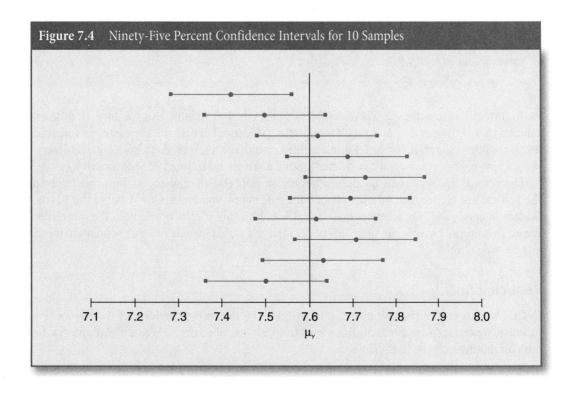

Figure 7.4   Ninety-Five Percent Confidence Intervals for 10 Samples

Table 7.1   Confidence Levels and Corresponding *Z* Values

Confidence Level	Z Value
90%	1.65
95%	1.96
99%	2.58

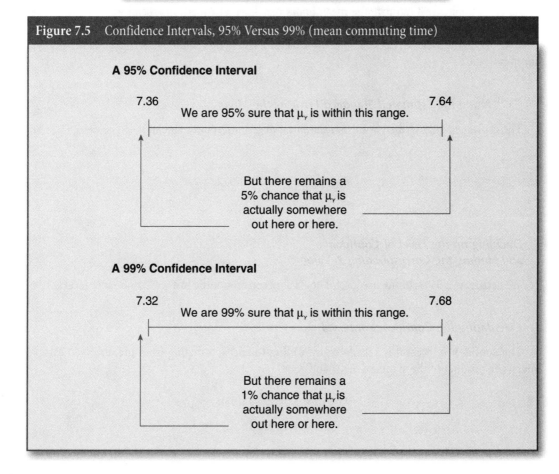

Figure 7.5   Confidence Intervals, 95% Versus 99% (mean commuting time)

A 95% Confidence Interval

7.36   7.64

We are 95% sure that μ_Y is within this range.

But there remains a 5% chance that μ_Y is actually somewhere out here or here.

A 99% Confidence Interval

7.32   7.68

We are 99% sure that μ_Y is within this range.

But there remains a 1% chance that μ_Y is actually somewhere out here or here.

Estimating Sigma

To calculate confidence intervals, we need to know the standard error of the sampling distribution. The standard error is a function of the population standard deviation and the sample size:

$$\sigma_{\bar{Y}} = \frac{\sigma_Y}{\sqrt{N}}$$

In our commuting example, we have been using a hypothetical value, $\sigma_Y = 1.5$, for the population standard deviation. Typically, both the mean (μ_Y) and the standard deviation (σ_Y) of the population are unknown to us. When $N \geq 50$, however, the sample standard deviation, S_Y, is a good estimate of $\sigma_{\bar{Y}}$. The standard error is then calculated as follows:

$$S_{\bar{Y}} = \frac{S_Y}{\sqrt{N}}$$ (7.3)

As an example, we'll estimate the mean hours per day that Americans spend watching television based on the 2010 GSS. The mean hours per day spent watching television for a sample of $N = 1,013$ is $\bar{Y} = 3.01$, and the standard deviation $S_Y = 2.65$ hours. Let's determine the 95% confidence interval for these data.

Calculating the Estimated Standard Error of the Mean

The estimated standard error for the sampling distribution of the mean is

$$S_{\bar{Y}} = \frac{S_Y}{\sqrt{N}} = \frac{2.65}{\sqrt{1,013}} = 0.08$$

Deciding on the Level of Confidence and Finding the Corresponding Z Value

We decide on a 95% confidence level. The Z value corresponding to a 95% confidence level is 1.96.

Calculating the Confidence Interval

The confidence interval is calculated by adding to and subtracting from the observed sample mean the product of the standard error and Z:

$$95\% \ CI = 3.01 \pm 1.96(0.08)$$
$$= 3.01 \pm 0.16$$
$$= 2.85 \text{ to } 3.17$$

Interpreting the Results

We can be 95% confident that the actual mean hours spent watching television by Americans from which the GSS sample was taken is not less than 2.85 hours and not greater than 3.17 hours. In other words, if we drew a large number of samples ($N = 1,013$) from this population, then 95 times out of 100, the true population mean would be included within our computed interval.

Sample Size and Confidence Intervals

Researchers can increase the precision of their estimate by increasing the sample size. In Chapter 6, we learned that larger samples result in smaller standard errors, and therefore, sampling distributions are more clustered around the population mean (Figure 6.3). A more tightly clustered sampling distribution means that our confidence intervals will be narrower and more precise. To illustrate the relationship between sample size and the standard error, and thus the confidence interval, let's calculate the 95% confidence interval for our GSS data with (1) a sample of $N = 195$ and (2) a sample of $N = 1,987$.

With a sample size $N = 195$, the estimated standard error for the sampling distribution is

$$S_{\bar{Y}} = \frac{S_Y}{\sqrt{N}} = \frac{2.65}{\sqrt{195}} = 0.19$$

and the 95% confidence interval is

$$95\% \text{ CI} = 3.01 \pm 1.96(0.19)$$

$$= 3.01 \pm 0.37$$

$$= 2.64 \text{ to } 3.38$$

With a sample size $N = 1,987$, the estimated standard error for the sampling distribution is

$$S_{\bar{Y}} = \frac{S_Y}{\sqrt{N}} = \frac{2.65}{\sqrt{1,987}} = 0.06$$

and the 95% confidence interval is

$$95\% \text{ CI} = 3.01 \pm 1.96(0.06)$$

$$= 3.01 \pm 0.12$$

$$= 2.89 \text{ to } 3.13$$

In Table 7.2, we summarize the 95% confidence intervals for the mean number of hours watching television for these three sample sizes: $N = 195$, $N = 1,013$, and $N = 1,987$.

Note that there is an inverse relationship between sample size and the width of the confidence interval. The increase in sample size is linked with increased precision of the confidence interval. The 95% confidence interval for the GSS sample of 195 cases is 0.74 hours. But the interval widths decrease to 0.32 and 0.24 hours, respectively, as the sample sizes increase to $N = 1,013$ and then to $N = 1,987$. We had to nearly double the size of the sample (from 1,013 to 1,987) to reduce the confidence interval by about one-fourth (from 0.32 to 0.24 hours). Researchers have to consider at what point the increase in precision is too small to justify the additional cost associated with a larger sample.

Table 7.2 Ninety-Five Percent Confidence Interval and Width for Mean Number of Hours per Day Watching Television for Three Different Sample Sizes

Sample Size (N)	Confidence Interval	Interval Width	S_Y	$S_{\bar{Y}}$
195	2.64–3.38	0.74	2.65	0.19
1,013	2.85–3.17	0.32	2.65	0.08
1,987	2.89–3.13	0.24	2.65	0.06

▣ CONFIDENCE INTERVALS FOR PROPORTIONS

Confidence intervals can also be computed for sample proportions or percentages to estimate population proportions or percentages. The procedures for estimating proportions and percentages are identical. Any of the formulas presented for proportions can be applied to percentages, and vice versa. We can obtain a confidence interval for a percentage by calculating the confidence interval for a proportion and then multiplying the result by 100.

Earlier, we saw that the sampling distribution of the means underlies the process of estimating population means from sample means. Similarly, the *sampling distribution of proportions* underlies the estimation of population proportions from sample proportions. Based on the central limit theorem, we know that with sufficient sample size the sampling distribution of proportions is approximately normal, with mean μ_p equal to the population proportion π and with a standard error of proportions (the standard deviation of the sampling distribution of proportions) equal to

$$\sigma_p = \sqrt{\frac{(\pi)(1-\pi)}{N}} \qquad (7.4)$$

where

σ_p = the standard error of proportions

π = the population proportion

N = the population size

However, since the population proportion, π, is unknown to us (that is what we are trying to estimate), we can use the sample proportion, p, as an estimate of π. The estimated standard error then becomes

$$S_p = \sqrt{\frac{(p)(1-p)}{N}} \qquad (7.5)$$

where

S_p = the estimated standard error of proportions

p = the sample proportion

N = the sample size

As an example, let's calculate the estimated standard error for the survey by Gallup. Based on a random sample of 1,535 adults, the percentage who support gay and lesbian relations was estimated to be 59%. Based on Formula 7.5, with $p = 0.59$, $1 - p = (1 - 0.59) = 0.41$, and $N = 1,535$, the standard error is $S_p = \sqrt{(0.59)(1-0.59)/1,535} = 0.013$. We will have to consider two factors to meet the assumption of normality with the sampling distribution of proportions: (1) the sample size N and (2) the sample proportions p and $1 - p$. When p and $1 - p$ are about 0.50, a sample size of at least 50 is sufficient. But when $p > 0.50$ (or $1 - p < 0.50$), a larger sample is required to meet the assumption of normality. Usually, a sample of 100 or more is adequate for any single estimate of a population proportion.

Procedures for Estimating Proportions

Because the sampling distribution of proportions is approximately normal, we can use the normal distribution to establish confidence intervals for proportions in the same manner that we used the normal distribution to establish confidence intervals or means.

The general formula for constructing confidence intervals for proportions for any level of confidence is

$$CI = p \pm Z(S_p) \tag{7.6}$$

where

CI = the confidence interval

p = the observed sample proportion

Z = the Z score corresponding to the confidence level

S_p = the estimated standard error of proportions

To determine the confidence interval for a proportion, we follow the same steps that were used to find confidence intervals for means.

To illustrate these steps, we use the results of a Gallup survey on the percentage of Americans who support gay and lesbian relations.

Calculating the Estimated Standard Error of the Proportion

The standard error of the proportion 0.59 (59%) with a sample $N = 1,535$ is 0.013.

*Deciding on the Desired Level of Confidence
and Finding the Corresponding Z Value*

We choose the 95% confidence level. The *Z* score corresponding to a 95% confidence level is 1.96.

Calculating the Confidence Interval

We calculate the confidence interval by adding to and subtracting from the observed sample proportion the product of the standard error and *Z*:

$$95\% \text{ CI} = 0.59 \pm 1.96(0.013)$$

$$= 0.59 \pm 0.025$$

$$= 0.565 \text{ to } 0.615$$

Interpreting the Results

We are 95% confident that the true population proportion is somewhere between 0.565 and 0.615. In other words, if we drew a large number of samples from the population of adults, then 95 times out of 100, the confidence interval we obtained would contain the true population proportion. We can also express this result in percentages and say that we are 95% confident that the true population percentage of Americans who support gay and lesbian relations is included somewhere within our computed interval of 56.5% to 61.5%.

Note that with a 95% confidence level, there is a 5% risk that we are wrong. If we continued to draw large samples from this population, in 5 out of 100 samples the true population proportion would not be included in the specified interval.

We can decrease our risk by increasing the confidence level from 95% to 99%.

$$99\% \text{ CI} = 0.59 \pm 2.58(0.013)$$

$$= 0.59 \pm 0.034$$

$$= 0.556 \text{ to } 0.624$$

When using the 99% confidence interval, we can be almost certain (99 times out of 100) that the true population proportion is included in the interval ranging from 0.556 (55.6%) to 0.624 (62.4%). However, as we saw earlier, there is a trade-off between achieving greater confidence in making an estimate and the precision of that estimate.[1,2]

▣ STATISTICS IN PRACTICE: THE 2012 BENGHAZI TERRORIST ATTACK INVESTIGATION

Poll or survey results may be limited to a single estimate of a parameter. For instance, political pollsters could estimate the percentage of Americans who closely follow controversial news stories. Most survey studies, however, are not limited to single estimates for the overall population. Often, separate estimates are reported for subgroups within the overall population of interest. In a report released on May 20, 2013, Gallup compared the percentage of Democrats and Republicans who were closely following the September 11, 2012, Benghazi terrorist attacks. They were interested in exploring whether or not there were differences across groups with different political affiliations.[3]

When estimates are reported for subgroups, the confidence intervals are likely to vary from subgroup to subgroup. Each confidence interval is based on the confidence level, the standard error of the proportion (which can be estimated from p), and the sample size. Even when a confidence interval is reported only for the overall sample, we can easily compute separate confidence intervals for each of the subgroups if the confidence level and the size of each of the subgroups are included.

To illustrate this, let's calculate the 95% confidence intervals for the proportions of Democrats and Republicans who were closely following the Benghazi terrorist attack investigation. Out of 317 Democrats in the sample, 0.18 (or 18%) were closely following the investigation. In contrast, of the 247 Republicans surveyed, 0.34 (or 34%) were closely following the same investigation.

Calculating the Estimated Standard Error of the Proportion

The estimated standard error for the proportion of Democrats is

$$S_p = \sqrt{\frac{(0.18)(1-0.18)}{317}} = 0.02$$

The estimated standard error for the proportion of Republicans is

$$S_p = \sqrt{\frac{(0.34)(1-0.34)}{247}} = 0.03$$

Deciding on the Desired Level of Confidence and Finding the Corresponding *Z* Value

We choose the 95% confidence level, with a corresponding Z value of 1.96.

Calculating the Confidence Interval

For Democrats,

$$95\% \text{ CI} = 0.18 \pm 1.96(0.02)$$
$$= 0.18 \pm 0.04$$
$$= 0.14 \text{ to } 0.22$$

and for Republicans,

$$95\% \text{ CI} = 0.34 \pm 1.96(0.03)$$
$$= 0.34 \pm 0.06$$
$$= 0.28 \text{ to } 0.40$$

The 95% confidence intervals for the proportions of Democrats and Republicans surveyed who closely followed the Benghazi terrorist attack investigation are illustrated in Figure 7.6.

Interpreting the Results

We are 95% confident that the true population proportion who closely followed the Benghazi terrorist attack investigation was between 0.14 and 0.22 (or between 14% and 22%) for Democrats, and between 0.28 and 0.40 (or between 28% and 40%) for Republicans. Based on the sample, it is clear that among Americans there were partisan differences in how closely persons followed this investigation. Republicans were much more likely than Democrats to closely follow the Benghazi terrorist attack investigation.

◙ THE MARGIN OF ERROR

The most common application of estimation using confidence intervals (also called the margin of error) is demonstrated in opinion and election polls. Pollsters usually interview a random sample representative of a defined population to assess their opinions on a certain issue or their voting preferences in a particular election. For example, a 2013 Pew Research Center poll of 1,504 adults reported that 75% believe U.S. immigration policy needs at least a major overhaul.[4] Also reported was the poll's margin of error of plus or minus 2.9 percentage points. The margin of error in a poll tells us how well a randomly selected sample represents the population from which it was

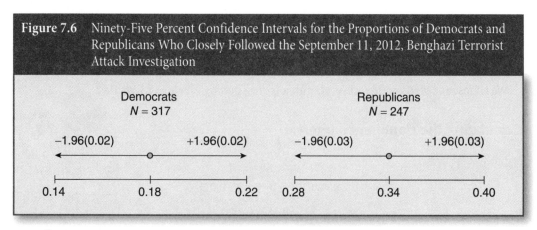

Figure 7.6   Ninety-Five Percent Confidence Intervals for the Proportions of Democrats and Republicans Who Closely Followed the September 11, 2012, Benghazi Terrorist Attack Investigation

Source: "Partisan Interest, Reactions to IRS and AP Controversies," *Pew Research Center,* May 20, 2013.

selected. The Pew Research Center poll results indicate that we can be 95% confident that the true percentage of Americans who feel U.S. immigration policy is in need of at least a major overhaul is somewhere between 72.1% (75% − 2.9%) and 77.9% (75% + 2.9%).

A margin of error of 2.9% means that with a sample of 1,504 individuals, we can be 95% confident that the true value of 75% is within ±2.9 percentage points of what it would be if the entire American adult population had been polled.

MAIN POINTS

- The goal of most research is to find population parameters. The major objective of sampling theory and statistical inference is to provide estimates of unknown parameters from sample statistics.

- Researchers make point estimates and interval estimates. Point estimates are sample statistics used to estimate the exact value of a population parameter. Interval estimates are ranges of values within which the population parameter may fall.

- Confidence intervals can be used to estimate population parameters such as means or proportions. Their accuracy is defined with the confidence level. The most common confidence levels are 90%, 95%, and 99%.

- To establish a confidence interval for a mean or a proportion, add or subtract from the mean or the proportion the product of the standard error and the Z value corresponding to the confidence level.

KEY TERMS

confidence interval (CI)

confidence level estimation

margin of error point estimate

$SAGE edge™

Sharpen your skills with SAGE edge at **edge.sagepub.com/ssdsess2e**. **SAGE edge for students** provides a personalized approach to help you accomplish your coursework goals in an easy-to-use learning environment.

CHAPTER EXERCISES

1. In the 2011 National Crime Victimization Study, the Federal Bureau of Investigation (FBI) found that 16.2% of Americans age 12 or older had been victims of crime during a 1-year period. This result was based on a sample of 143,120 persons.
   a. Estimate the percentage of U.S. adults who were victims at the 90% confidence level. State in words the meaning of the result.
   b. Estimate the percentage of victims at the 99% confidence level.

   c. Imagine that the FBI cuts the sample size in half but finds the same value of 16.2% for the percentage of victims in the second sample. By how much would the 90% confidence interval increase? By how much would the 99% confidence interval increase?

   d. Considering your answers to (a), (b), and (c), can you suggest why national surveys, such as those by Gallup, Roper, or *The New York Times*, typically take samples of size 1,000 to 1,500?

2. Use the data on education from Chapter 5, Exercise 4.

	Mean	Standard Deviation	N
Lower class	11.61	2.67	123
Working class	12.80	2.85	697
Middle class	14.45	3.08	626
Upper class	15.45	2.98	38

   a. Construct the 95% confidence interval for the mean number of years of education for lower-class and middle-class respondents.

   b. Construct the 99% confidence interval for the mean number of years of education for lower-class and middle-class respondents.

   c. As our confidence in the result increases, how does the size of the confidence interval change? Explain why this is so.

3. There has been a great deal of discussion about global warming in recent years. In 2012, the Pew Research Center conducted a survey of 1,511 Americans to assess their opinions of global warming.[5] The data show that 589 respondents of the 1,511 surveyed felt global warming is a very serious problem.

   a. Estimate the proportion of all adult Americans who felt global warming is a very serious problem at the 95% confidence interval.

   b. Estimate the proportion of all adult Americans who felt global warming is a very serious problem at the 99% confidence interval.

   c. If you were going to write a report on this poll result, would you prefer to use the 99% or 95% confidence interval? Explain why.

4. Use the data in Chapter 4, Exercise 3, about occupational prestige and education.

PRESTG80			Statistic
RS OCCUPATIONAL PRESTIGE SCORE (1980)	High School Diploma	Mean	40.59
		Median	40.00
		Std. Deviation	11.419

PRESTG80			*Statistic*
		Minimum	17
		Maximum	75
		Range	58
		Interquartile Range	17
	Bachelor's Degree	Mean	50.95
		Median	51.00
		Std. Deviation	12.930
		Minimum	23
		Maximum	75
		Range	52
		Interquartile Range	23

    a. Construct the 90% confidence interval for occupational prestige for respondents with only high school diplomas ($N = 702$).

    b. Construct the 90% confidence interval for occupational prestige for respondents with bachelor's degrees ($N = 270$). State in words the meaning of the result.

    c. Use these statistics to discuss differences in occupational prestige scores by educational attainment.

5. Gallup conducted a survey in April 2010 to determine the congressional voting preferences of American voters.[6] They found that 51% of the male voters preferred a Republican candidate to a Democratic candidate in a sample of 5,490 registered voters. Gallup asks you, their statistical consultant, to tell them whether you could declare the Republican candidate as the likely winner of men's votes if there were an election today. What is your advice? Why?

6. You have been doing research for your statistics class on the prevalence of severe binge drinking among teens. You have decided to use 2011 Monitoring the Future (MTF) data that have a scale (from 0 to 14) measuring the number of times teens drank 10 or more alcoholic beverages in a single sitting in the past 2 weeks.

    a. According to 2011 MTF data, the average severe binge drinking score, for this sample of 914 teens, is 1.27, with a standard deviation of 0.80. Construct the 95% confidence interval for the true average severe binge drinking score.

    b. One of your classmates, who claims to be good at statistics, complains about your confidence interval calculation. She or he asserts that the severe binge drinking scores are not normally distributed, which in turn makes the confidence interval calculation meaningless. Assume that she or he is correct about the distribution of severe binge drinking scores. Does that imply that the calculation of a confidence interval is not appropriate? Why or why not?

7. From the 2010 GSS subsample, we find that 72.7% of respondents believe in some form of life after death ($N = 1,500$).

   a. What is the 95% confidence interval for the percentage of the U.S. population who believe in life after death?

   b. Without doing any calculations, make an educated guess at the lower and upper bounds of 90% and 99% confidence intervals.

8. A social service agency plans to conduct a survey to determine the mean income of its clients. The director of the agency prefers that you measure the mean income very accurately, to within ±$500. From a sample taken 2 years ago, you estimate that the standard deviation of income for this population is about $5,000. Your job is to figure out the necessary sample size to reduce sampling error to ±$500.

   a. Do you need to have an estimate of the current mean income to answer this question? Why or why not?

   b. What sample size should be drawn to meet the director's requirement at the 95% level of confidence? (*Hint:* Use the formula for a confidence interval and solve for N, the sample size.)

   c. What sample size should be drawn to meet the director's requirement at the 99% level of confidence?

9. Data from a 2010 GSS subsample show that the mean number of children per respondent was 1.97, with a standard deviation of 1.73. A total of 1,496 people answered this question. Estimate the population mean number of children per adult using a 90% confidence interval.

10. A sample of the 2011 MTF survey suggests that adolescents are divided in terms of their attitudes toward others trying marijuana at least once. In fact, 49.3% of the 1,202 respondents who answered the question reported that they don't disapprove of others trying marijuana at least once. Estimate at the 95% and 99% confidence levels the proportion of all adolescents who don't disapprove of others trying marijuana at least once.

11. According to a 2010 survey by the Pew Research Center, 61% of adult Americans use social networking websites such as Twitter and Facebook.[7] Interestingly, 21% of the 2,257 adults surveyed said that they used social network websites to access the 2010 midterm elections. What is the 95% confidence interval for the percentage of American adults who use social networking websites to connect to elections?

12. According to a report published by the Pew Research Center in February 2010, 61% of millennials (Americans in their teens and 20s) think that their generation has a unique and distinctive identity ($N = 527$).[8]

   a. Calculate the 95% confidence interval to estimate the percentage of millennials who believe that their generation has a distinctive identity as compared with other generations (generation X, baby boomers, or the silent generation).

   b. Calculate the 99% confidence interval.

   c. Are both these results compatible with the conclusion that the majority of millennials believe that they have a unique identity that separates them from the previous generations?

13. Whether one views homosexual relations as wrong is closely related to whether one views homosexuality as a biological trait or the outcome of one's environment and/or socialization. Thus, it is not surprising that several religious groups that condemn homosexual relations have proclaimed their ability to "cure" gays of their sexual orientation. After all, their assumption is that homosexuality is not a trait that a person is born with. In 2010, GSS respondents ($N = 930$) were asked what they

thought about homosexual relations. The data show that 50.2% believed that homosexual relations are always wrong, while 37.2% believed that homosexual relations are not wrong at all.

a. For each reported percentage, calculate the 95% confidence interval.

b. Approximately 13% of GSS respondents were in the middle, some saying that homosexual relations are almost always wrong or sometimes wrong. Calculate the 95% confidence interval.

c. What conclusions can you draw about the public's opinions of homosexual behavior based on your calculations?

14. Provided below is SPSS output from the 2010 GSS. Two sets of respondents, those who found life exciting and those who found life dull, were polled on how many hours of television they watch daily.

Descriptives

IS LIFE EXCITING OR DULL				Statistic	Std. Error
HOURS PER DAY WATCHING TV	EXCITING	Mean		2.34	.129
		99% Confidence Interval for Mean	Lower Bound	2.01	
			Upper Bound	2.68	
		5% Trimmed Mean		2.16	
		Median		2.00	
		Variance		3.983	
		Std. Deviation		1.996	
		Minimum		0	
		Maximum		21	
		Range		21	
		Interquartile Range		2	
		Skewness		4.049	.157
		Kurtosis		32.025	.313
	DULL	Mean		6.14	1.048
		99% Confidence Interval for Mean	Lower Bound	3.24	
			Upper Bound	9.03	
		5% Trimmed Mean		5.49	
		Median		4.00	
		Variance		31.837	
		Std. Deviation		5.642	
		Minimum		0	
		Maximum		24	
		Range		24	
		Interquartile Range		6	
		Skewness		2.140	.434
		Kurtosis		4.886	.845

a. Can we say, with 99% confidence, that people who find life exciting watch less television than those who find it dull?

b. What might account for the fact that the interval for "dull" is much wider?

15. Provided below is SPSS output from the 2010 GSS. Eight hundred sixty-five respondents were polled on what they believe is the ideal number of children.

Descriptives

			Statistic	Std. Error
IDEAL NUMBER OF CHILDREN	Mean		2.54	.032
	95% Confidence Interval for Mean	Lower Bound	2.47	
		Upper Bound	2.60	
	5% Trimmed Mean		2.48	
	Median		2.00	
	Variance		.883	
	Std. Deviation		.940	
	Minimum		0	
	Maximum		7	
	Range		7	
	Interquartile Range		1	
	Skewness		1.236	.083
	Kurtosis		3.075	.166

a. Looking at these data, is it possible that 2.3 children is actually the American ideal? Answer for a 95% confidence level.

b. Looking at these data, is it possible that 2.3 children is actually the American ideal? Answer for a 99% confidence level. Please show your work for this part of the problem.

Chapter 8

Testing Hypotheses

- ❖ Understanding the assumptions of statistical hypothesis testing
- ❖ Defining and applying the components of hypothesis testing: the research and null hypotheses, sampling distribution, and test statistic
- ❖ Understanding what it means to reject or fail to reject a null hypothesis
- ❖ Applying hypothesis testing to two sample cases, with means or proportions

According to economist Ethan Harris, "People may not remember too many numbers about the economy, but there are certain signposts they do pay attention to. As a short hard way to assess how the economy is doing, everybody notices the price of gas."[1] In July 2008, the national record for the price of gasoline was set at $4.11 per gallon. The impact of high and volatile fuel prices is felt across the nation, affecting consumer spending and the economy, but the burden remains greater among distinct social economic groups and geographic areas.

Lower income Americans spend eight times more of their disposable income on gasoline than wealthier Americans do.[2] For example, in Wilcox, Alabama, individuals spend 12.72% of their income to fuel one vehicle, while in Hunterdon County, New Jersey, people spend 1.52%. Nationally, Americans spend 3.8% of their income fueling one vehicle. The first state to reach the $5 per gallon milestone was California in 2012. California's drivers were hit especially hard by the rising price of gas, in part because of their reliance on automobiles, especially for work commuters. Declines in consumer spending and confidence in the economy have been attributed in part to the high (and rising) cost of gasoline.

In 2013, gasoline prices remained higher for states along the West Coast, particularly in Alaska, California, and Hawaii. Let's say we drew a random sample of California gas stations ($N = 100$) and calculated the mean price for a gallon of regular gas. Based on consumer information,[3] we also know that nationally the mean price of a gallon was $3.53 with a standard deviation of 0.21 for the same week. We can thus compare the mean price of gas in California with the mean price of all gas stations in May 2013. By comparing these means, we are asking whether it is reasonable to consider our random sample of California gas as representative of the population of gas stations in the United States. Actually, we expect to find that the average price of gas from a sample of California gas stations will be unrepresentative of the population of gas stations because we assume higher gas prices in the state.

The mean price for our sample is $3.90. This figure is higher than $3.53, the mean price per gallon across the nation. But is the observed gap of 37 cents ($3.90 – $3.53) large enough to convince us that the sample of California gas stations is not representative of the population?

The sample mean of $3.90 is higher than the population mean, but it is an estimate based on a single sample. Thus, it could mean one of two things: (1) the average price of gas in California is indeed higher than the national average, or (2) the average price of gas in California is about the same as the national average, and this sample happens to show a particularly high mean.

How can we decide which of these explanations makes more sense? Because most estimates are based on single samples and different samples may result in different estimates, sampling results cannot be used directly to make statements about a population. We need a procedure that allows us to evaluate hypotheses about population parameters based on sample statistics. In Chapter 7, we saw that population parameters can be estimated from sample statistics. In this chapter, we will learn how to use sample statistics to make decisions about population parameters. This procedure is called **statistical hypothesis testing**.

Statistical hypothesis testing A procedure that allows us to evaluate hypotheses about population parameters based on sample statistics.

▣ ASSUMPTIONS OF STATISTICAL HYPOTHESIS TESTING

Statistical hypothesis testing requires several assumptions. These assumptions include considerations of the level of measurement of the variable, the method of sampling, the shape of the population distribution, and the sample size. The specific assumptions may vary, depending on the test or the conditions of testing. However, without exception, *all* statistical tests assume random sampling. Tests of hypotheses about means also assume the interval-ratio level of measurement and require that the population under consideration be normally distributed or that the sample size be larger than 50.

Based on our data, we can test the hypothesis that the average price of gas in California is higher than the average national price of gas. The test we are considering meets these conditions:

1. The sample of California gas stations was randomly selected.

2. The variable *price per gallon* is measured at the interval-ratio level.

3. We cannot assume that the population is normally distributed. However, because our sample size is sufficiently large ($N > 50$), we know, based on the central limit theorem, that the sampling distribution of the mean will be approximately normal.

▣ STATING THE RESEARCH AND NULL HYPOTHESES

Hypotheses are usually defined in terms of interrelations between variables and are often based on a substantive theory. Earlier, we defined *hypotheses* as tentative answers to research questions. They are tentative because we can find evidence for them only after being empirically tested. The testing of hypotheses is an important step in this evidence-gathering process.

The Research Hypothesis (*H₁*)

Our first step is to formally express the hypothesis in a way that makes it amenable to a statistical test. The substantive hypothesis is called the **research hypothesis** and is symbolized as H_1. Research hypotheses are always expressed in terms of population parameters because we are interested in making statements about population parameters based on our sample statistics.

Research hypothesis (H_1*)* A statement reflecting the substantive hypothesis. It is always expressed in terms of population parameters, but its specific form varies from test to test.

In our research hypothesis (H_1), we state that the average price of gas in California is higher than the average price of gas nationally. Symbolically, we use μ_Y to represent the population mean; our hypothesis can be expressed as

$$H_1: \mu_Y > \$3.53$$

In general, the research hypothesis (H_1) specifies that the population parameter is one of the following:

1. Not equal to some specified value: $\mu_Y \neq$ some specified value

2. Greater than some specified value: $\mu_Y >$ some specified value

3. Less than some specified value: $\mu_Y <$ some specified value

The Null Hypothesis (*H₀*)

Is it possible that in the population there is no real difference between the mean price of gas in California and the mean price of gas in the nation and that the observed difference of $0.37 is actually due to the fact that this particular sample happened to contain California gas stations with higher prices? Since statistical inference is based on probability theory, it is not possible to prove or disprove the research hypothesis directly. We can, at best, estimate the *likelihood* that it is true or false.

To assess this likelihood, statisticians set up a hypothesis that is counter to the research hypothesis. The **null hypothesis**, symbolized as H_0, contradicts the research hypothesis and usually states that there is no difference between the population mean and some specified value. It is also referred to as the hypothesis of "no difference." Our null hypothesis can be stated symbolically as

$$H_0: \mu_Y = \$3.53$$

Rather than directly testing the substantive hypothesis (H_1) that there is a difference between the mean price of gas in California and the mean price nationally, we test the null hypothesis (H_0) that there is no difference in prices. In hypothesis testing, we hope to reject the null hypothesis to provide support for the research hypothesis. Rejection of the null hypothesis will strengthen our belief in the research hypothesis and increase our confidence in the importance and utility of the broader theory from which the research hypothesis was derived.

Null hypothesis (H_0)   A statement of "no difference" that contradicts the research hypothesis and is always expressed in terms of population parameters.

More About Research Hypotheses: One- and Two-Tailed Tests

In a **one-tailed test**, the research hypothesis is directional; that is, it specifies that a population mean is either less than ($<$) or greater than ($>$) some specified value. We can express our research hypothesis as either

$$H_1: \mu_Y < \text{some specified value}$$

or

$$H_1: \mu_Y > \text{some specified value}$$

The research hypothesis we've stated for the average price of a gallon of regular gas in California is a one-tailed test.

When a one-tailed test specifies that the population mean is *greater than* some specified value, we call it a **right-tailed test** because we will evaluate the outcome at the right tail of the sampling distribution. If the research hypothesis specifies that the population mean is *less than* some specified value, it is called a **left-tailed test** because the outcome will be evaluated at the left tail of the sampling distribution. Our example is a right-tailed test because the research hypothesis states that the mean gas prices in California are higher than $3.53. (Refer to Figure 8.1.)

Sometimes, we have some theoretical basis to believe that there is a difference between groups, but we cannot anticipate the direction of that difference. For example, we may have reason to believe that the average price of California gas is *different* from that of the general population, but we may not have enough research or support to predict whether it is *higher* or *lower*. When we have no theoretical reason for specifying a direction in the research hypothesis, we conduct a **two-tailed test**. The research hypothesis specifies that the population mean is not equal to some specified value. For example, we can express the research hypothesis about the mean price of gas as

$$H_1: \mu_Y \neq \$3.53$$

With both one- and two-tailed tests, our null hypothesis of no difference remains the same. It can be expressed as

$$H_0: \mu_Y = \text{some specified value}$$

One-tailed test   A type of hypothesis test that involves a directional research hypothesis. It specifies that the values of one group are either larger or smaller than some specified population value.

Right-tailed test   A one-tailed test in which the sample outcome is hypothesized to be at the right tail of the sampling distribution.

Left-tailed test   A one-tailed test in which the sample outcome is hypothesized to be at the left tail of the sampling distribution.

Two-tailed test   A type of hypothesis test that involves a nondirectional research hypothesis. We are equally interested in whether the values are less than or greater than one another. The sample outcome may be located at both the low and the high ends of the sampling distribution.

▣ DETERMINING WHAT IS SUFFICIENTLY IMPROBABLE: PROBABILITY VALUES AND ALPHA

Now let's put all our information together. We're assuming that our null hypothesis ($\mu_Y = \$3.53$) is true, and we want to determine whether our sample evidence casts doubt on that assumption, suggesting that there is evidence for our research hypothesis, $\mu_Y > \$3.53$. What are the chances that we would have randomly selected a sample of California gas stations such that the average price per gallon is higher than \$3.53, the average for the nation? We can determine the chances or probability because of what we know about the sampling distribution and its properties. We know, based on the central limit theorem, that if our sample size is larger than 50, the sampling distribution of the mean is approximately normal, with a mean and a standard deviation (standard error) of

$$\sigma_{\bar{Y}} = \frac{\sigma_Y}{\sqrt{N}}$$

We will assume that the null hypothesis is true and then see if our sample evidence casts doubt on that assumption. We have a population mean $\mu_Y = \$3.53$ and a standard deviation $\sigma_Y = 0.21$. Our sample size is $N = 100$, and the sample mean is 3.90. We can assume that the distribution of means of all possible samples of size $N = 100$ drawn from this distribution would be approximately normal, with a mean of 3.53 and a standard deviation of

$$\sigma_{\bar{Y}} = \frac{.21}{\sqrt{100}} = .02$$

This sampling distribution is shown in Figure 8.1. Also shown in Figure 8.1 is the mean gas price we observed for our sample of California gas stations.

Because this distribution of sample means is normal, we can use Appendix A to determine the probability of drawing a sample mean of 3.90 or higher from this population. We will

Figure 8.1   Sampling Distribution of Sample Means Assuming H_0 Is True for a Sample $N = 100$

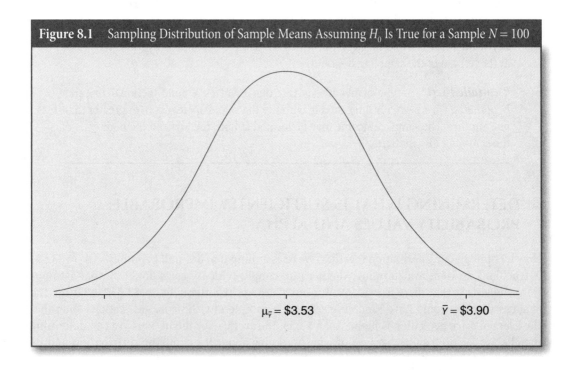

$\mu_{\bar{Y}} = \$3.53$ $\bar{Y} = \$3.90$

translate our sample mean into a Z score so that we can determine its location relative to the population mean. In Chapter 5, we learned how to translate a raw score into a Z score by using Formula 5.1:

$$Z = \frac{Y - \bar{Y}}{S_Y}$$

Because we are dealing with a sampling distribution in which our raw score is $\bar{Y}$, the mean, and the standard deviation (standard error) is $\sigma_Y / \sqrt{N}$, we need to modify the formula somewhat:

$$Z = \frac{\bar{Y} - \mu_{\bar{Y}}}{\sigma_Y / \sqrt{N}} \tag{8.1}$$

Converting the sample mean to a Z-score equivalent is called computing the *test statistic*. The Z value we obtain is called the **Z statistic (obtained)**. The obtained Z gives us the number of standard deviations (standard errors) that our sample is from the hypothesized value (μ_Y or $\mu_{\bar{Y}}$), assuming the null hypothesis is true. For our example, the obtained Z is

$$Z = \frac{3.90 - 3.53}{.21 / \sqrt{100}} = 18.50$$

Z statistic (obtained)   The test statistic computed by converting a sample statistic (such as the mean) to a Z score. The formula for obtaining Z varies from test to test.

Before we determine the probability of our obtained Z statistic, let's determine whether it is consistent with our research hypothesis. Recall that we defined our research hypothesis as a right-tailed test ($\mu_Y > \$3.53$), predicting that the difference would be assessed on the right tail of the sampling distribution. The positive value of our obtained Z statistic confirms that we will be evaluating the difference on the right tail. (If we had a negative obtained Z, it would mean the difference would have to be evaluated at the left tail of the distribution, contrary to our research hypothesis.)

To determine the probability of observing a Z value of 18.50, assuming that the null hypothesis is true, look up the value in Appendix A to find the area to the right of (above) the Z of 18.50. Our calculated Z value is not listed in Appendix A, so we'll need to rely on the last Z value reported in the table, 4.00. Recall from Chapter 5, in which we calculated Z scores and their probability, that the Z values are located in Column A. The P value is the probability to the right of the obtained Z, or the "area beyond Z" in Column C. This area includes the proportion of all sample means that are \$3.90 or higher. The proportion is less than 0.0001 (Figure 8.2). This value is the probability of getting a result as extreme as the sample result if the null hypothesis is true; it is symbolized as P. Thus, for our example, $P \leq .0001$.

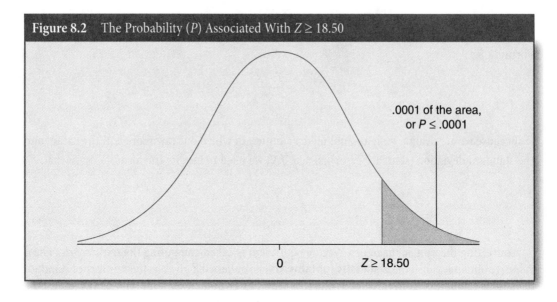

Figure 8.2 The Probability (*P*) Associated With $Z \geq 18.50$

A *P* **value** can be defined as the actual probability associated with the obtained value of *Z*. It is a measure of how unusual or rare our obtained statistic is compared with what is stated in our null hypothesis. The smaller the *P* value, the more evidence we have that the null hypothesis should be rejected in favor of the research hypothesis.

P *value*   The probability associated with the obtained value of *Z*.

Researchers usually define in advance what a sufficiently improbable *Z* value is by specifying a cutoff point below which *P* must fall to reject the null hypothesis. This cutoff point, called **alpha** and denoted by the Greek letter α, is customarily set at the .05, .01, or .001 level. Let's say that we decide to reject the null hypothesis if $P \leq .05$. The value .05 is referred to as alpha (α); it defines for us what result is sufficiently improbable to allow us to take the risk and reject the null hypothesis. An alpha (α) of .05 means that even if the obtained *Z* statistic is due to sampling error, so that the null hypothesis is true, we would allow a 5% risk of rejecting it. Alpha values of .01 and .001 are more cautionary levels of risk. The difference between *P* and alpha is that *P* is the *actual probability* associated with the obtained value of *Z*, whereas alpha is the level of probability *determined in advance* at which the null hypothesis is rejected. The null hypothesis is rejected when $P \leq \alpha$.

Alpha (α)   The level of probability at which the null hypothesis is rejected. It is customary to set alpha at the .05, .01, or .001 level.

We have already determined that our obtained *Z* has a probability value less than .0001. Since our observed *P* is less than .05 ($P = .0001 < \alpha = .05$), we reject the null hypothesis. The value of .0001 means that fewer than 1 out of 10,000 samples drawn from this population are likely to have a mean that is 18.50 Z scores above the hypothesized mean of \$3.53. Another way to say it is as follows: There is only 1 chance out of 10,000 (or 0.0001%) that we would draw a random sample with $Z \geq 18.50$ if the mean price of California gas were equal to the national mean price.

Based on the P value, we can also make a statement regarding the "significance" of the results. If the P value is equal to or less than our alpha level, our obtained Z statistic is considered *statistically significant*—that is to say, it is very unlikely to have occurred by random chance or sampling error. We can state that the difference between the average price of gas in California and nationally is significantly different at the .05 level, or we can specify the actual level of significance by saying that the level of significance is less than .0001.

Recall that our hypothesis was a one-tailed test ($\mu_Y > \$3.53$). In a two-tailed test, sample outcomes may be located at both the higher and the lower ends of the sampling distribution. Thus, the null hypothesis will be rejected if our sample outcome falls either at the left or right tail of the sampling distribution. For instance, a .05 alpha or P level means that H_0 will be rejected if our sample outcome falls among either the lowest or the highest 5% of the sampling distribution.

Suppose we had expressed our research hypothesis about the mean price of gas as

$$H_1: \mu_Y \neq \$3.53$$

The null hypothesis to be directly tested still takes the form $H_0: \mu_Y = \$3.53$ and our obtained Z is calculated using the same formula (8.1) as was used with a one-tailed test. To find P for a two-tailed test, look up the area in Column C of Appendix A that corresponds to your obtained Z (as we did earlier) and then multiply it by 2 to obtain the two-tailed probability. Thus, the two-tailed P value for $Z = 18.50$ is $.0001 \times 2 = .0002$. This probability is less than our stated alpha (.05), and thus, we reject the null hypothesis.

▣ THE FIVE STEPS IN HYPOTHESIS TESTING: A SUMMARY

Regardless of the particular application or problem, statistical hypothesis testing can be organized into five basic steps. Let's summarize these steps:

1. Making assumptions

2. Stating the research and null hypotheses and selecting alpha

3. Selecting the sampling distribution and specifying the test statistic

4. Computing the test statistic

5. Making a decision and interpreting the results

Making Assumptions. Statistical hypothesis testing involves making several assumptions regarding the level of measurement of the variable, the method of sampling, the shape of the population distribution, and the sample size. In our example, we made the following assumptions:

1. A random sample was used.

2. The variable *price per gallon* is measured on an interval-ratio level of measurement.

3. Because $N > 50$, the assumption of normal population is not required.

Stating the Research and Null Hypotheses and Selecting Alpha. The substantive hypothesis is called the *research hypothesis* and is symbolized as H_1. Research hypotheses are always expressed in terms of population parameters because we are interested in making statements about population parameters based on sample statistics. Our research hypothesis was

$$H_1: \mu_Y > \$3.53$$

The *null hypothesis*, symbolized as H_0, contradicts the research hypothesis in a statement of no difference between the population mean and our hypothesized value. For our example, the null hypothesis was stated symbolically as

$$H_0: \mu_Y = \$3.53$$

We set alpha at .05, meaning that we would reject the null hypothesis if the probability of our obtained Z was less than or equal to .05.

Selecting the Sampling Distribution and Specifying the Test Statistic. The normal distribution and the Z statistic are used to test the null hypothesis.

Computing the Test Statistic. Based on Formula 8.1, our Z statistic is 18.50.

Making a Decision and Interpreting the Results. We confirm that our obtained Z is on the right tail of the distribution, consistent with our research hypothesis. We determine that the P value of 18.50 is less than .0001, less than our .05 alpha level. We have evidence to reject the null hypothesis of no difference between the mean price of California gas and the mean price of gas nationally. We thus conclude that the average price of California gas is significantly higher than the national average.

▣ ERRORS IN HYPOTHESIS TESTING

We should emphasize that because our conclusion is based on sample data, we will never really know if the null hypothesis is true or false. In fact, as we have seen, there is a 0.01% chance that the null hypothesis is true and that we are making an error by rejecting it.

The null hypothesis can be either true or false, and in either case, it can be rejected or not rejected. If the null hypothesis is true and we reject it nonetheless, we are making an incorrect decision. This type of error is called a **Type I error**. Conversely, if the null hypothesis is false but we fail to reject it, this incorrect decision is a **Type II error**.

Type I error   The probability associated with rejecting a null hypothesis when it is true.

Type II error   The probability associated with failing to reject a null hypothesis when it is false.

In Table 8.1, we show the relationship between the two types of errors and the decisions we make regarding the null hypothesis. The probability of a Type I error—rejecting a true hypothesis—is equal to the chosen alpha level. For example, when we set alpha at the .05 level, we know that the probability that the null hypothesis is in fact true is .05 (or 5%).

Table 8.1   Type I and Type II Errors

	True State of Affairs	
Decision Made	H_0 *Is True*	H_0 *Is False*
Reject H_0	Type I error (α)	Correct decision
Do not reject H_0	Correct decision	Type II error

We can control the risk of rejecting a true hypothesis by manipulating alpha. For example, by setting alpha at .01, we are reducing the risk of making a Type I error to 1%. Unfortunately, however, Type I and Type II errors are inversely related; thus, by reducing alpha and lowering the risk of making a Type I error, we are increasing the risk of making a Type II error (Table 8.1).

As long as we base our decisions on sample statistics and not population parameters, we have to accept a degree of uncertainty as part of the process of statistical inference.

✓ *Learning Check*

The implications of research findings are not created equal. For example, researchers might hypothesize that eating spinach increases the strength of weight lifters. Little harm will be done if the null hypothesis that eating spinach has no effect on the strength of weight lifters is rejected in error. The researchers would most likely be willing to risk a high probability of a Type I error, and all weight lifters would eat spinach. However, when the implications of research have important consequences (funding of social programs or medical testing), the balancing act between Type I and Type II errors becomes more important. Can you think of some examples where researchers would want to minimize Type I errors? When might they want to minimize Type II errors?

The *t* Statistic and Estimating the Standard Error

The Z statistic we have calculated (Formula 8.1) to test the hypothesis involving a sample of California gas stations assumes that the population standard deviation σ_Y is known. The value of σ_Y is required to calculate the standard error

$$\sigma_Y / \sqrt{N}$$

In most situations, σ_Y will not be known, and we will need to estimate it using the sample standard deviation S_Y. We then use the t statistic instead of the Z statistic to test the null hypothesis. The formula for computing the t statistic is

$$t = \frac{\overline{Y} - \mu_Y}{S_Y / \sqrt{N}} \tag{8.2}$$

The t value we calculate is called the **t statistic (obtained)**. The obtained t represents the number of standard deviation units (or standard error units) that our sample mean is from the hypothesized value of μ_Y, assuming that the null hypothesis is true.

---

t *statistic (obtained)*   The test statistic computed to test the null hypothesis about a population mean when the population standard deviation is unknown and is estimated using the sample standard deviation.

The *t* Distribution and Degrees of Freedom

To understand the t statistic, we should first be familiar with its distribution. The **t distribution** is actually a family of curves, each determined by its *degrees of freedom*. The concept of degrees of freedom is used in calculating several statistics, including the t statistic. The **degrees of freedom (*df*)** represent the number of scores that are free to vary in calculating each statistic.

---

t *distribution*   A family of curves, each determined by its degrees of freedom *(df)*. It is used when the population standard deviation is unknown and the standard error is estimated from the sample standard deviation.

Degrees of freedom (**df**)   The number of scores that are free to vary in calculating a statistic.

To calculate the degrees of freedom, we must know the sample size and whether there are any restrictions in calculating that statistic. The number of restrictions is then subtracted from the sample size to determine the degrees of freedom. When calculating the t statistic for a one-sample test, we start with the sample size N and lose 1 degree of freedom for the population standard deviation we estimate.[4] Note that the degrees of freedom will increase as the sample size increases. In the case of a single-sample mean, the *df* is calculated as follows:

$$df = N - 1 \qquad (8.3)$$

Comparing the *t* and *Z* Statistics

Notice the similarities between the formulas for the *t* and *Z* statistics. The only apparent difference is in the denominator. The denominator of *Z* is the standard error based on the population standard deviation σ_Y. For the denominator of *t*, we replace $\sigma_Y/\sqrt{N}$ with $S_Y/\sqrt{N}$, the estimated standard error based on the sample standard deviation.

However, there is another important difference between the *Z* and *t* statistics: Because it is estimated from sample data, the denominator of the *t* statistic is subject to sampling error. The sampling distribution of the test statistic is not normal, and the standard normal distribution cannot be used to determine probabilities associated with it.

In Figure 8.3, we present the *t* distribution for several *df*s. Like the standard normal distribution, the *t* distribution is bell shaped. The *t* statistic, similar to the *Z* statistic, can have positive and negative values. A positive *t* statistic corresponds to the right tail of the distribution; a negative value corresponds to the left tail. Note that when the *df* is small, the *t* distribution is much flatter than the normal curve. But as the degrees of freedom increases, the shape of the *t* distribution gets closer to the normal distribution, until the two are almost identical when *df* is greater than 120.

Appendix B summarizes the *t* distribution. Note that the *t* table differs from the normal (*Z*) table in several ways. First, the column on the left side of the table shows the degrees of freedom. The *t*

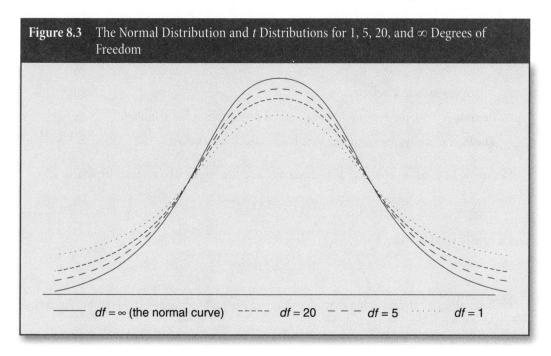

Figure 8.3 The Normal Distribution and *t* Distributions for 1, 5, 20, and ∞ Degrees of Freedom

——— *df* = ∞ (the normal curve)   - - - - - *df* = 20   - - - *df* = 5   · · · · · · *df* = 1

statistic will vary depending on the degrees of freedom, which must first be computed ($df = N - 1$). Second, the probabilities or alpha, denoted as significance levels, are arrayed across the top of the table in two rows, the first for a one-tailed and the second for a two-tailed test. Finally, the values of t, listed as the entries of this table, are a function of (1) the degrees of freedom, (2) the level of significance (or probability), and (3) whether the test is a one- or a two-tailed test.

To illustrate the use of this table, let's determine the probability of observing a t value of 2.021 with 40 degrees of freedom and a two-tailed test. Locating the proper row ($df = 40$) and column (two-tailed test), we find the t statistic of 2.021 corresponding to the .05 level of significance. Restated, we can say that the probability of obtaining a t statistic of 2.021 is .05, or that there are fewer than 5 chances out of 100 that we would have drawn a random sample with an obtained t of 2.021 if the null hypothesis were correct.

▣ STATISTICS IN PRACTICE: THE EARNINGS OF WHITE WOMEN

To illustrate the application of the t statistic, let's test a two-tailed hypothesis about a population mean μ_Y. Let's say we drew a random sample of 320 white women who worked full-time in 2009. We found their mean earnings to be $36,471, with a standard deviation $S_Y = \$28,563$. Based on data from the U.S. Census Bureau,[5] we also know that the 2009 mean earnings nationally for all women was $\mu_Y = \$33,797$. However, we do not know the value of the population standard deviation. We want to determine whether the sample of white women was representative of the population of all women working full-time in 2009. Although we suspect that white American women experienced a relative advantage in earnings, we are not sure enough to predict that their earnings were indeed higher than the earnings of all women nationally. Therefore, the statistical test is two tailed.

Let's apply the five-step model to test the hypothesis that the average earnings of white women differed from the average earnings of all women working full-time in the United States in 2009.

Making Assumptions. Our assumptions are as follows:

1. A random sample is selected.

2. Because $N > 50$, the assumption of a normal population is not required.

3. The level of measurement of the variable *income* is interval-ratio.

Stating the Research and the Null Hypotheses and Selecting Alpha. The research hypothesis is

$$H_1: \mu_Y \neq \$33,797$$

and the null hypothesis is

$$H_0: \mu_Y = \$33,797$$

We'll set alpha at .05, meaning that we will reject the null hypothesis if the probability of our obtained statistic is less than or equal to .05.

Selecting the Sampling Distribution and Specifying the Test Statistic. We use the *t* distribution and the *t* statistic to test the null hypothesis.

Computing the Test Statistic. We first calculate the *df* associated with our test:

$$df = (N - 1) = (320 - 1) = 319$$

To evaluate the probability of obtaining a sample mean of $36,471, assuming the average earnings of white women were equal to the national average of $33,797, we need to calculate the obtained *t* statistic by using Formula 8.2:

$$t = \frac{\overline{Y} - \mu_Y}{S_Y / \sqrt{N}} = \frac{36,471 - 33,797}{28,563 / \sqrt{320}} = 1.67$$

Making a Decision and Interpreting the Results. Given our research hypothesis, we will conduct a two-tailed test. To determine the probability of observing a *t* value of 1.67 with 319 degrees of freedom, let's refer to Appendix B. From the first column, we can see that 319 degrees of freedom is not listed, so we'll have to use the last row, $df = \infty$, to assess the significance of our obtained *t* statistic.

Though our obtained *t* statistic of 1.67 is not listed in the last row, we can see that it lies somewhere between 1.645 (for .05 one-tailed test) and 1.960 (for .025 one-tailed test). The probability of 1.67 can be estimated as $.05 > p > .025$, allowing us to determine that we can reject the null hypothesis. We can also note how the obtained *t* statistic of 1.67 is greater than the *t* critical of 1.645 and make the same decision. We have sufficient evidence to reject the null hypothesis and conclude that the 2009 average earnings of white women were significantly higher than the average earnings of all women. The difference of $2,674 is significant at the .05 level.

▣ TESTING HYPOTHESES ABOUT TWO SAMPLES

The two examples that we reviewed at the beginning of this chapter dealt with data from one sample compared with data from the population. In practice, social scientists are often more interested in situations involving two (sample) parameters than those involving one, such as the differences between men and women, Democrats and Republicans, whites and nonwhites, or high school or college graduates. Specifically, we may be interested in finding out whether the average years of education for one racial/ethnic group is the same, lower, or higher than another group.

U.S. data on educational attainment reveal that Asian and Pacific Islanders have more years of education than any other racial/ethnic group; this includes the percentage of those earning high school degrees or higher or college degrees or higher. Though years of education have steadily increased for blacks and Hispanics since 1990, their numbers remain behind Asian and Pacific Islanders and whites.

Using data from the 2010 General Social Survey (GSS), we examine the difference in white and black educational attainment. From the GSS sample, white respondents reported an average of 13.66 years of education and blacks an average of 12.89 years, as shown in Table 8.2. These sample averages could mean either (1) the average number of years of education for whites is higher than the average for blacks, or (2) the average for whites is actually about the same as for blacks, but our sample just happens to indicate a higher average for whites. What we are applying here is a bivariate analysis (for more information, refer to Chapter 9), a method to detect and describe the relationship between two variables—race/ethnicity and educational attainment.

Table 8.2   Years of Education for White and Black Men and Women, GSS 2010

	Whites (Sample 1)	Blacks (Sample 2)
Mean	13.66	12.89
Standard deviation	3.16	3.03
Variance	9.99	9.18
N	584	118

The statistical procedures discussed in the following sections allow us to test whether the differences that we observe between two samples are large enough for us to conclude that the populations from which these samples are drawn are different as well. We present tests for the significance of the differences between two groups. Primarily, we consider differences between sample means and differences between sample proportions.

Hypothesis testing with two samples follows the same structure as for one-sample tests: The assumptions of the test are stated, the research and null hypotheses are formulated and the alpha level is selected, the sampling distribution and the test statistic are specified, the test statistic is computed, and a decision is made whether to reject the null hypothesis.

The Assumption of Independent Samples

One important difference between one- and two-sample hypothesis testing involves sampling procedures. With a two-sample case, we assume that the samples are independent of each other. The choice of sample members from one population has no effect on the choice of sample members from the second population. In our comparison of whites and blacks, we are assuming that the selection of whites is independent of the selection of black individuals. (The requirement of independence is also satisfied by selecting one sample randomly and then dividing the sample into appropriate subgroups. For example, we could randomly select a sample and then divide it into groups based on gender, religion, income, or any other attribute that we are interested in.)

Stating the Research and Null Hypotheses

The second difference between one- and two-sample tests is in the form taken by the research and the null hypotheses. In one-sample tests, both the null and the research hypotheses are statements about a single population parameter, μ_Y. In contrast, with two-sample tests, we compare two population parameters.

Our research hypothesis (H_1) is that the average years of education for whites is not equal to the average years of education for black respondents. We are stating a hypothesis about the relationship between race/ethnicity and education in the general population by comparing the mean educational attainment of whites with the mean educational attainment of blacks. Symbolically, we use μ to represent the population mean; the subscript 1 refers to our first sample (whites) and the subscript 2 to our second sample (blacks). Our research hypothesis can then be expressed as

$$H_1: \mu_1 \neq \mu_2$$

Because H_1 specifies that the mean education for whites is not equal to the mean education for blacks, it is a nondirectional hypothesis. Thus, our test will be a two-tailed test. Alternatively, if there were sufficient basis for deciding which population mean score is larger (or smaller), the research hypothesis for our test would be a one-tailed test:

$$H_1: \mu_1 < \mu_2 \text{ or } H_1: \mu_1 > \mu_2$$

In either case, the null hypothesis states that there are no differences between the two population means:

$$H_0: \mu_1 = \mu_2$$

We are interested in finding evidence to reject the null hypothesis of no difference so that we have sufficient support for our research hypothesis.

✓ *Learning Check*

For the following research situations, state your research and null hypotheses:

- *There is a difference between the mean statistics grades of social science majors and the mean statistics grades of business majors.*
- *The average number of children in two-parent black families is lower than the average number of children in two-parent nonblack families.*
- *Grade point averages are higher among girls who participate in organized sports than among girls who do not.*

▣ THE SAMPLING DISTRIBUTION OF THE DIFFERENCE BETWEEN MEANS

The sampling distribution allows us to compare our sample results with all possible sample outcomes and estimate the likelihood of their occurrence. Tests about differences between two sample means are based on the **sampling distribution of the difference between means**. The sampling distribution of the difference between two sample means is a theoretical probability distribution that would be obtained by calculating all the possible mean differences by drawing all possible independent random samples of size N_1 and N_2 from two populations.

Sampling distribution of the difference between means   A theoretical probability distribution that would be obtained by calculating all the possible mean differences that would be obtained by drawing all the possible independent random samples of size N_1 and N_2 from two populations where N_1 and N_2 are both greater than 50.

The properties of the sampling distribution of the difference between two sample means are determined by a corollary to the central limit theorem. This theorem assumes that our samples are independently drawn from normal populations, but that with sufficient sample sizes ($N_1 > 50$, $N_2 > 50$) the sampling distribution of the difference between means will be approximately normal, even if the original populations are not normal. This sampling distribution has a mean $\mu_{\bar{Y}_1} - \mu_{\bar{Y}_2}$ and a standard deviation (standard error)

$$\sigma_{\bar{Y}_1 - \bar{Y}_2} = \sqrt{\frac{\sigma_{Y_1}^2}{N_1} + \frac{\sigma_{Y_2}^2}{N_2}} \qquad (8.4)$$

which is based on the variances in each of the two populations ($\sigma_{Y_1}^2$ and $\sigma_{Y_2}^2$).

Estimating the Standard Error

Formula 8.4 assumes that the population variances are known and that we can calculate the standard error $\sigma_{\bar{Y}_1 - \bar{Y}_2}$ (the standard deviation of the sampling distribution). However, in most situations, the only data we have are based on sample data, and we do not know the true value of the population variances, $\sigma_{Y_1}^2$ and $\sigma_{Y_2}^2$. Thus, we need to estimate the standard error from the sample variances, $S_{Y_1}^2$ and $S_{Y_2}^2$. The estimated standard error of the difference between means is symbolized as $S_{\bar{Y}_1 - \bar{Y}_2}$ (instead of $\sigma_{\bar{Y}_1 - \bar{Y}_2}$).

Calculating the Estimated Standard Error

When we can assume that the two population variances are equal, we combine information from the two sample variances to calculate the estimated standard error.

$$S_{\bar{Y}_1 - \bar{Y}_2} = \sqrt{\frac{(N_1 - 1)S_{Y_1}^2 + (N_2 - 1)S_{Y_2}^2}{(N_1 + N_2) - 2}} \sqrt{\frac{N_1 + N_2}{N_1 N_2}} \qquad (8.5)$$

where $S_{\bar{Y}_1 - \bar{Y}_2}$ is the estimated standard error of the difference between means, and $S_{Y_1}^2$ and $S_{Y_2}^2$ are the variances of the two samples. As a rule of thumb, when either sample variance is more than *twice* as large as the other, we can no longer assume that the two population variances are equal and would need to use Formula 8.8.

The *t* Statistic

As with single sample means, we use the *t* distribution and the *t* statistic whenever we estimate the standard error for a difference between means test. The *t* value we calculate is the obtained *t*. It represents the number of standard deviation units (or standard error units) that our mean difference $\left(\overline{Y}_1 - \overline{Y}_2\right)$ is from the hypothesized value of $\mu_1 - \mu_2$, assuming that the null hypothesis is true.

The formula for computing the *t* statistic for a difference between means test is

$$t = \frac{\overline{Y}_1 - \overline{Y}_2}{S_{\overline{Y}_1 - \overline{Y}_2}} \tag{8.6}$$

where $S_{\overline{Y}_1 - \overline{Y}_2}$ is the estimated standard error.

Calculating the Degrees of Freedom for a Difference Between Means Test

To use the *t* distribution for testing the difference between two sample means, we need to calculate the degrees of freedom. As we saw earlier, the degrees of freedom (*df*) represent the number of scores that are free to vary in calculating each statistic. When calculating the *t* statistic for the two-sample test, we lose 2 degrees of freedom, one for every population variance we estimate. When population variances are assumed to be equal or if the size of both samples is greater than 50, the *df* is calculated as follows:

$$df = (N_1 + N_2) - 2 \tag{8.7}$$

When we cannot assume that the population variances are equal and when the size of one or both samples is equal to or less than 50, we use Formula 8.9 to calculate the degrees of freedom.

▣ POPULATION VARIANCES ARE ASSUMED TO BE UNEQUAL

If the variances of the two samples ($s^2_{Y_1}$ and $s^2_{Y_2}$) are very different (one variance is twice as large as the other), the formula for the estimated standard error becomes

$$S_{\overline{Y}_1 - \overline{Y}_2} = \sqrt{\frac{S^2_{Y_1}}{N_1} + \frac{S^2_{Y_2}}{N_2}} \tag{8.8}$$

When the population variances are unequal and the size of one or both samples is equal to or less than 50, we use another formula to calculate the degrees of freedom associated with the *t* statistic:[6]

$$df = \frac{(S^2_{Y_1}/N_1 + S^2_{Y_2}/N_2)^2}{(S^2_{Y_1}/N_1)^2/(N_1 - 1) + (S^2_{Y_2}/N_2)^2/(N_2 - 1)} \tag{8.9}$$

▣ THE FIVE STEPS IN HYPOTHESIS TESTING ABOUT DIFFERENCE BETWEEN MEANS: A SUMMARY

As with single-sample tests, statistical hypothesis testing involving two sample means can be organized into five basic steps. Let's summarize these steps:

1. Making assumptions

2. Stating the research and null hypotheses and selecting alpha

3. Selecting the sampling distribution and specifying the test statistic

4. Computing the test statistic

5. Making a decision and interpreting the results

Making Assumptions. In our example, we made the following assumptions:

1. Independent random samples are used.

2. The variable *years of education* is measured at an interval-ratio level of measurement.

3. Because $N_1 > 50$ and $N_2 > 50$, the assumption of normal populations is not required.

4. The population variances are assumed to be equal.

Stating the Research and Null Hypotheses and Selecting Alpha. Our research hypothesis is that the mean education of whites is different from the mean education of blacks, indicating a two-tailed test. Symbolically, the research hypothesis is expressed as

$$H_1: \mu_1 \neq \mu_2$$

with μ_1 representing the mean education of whites and μ_2 the mean education of blacks.

The null hypothesis states that there are no differences between the two population means, or

$$H_0: \mu_1 = \mu_2$$

We are interested in finding evidence to reject the null hypothesis of no difference so that we have sufficient support for our research hypothesis. We will reject the null hypothesis if the probability of *t* (obtained) is less than or equal to .05 (our alpha value).

Selecting the Sampling Distribution and Specifying the Test Statistic. The *t* distribution and the *t* statistic are used to test the significance of the difference between the two sample means.

Computing the Test Statistic. To test the null hypothesis about the differences between the mean education of whites and blacks, we need to translate the ratio of the observed differences to its standard error into a *t* statistic (based on data presented in Table 8.2). The obtained *t* statistic is calculated using Formula 8.6:

$$t = \frac{\overline{Y}_1 - \overline{Y}_2}{S_{\overline{Y}_1 - \overline{Y}_2}}$$

where $S_{\overline{Y}_1 - \overline{Y}_2}$ is the estimated standard error of the sampling distribution. Because the population variances are assumed to be equal, $df = (N_1 + N_2) - 2 = (584 + 118) - 2 = 700$, and we can combine information from the two sample variances to estimate the standard error (Formula 8.5):

$$S_{\overline{Y}_1 - \overline{Y}_2} = \sqrt{\frac{(584-1)(3.16)^2 + (118-1)(3.03)^2}{(584+118)-2}} \sqrt{\frac{584+118}{584(118)}} = 3.14(.10) = .31$$

We substitute this value into the denominator for the t statistic (Formula 8.6):

$$t = \frac{13.66 - 12.89}{.31} = \frac{.77}{.31} = 2.48$$

Making a Decision and Interpreting the Results. We confirm that our obtained t is on the right tail of the distribution. Since our obtained t statistic of 2.48 is greater than $t = 2.326$ ($df = \infty$, two tailed; see Appendix B), we can state that its probability is less than .02. This is less than our .05 alpha level, and we can reject the null hypothesis of no difference between the educational attainment of whites and blacks. We conclude that white men and women, on average, have significantly higher years of education than black men and women do.

▣ FOCUS ON INTERPRETATION: CIGARETTE USE AMONG TEENS

Administered annually since 1975, the Monitoring the Future (MTF) survey measures the extent of and beliefs regarding drug use among 8th, 10th, and 12th graders. In recent years, data collected from the MTF surveys revealed decreases or stability in drug use among youths, particularly for cigarettes, alcohol, marijuana, cocaine, and methamphetamine.[7]

Let's examine data from the MTF 2011 survey, comparing first-time cigarette use between black and white students. The survey results (Table 8.3) indicate that black students are more likely to smoke cigarettes later (in later grades) than white students. The mean grade of first use of cigarettes is 6.38 for white students and 7.15 for black students.

We will rely on SPSS to calculate the t obtained for the data. We will not present the complete five-step model and t-test calculation, because we want to focus on interpreting the SPSS output. However, we will need a research hypothesis and an alpha level to guide our interpretation. SPSS always estimates a two-tailed test, namely, does the gap of 0.77 (7.15 − 6.38) indicate a difference in when black and white adolescents first smoke cigarettes? We'll set alpha at .05.

The output includes two tables. The Group Statistics table (Figure 8.4) presents descriptive statistics for each group. In the second table (Figure 8.5), labeled Independent-Samples Test, t statistics are presented for equal variances assumed (3.891) and equal variances not assumed (4.266). In order to determine which t statistic to use, review the results of the Levene's test for equality of variances. Levene's test (a calculation that we will not cover in this text) tests the null hypothesis that the population variances are equal. If the significance of the reported F statistic is equal to or

Table 8.3   Grade When First Smoked Cigarettes by Race, MTF 2011

	Black Students	White Students
Mean	7.15	6.38
Standard deviation	1.99	2.29
N	156	748

Figure 8.4   Group Statistics

Group Statistics

	race Respondent's race (trichotomized B/W/H)	N	Mean	Std. Deviation	Std. Error Mean
grsmoke What grade when first smoked cigarettes?	1 BLACK: (1)	156	7.15	1.990	.159
	2 WHITE: (2)	748	6.38	2.291	.084

Figure 8.5   Independent-Samples Test

Independent Samples Test

		Levene's Test for Equality of Variances	
		F	Sig.
grsmoke What grade when first smoked cigarettes?	Equal variances assumed	34.238	.000
	Equal variances not assumed		

	t-test for Equality of Means						
						95% Confidence interval of the Difference	
	t	df	Sig. (2-tailed)	Mean Difference	Std. Error Difference	Lower	Upper
Equal variances assumed	3.891	902	.000	.768	.197	.380	1.155
Equal variances not assumed	4.266	248.610	.000	.768	.180	.413	1.122

less than .05 (the baseline alpha for Levene's test), we can reject the null hypothesis that the variances are equal; if the significance is greater than .05, we fail to reject the null hypothesis. (To say it another way: if the significance for Levene's test is greater than .05, refer to the t obtained for equal variances assumed; if the significance is less than .05, refer to the t obtained for equal variances not assumed.) Since the significance of F is .000 < .05, we reject the null hypothesis and conclude that the variances are unequal. Thus, the *t* obtained that we will use for this model is 4.266 (the one corresponding to equal variances not assumed).

SPSS calculates the probability of the *t* obtained for a two-tailed test. There is no need to estimate it based on Appendix B. The significance of 4.266 is .000, which is less than our alpha level of .05. We reject the null hypothesis of no difference for grade of first-time cigarette use between white and black students. On average, black students first use cigarettes at a later grade (0.77 grades later) than white students.

Would you change your decision in the previous example if alpha were .01? Why or why not?

✓ *Learning* *Check*

◙ TESTING THE SIGNIFICANCE OF THE DIFFERENCE BETWEEN TWO SAMPLE PROPORTIONS

In the preceding sections, we have learned how to test for the significance of the difference between two population means when the variable is measured at an interval-ratio level. Yet numerous variables in the social sciences are measured at a nominal or an ordinal level. These variables are often described in terms of proportions or percentages. For example, we might be interested in comparing the proportion of those who support immigrant policy reform among Hispanics and non-Hispanics or the proportion of men and women who supported the Democratic candidate during the last presidential election. In this section, we present statistical inference techniques to test for significant differences between two sample proportions.

Hypothesis testing with two sample proportions follows the same structure as the statistical tests presented earlier: The assumptions of the test are stated, the research and null hypotheses are formulated, the sampling distribution and the test statistic are specified, the test statistic is calculated, and a decision is made whether or not to reject the null hypothesis.

◙ STATISTICS IN PRACTICE: COMPARING FIRST- AND SECOND-GENERATION HISPANIC AMERICANS

In 2013 the Pew Research Center[8] presented a comparison of first-generation Americans (immigrants who were foreign born) and second-generation Americans (adults who have at least one immigrant parent) on several key demographic variables. Based on several measures of success, the center documented social mobility between the generations, confirming that second-generation Americans were doing better than the first-generation Americans. The statistical question we examine here is whether the difference between the generations is significant.

For example, according to the Pew report, the proportion of first-generation Hispanic Americans who earned a bachelor's degree or higher was 0.11 (p_1); the proportion of second-generation Hispanic Americans with the same response was 0.21 (p_2). A total of 899 first-generation Hispanic Americans (N_1) and 351 second-generation Hispanic Americans (N_2) answered this question. We use the five-step model to determine whether the difference between the two proportions is significant.

Making Assumptions. Our assumptions are as follows:

1. Independent random samples of $N_1 > 50$ and $N_2 > 50$ are used.
2. The level of measurement of the variable is nominal.

Stating the Research and Null Hypotheses and Selecting Alpha. We propose a two-tailed test that the population proportions for first-generation and second-generation Hispanic Americans are not equal.

$$H_1: \pi_1 \neq \pi_2$$
$$H_0: \pi_1 = \pi_2$$

We decide to set alpha at .05.

Selecting the Sampling Distribution and Specifying the Test Statistic. The population distributions of dichotomies are not normal. However, based on the central limit theorem, we know that the sampling distribution of the difference between sample proportions is normally distributed when the sample size is large (when $N_1 > 50$ and $N_2 > 50$), with mean μ_{p1-p2} and the estimated standard error S_{p1-p2}. Therefore, we can use the normal distribution as the sampling distribution, and we can calculate Z as the test statistic.[9]
   The formula for computing the Z statistic for a difference between proportions test is

$$Z = \frac{p^1 - p^2}{S_{p^1 - p^2}} \tag{8.10}$$

where p_1 and p_2 are the sample proportions for first- and second-generation Hispanic Americans, and S_{p1-p2} is the estimated standard error of the sampling distribution of the difference between sample proportions.
   The estimated standard error is calculated using the following formula:

$$S_{p_1-p_2} = \sqrt{\frac{p_1(1-p_1)}{N_1} + \frac{p_2(1-p_2)}{N_2}} \tag{8.11}$$

Calculating the Test Statistic. We calculate the standard error using Formula 8.11:

$$S_{p_1-p_2} = \sqrt{\frac{.11(1-.11)}{899} + \frac{.21(1-.21)}{351}} = \sqrt{.000581547} = .02$$

Substituting this value into the denominator of Formula 8.10, we get

$$Z = \frac{.11 - .21}{.02} = -5.00$$

Making a Decision and Interpreting the Results. Our obtained Z of -5.00 indicates that the difference between the two proportions will be evaluated at the left tail (the negative side) of the Z distribution. To determine the probability of observing a Z value of -5.00 if the null hypothesis is true, look up the value in Appendix A (Column C) to find the area to the right of (above) the obtained Z.

Note that a Z score of 5.00 is not listed in Appendix A; however, the value exceeds the largest Z reported in the table, 4.00. The P value corresponding to a Z score of -5.00 would be less than .0001. For a two-tailed test, we'll have to multiply P by 2 (.0001 $\times$ 2 = .0002). If this were a one-tailed test, we would not have to multiply the P value by 2. The probability of -5.00 for a two-tailed test is less than our alpha level of .05 (.0002 < .05).

Thus, we reject the null hypothesis of no difference and conclude that there is a significant difference in the proportion of college graduates among first- and second-generation Hispanic Americans. There is a significantly higher proportion of college graduates among second-generation Hispanic Americans compared with first-generation Hispanic Americans.

▣ IS THERE A SIGNIFICANT DIFFERENCE?

The news media made note of a 2010 Centers for Disease Control and Prevention (CDC) study that examined the difference in length of marriage between couples in 2005 who first cohabited before marriage and couples who did not cohabit before marriage. Several news services released stories noting the "troubles" associated with living together. As reported, the percentage of marriages surviving to the 10th anniversary, among those who cohabited before marriage, was lower than those who did not cohabit before their first marriage. A closer look at the report reveals important (overlooked) details.

CDC researchers Goodwin, Mosher, and Chandra (2010) reported that previous cohabitation experience was significantly associated with marriage survival probabilities for men. Conversely, though the probability that a woman's marriage would last at least 10 years was lower than for those who cohabited before marriage (60%) than for women who did not (66%), the researchers wrote, "in the 2002 data, the difference was not significant at the 5% level" (p. 13).[10]

Throughout this chapter, we've assessed the difference between two means and two proportions, attempting to determine whether the difference between them is due to real effects in the population or due to sampling error. A significant difference is one that confirms that effects of the independent variable, such as cohabiting before marriage, are real. As in the case of marriage survival rate, cohabitation before marriage makes a significant difference in marital outcomes for men, but not for women in the CDC sample. Take caution in accepting comparative statements that fail to mention significance. There may be a difference, but you have to ask, is it a significant difference?

▣ READING THE RESEARCH LITERATURE: REPORTING THE RESULTS OF STATISTICAL HYPOTHESIS TESTING

Robert Emmet Jones and Shirley A. Rainey (2006) examined the relationship between race, environmental attitudes, and perceptions about environmental health and justice.[11] Researchers have documented how people of color and the poor are more likely than whites and more affluent groups to live in areas with poor environmental quality and protection, exposing them to greater health risks. Yet little is known about how this disproportional exposure and risk are perceived by those affected. Jones and Rainey studied black and white residents from the Red River community in Tennessee, collecting data from interviews and a mail survey during 2001 to 2003.

They created a series of index scales measuring residents' attitudes pertaining to environmental problems and issues. The Environmental Concern (EC) Index measures public concern for specific environmental problems in the neighborhood. It includes questions on drinking water quality, landfills, loss of trees, lead paint and poisoning, the condition of green areas, and stream and river conditions. EC-II measures public concern (very unconcerned to very concerned) for the overall environmental quality in the neighborhood. EC-III measures the seriousness (not serious at all to very serious) of environmental problems in the neighborhood. Higher scores on all EC indicators indicate greater concern for environmental problems in their neighborhood. The Environmental Health (EH) Index measures public perceptions of certain physical side effects, such as headaches, nervous disorders, significant weight loss or gain, skin rashes, and breathing problems. The EH Index measures the likelihood (very unlikely to very likely) that a person believes that he or she or a household member experienced health problems due to exposure to environmental contaminants in his or her neighborhood. Higher EH scores reflect a greater likelihood that respondents believe that they have experienced health problems from exposure to environmental contaminants. Finally, the Environmental Justice (EJ) Index measures public perceptions about environmental justice, measuring the extent to which they agreed (or disagreed) that public officials had informed residents about environmental problems, enforced environmental laws, or held meetings to address residents' concerns. A higher mean EJ score indicates a greater likelihood that respondents think public officials failed to deal with environmental problems in their neighborhood. Index score comparisons between black and white respondents are presented in Table 8.4.

✓ *Learning Check*

Review the information provided in Table 8.4. What would be the t *critical at the .05 level for the first indicator, EC Index? Assume a two-tailed test.*

Let's examine the table carefully. Each row represents a single index measurement, reporting means and standard deviations separately for black and white residents. Obtained t-test statistics are reported in the second-to-last column. The probability of each t test is reported in the last column ($P < .001$), indicating a significant difference in responses between the two groups. All index score comparisons are significant at the .001 level.

While not referring to specific differences in index scores or to t-test results, Jones and Rainey use data from this table to summarize the differences between black and white residents on the three environmental index measurements:

Table 8.4   Environmental Concern (EC), Environmental Health (EH), and Environmental Justice (EJ)

Indicator	Group	Mean	Standard Deviation	t	Significance (one tailed)
EC Index	Blacks	56.2	13.7	6.2	<.001
	Whites	42.6	15.5		
EC-II	Blacks	4.4	1.0	5.6	<.001
	Whites	3.5	1.3		
EC-III	Blacks	3.4	1.1	6.7	<.001
	Whites	2.3	1.0		
EH Index	Blacks	23.0	10.5	5.1	<.001
	Whites	16.0	7.3		
EJ Index	Blacks	31.0	7.3	3.8	<.001
	Whites	27.2	6.3		

Source: Robert E. Jones and Shirley A. Rainey, "Examining Linkages Between Race, Environmental Concern, Health and Justice in a Highly Polluted Community of Color," *Journal of Black Studies* 36, no. 4 (2006): 473–496.

Note: N = 78 blacks, 113 whites.

The results presented [in Table 1] suggest that as a group, Blacks are significantly more concerned than Whites about local environmental conditions (EC Index). . . . The results . . . also indicate that as a group, Blacks believe they have suffered more health problems from exposure to poor environmental conditions in their neighborhood than Whites (EH Index). . . . [T]here is greater likelihood that Blacks feel local public agencies and officials failed to deal with environmental problems in their neighborhood in a fair, just, and effective manner (EJ Index). (p. 485)

MAIN POINTS

• Statistical hypothesis testing is a decision-making process that enables us to determine whether a particular sample result falls within a range that can occur by an acceptable level of chance. The process of statistical hypothesis testing consists of five steps: (1) making assumptions, (2) stating the research and null hypotheses and selecting alpha, (3) selecting a sampling distribution and a test statistic, (4) computing the test statistic, and (5) making a decision and interpreting the results.

• Statistical hypothesis testing may involve a comparison between a sample mean and a population mean or a comparison between two sample means. If we know the population variance(s) when testing for differences between means, we can use the Z statistic and the normal distribution. However, in practice, we are unlikely to have this information.

- When testing for differences between means when the population variance(s) are unknown, we use the t statistic and the t distribution.

- Tests involving differences between proportions follow the same procedure as tests for differences between means when population variances are known. The test statistic is Z, and the sampling distribution is approximated by the normal distribution.

KEY TERMS

alpha (α)
degrees of freedom (df)
left-tailed test
null hypothesis (H_0)
one-tailed test
P value
research hypothesis
   (H_1)

right-tailed test
sampling distribution
   of the difference
   between means
statistical hypothesis
   testing
t distribution
t statistic (obtained)

two-tailed test
Type I error
Type II error
Z statistic (obtained)

⑤SAGE edge™

Sharpen your skills with SAGE edge at **edge.sagepub.com/ssdsess2e**. **SAGE edge for students** provides a personalized approach to help you accomplish your coursework goals in an easy-to-use learning environment.

CHAPTER EXERCISES

1. It is known that, nationally, doctors working for health maintenance organizations (HMOs) average 13.5 years of experience in their specialties, with a standard deviation of 7.6 years. The executive director of an HMO in a western state is interested in determining whether or not its doctors have less experience than the national average. A random sample of 150 doctors from this HMO shows a mean of only 10.9 years of experience.
   a. State the research and the null hypotheses to test whether or not doctors in this HMO have less experience than the national average.
   b. Using an alpha level of .01, make this test.

2. We continue our analysis of the 2013 Pew Research Center data (first presented in this chapter), this time examining the difference in educational attainment between first- and second-generation Asian Americans.

   Our research hypothesis is that there is a lower proportion of college graduates among first-generation Asian Americans than second-generation Asian Americans.

   The data and results of our obtained Z statistic are presented.

First-Generation	Second-Generation	Obtained Z Statistic
$P_1 = .50$	$P_2 = .55$	−2.50
$N_1 = 2{,}684$	$N_2 = 566$	

   a. Is the research hypothesis a one- or two-tailed test?
   b. Based on the obtained Z statistic, what can you conclude about the difference in the proportion of college graduates between the two groups?

3. For each of the following situations, determine whether a one- or a two-tailed test is appropriate. Also, state the research and the null hypotheses.
   a. You are interested in finding out if the average household income of residents in your state is different from the national average household. According to the U.S. census, for 2011, the national average household income is $50,054.
   b. You believe that students at small liberal arts colleges attend more parties per month than students nationwide. It is known that nationally undergraduate students attend an average of 3.2 parties per month. The average number of parties per month will be calculated from a random sample of students from small liberal arts colleges.
   c. A sociologist believes that the average income of elderly women is lower than the average income of elderly men.
   d. Is there a difference in the amount of study time on-campus and off-campus students devote to their schoolwork during an average week? You prepare a survey to determine the average number of study hours for each group of students.
   e. Reading scores for a group of third graders enrolled in an accelerated reading program are predicted to be higher than the scores for nonenrolled third graders.
   f. Stress (measured on an ordinal scale) is predicted to be lower for adults who own dogs (or other pets) than for non–pet owners.

4. a. For each situation in Exercise 3, describe the Type I and Type II errors that could occur.
   b. What are the general implications of making a Type I error? Of making a Type II error?
   c. When would you want to minimize Type I error? Type II error?

5. One way to check on how representative a survey is of the population from which it was drawn is to compare various characteristics of the sample with the population characteristics. A typical variable used for this purpose is age. The 2010 GSS of the American adult population found a mean age of 49.28 years and a standard deviation of 17.21 for its sample of 4,857 adults. Assume that we know from census data that the mean age of all American adults is 37.2 years. Use this information to answer these questions.
   a. State the research and the null hypotheses for a two-tailed test.
   b. Calculate the t statistic and test the null hypothesis setting alpha at .01. What did you find?
   c. What is your decision about the null hypothesis? What does this tell us about how representative the sample is of the American adult population?

6. In the chapter, we examined the difference between white and black students in the timing of when they first tried cigarettes. In this SPSS output based on MTF11SSDS, we examine the grade when black and Hispanic students first tried alcohol. Present step 5 (final decision) for this data. Assume alpha = .05 and a two-tailed test.

Figure 8.6 *t*-Test Output for RACE and GRDRINK

Group Statistics

	race Respondent's race (trichoromized B/W/H)	N	Mean	Std. Deviation	Std. Error Mean
grdrink What grade when first tried alcohol, even a few sips?	1 BLACK:(1)	148	5.74	2.269	.187
	3 HISPANIC:(3)	166	4.78	2.390	.185

Independent Samples Test

		Levene's Test for Equality of Variances		t-test for Equality of Means						
		F	sig.	t	df	Sig. (2–Tailed)	Mean Difference	Std. Error Difference	95% Confidence Interval of the Difference Lower	Upper
grdrink What grade when first tried alcohol, even a few sips?	Equal variances assumed	.590	.443	3.614	312	.000	.953	.264	.434	1.472
	Equal variances not assumed			3.625	310.753	.000	.953	.263	.436	1.471

7. In this exercise, we will examine the attitudes of liberals and conservatives toward affirmative action policies in the workplace. Data from the 2010 GSS reveal that 12% of conservatives ($N = 336$) and 27% of liberals ($N = 267$) indicate that they "strongly support" or "support" affirmative action policies for African Americans in the workplace.
   a. What is the appropriate test statistic? Why?
   b. Test the null hypothesis with a one-tailed test (conservatives are less likely to support affirmative action policies than liberals); $\alpha = .05$. What do you conclude about the difference in attitudes between conservatives and liberals?
   c. Test the null hypothesis with a one-tailed test, alpha = .01. What do you conclude?

8. The GSS 2010 measures the number of hours individuals spend on the Internet per week. Data for male and female respondents are presented, along with the obtained *t* statistic.

   We tested the research hypothesis that men use the Internet more than women, setting alpha at .01.

Men	Women	Obtained t Statistic
Mean = 10.17	Mean = 9.08	.757
Standard deviation = 11.71	Standard deviation = 12.26	
N = 118	N = 157	

Based on the obtained *t* statistic, what can you conclude about the difference in Internet use between male and female GSS respondents?

9. During the 2012 presidential campaign, pollsters consistently reported how President Obama's supporters were mostly women, with men less likely to support his candidacy. In surveys conducted during March 2012 (months before the election), the Pew Research Center reported that among 762 men, 49% indicated that they supported President Obama. Among 741 women, 58% reported the same. Do these differences reflect a significant gender gap among Obama supporters?[12]
   a. If you wanted to test the research hypothesis that the proportion of male voters supporting President Obama is less than the proportion of female voters, would you conduct a one- or a two-tailed test?
   b. Test the research hypothesis at the .05 alpha level. What do you conclude?
   c. If alpha were changed to .01, would your decision remain the same?

10. Is there a significant difference in the level of community service participation between college and high school graduates? According to the 2010 GSS, 38% of 160 college graduates reported volunteering in the previous month compared with 29% of 210 high school graduates.
   a. What is the research hypothesis? Should you conduct a one- or a two-tailed test? Why?
   b. Present the five-step model, testing your hypothesis at the .05 level. What do you conclude?
   c. If alpha were changed to .01, would your decision change?

11. We will continue to examine media use in two additional categories: e-mail hours per week and hours per day watching television. We compare use among high school and college graduates from the GSS 2010.

Media Use Means and Standard Deviations for High School and College Graduates

	E-mail Hours per Week	*Television Watching Hours per Day*
High school graduates	Mean = 5.71 Standard deviation = 7.26 N = 84	Mean = 3.25 Standard deviation = 2.60 N = 246
College graduates	Mean = 7.32 Standard deviation = 7.23 N = 41	Mean = 2.41 Standard deviation = 1.74 N = 111

   a. Determine whether high school graduates have significantly fewer e-mail hours per week than college graduates. Test at the .05 alpha level.
   b. Test whether there is a significant difference in hours of television viewing between high school and college graduates. Assume alpha = .01.

12. GSS 2010 male and female respondents reported their ages when their first children were born. Based on the SPSS output, determine whether there is a significant difference between the two groups. Assume a two-tailed test and $\alpha = .05$.

Figure 8.7   *t*-Test Output for SEX and AGEKDBRN

Group Statistics

	sex RESPONDENT SEX	N	Mean	Std. Deviation	Std. Error Mean
agekdbrn R'S AGE WHEN 1ST CHILD BORN	1 MALE	379	25.78	6.561	.337
	2 FEMALE	583	22.42	5.155	.214

Independent Samples Test

		Levene's Test for Equality of Variances		t-test for Equality of Means					95% Confidence Interval of the Difference	
		F	sig.	t	df	Sig. (2–Tailed)	Mean Difference	Std. Error Difference	Lower	Upper
agekdbrn R'S AGE WHEN 1ST CHILD BORN	Equal variances assumed	15.334	.000	8.858	960	.000	3.361	.379	2.616	4.105
	Equal variances not assumed			8.424	672.029	.000	3.361	.399	2.577	4.144

13. We calculate the statistic for high school and college graduates, measuring their average ages when their first children were born. Results, based on the GSS10SSDS, are presented below.

Figure 8.8   *t*-Test Output for DEGREE and AGEKDBRN

Group Statistics

	degree RS HIGHEST DEGREE	N	Mean	Std. Deviation	Std. Error Mean
agekdbrn R'S AGE WHEN 1ST CHILD BORN	1 HIGH SCHOOL	485	22.66	5.111	.232
	3 BACHELOR	159	26.75	5.510	.437

Independent Samples Test

		Levene's Test for Equality of Variances		t-test for Equality of Means					95% Confidence Interval of the Difference	
		F	sig.	t	df	Sig. (2–Tailed)	Mean Difference	Std. Error Difference	Lower	Upper
agekdbrn R'S AGE WHEN 1ST CHILD BORN	Equal variances assumed	2.054	.152	–8.593	642	.000	–4.093	.476	–5.028	–3.158
	Equal variances not assumed			–8.272	253.116	.000	–4.093	.495	–5.067	–3.158

Based on a two-tailed test, $\alpha = .05$, present step 5 (final decision). What do you conclude?

14. In the chapter we examined the difference in educational attainment between first- and second-generation Hispanic and Asian Americans based on the proportion of each group with bachelor's degrees. We present additional data from the Pew Research Center's 2013 report, measuring the percentage of each group who are home owners.

	Percentage Owning a Home
First-generation Hispanic Americans	43
$N = 899$	
Second-generation Hispanic Americans	50
$N = 351$	
First-generation Asian Americans	58
$N = 2,684$	
Second-generation Asian Americans	51
$N = 566$	

Source: Pew Research Center, *Second-Generation Americans: A Portrait of the Adult Children of Immigrants*, February 7, 2013.

    a. Test whether there is a significant difference in the proportion of home owners between first- and second-generation Hispanic Americans. Set alpha at .05.

    b. Test whether there is a significant difference in the proportion of home owners between first- and second-generation Asian Americans. Set alpha at .01.

Chapter 9

Bivariate Tables

Chapter Learning Objectives

- ❖ Constructing a bivariate table
- ❖ Dealing with ambiguous relationships between variables
- ❖ Determining the properties of a bivariate relationship: existence, strength, and direction
- ❖ Understanding how to elaborate the relationship between variables: nonspuriousness, intervening, and conditional relationships
- ❖ Understanding hypothesis testing and statistical independence with chi-square
- ❖ Recognizing the limitations of the chi-square test—sample size and statistical significance
- ❖ Understanding the concept of PRE (proportional reduction of error) and how to interpret measures of association
- ❖ Calculating and interpreting lambda, gamma, and Kendall's tau-*b*
- ❖ Interpreting Cramer's *V*: a chi-square–related measure of association

One of the main objectives of social science is to make sense out of human and social experience by uncovering regular patterns among events. Therefore, the language of *relationships* is at the heart of social science inquiry. Consider the following examples from articles and research reports:

Example 1: Americans 50 years and older are more likely to oppose creating a path to citizenship for illegal immigrants than are younger Americans.[1] (This example indicates a relationship between age and immigration reform.)

Example 2: Contrary to the stereotype, whites use government safety net programs more than blacks or Latinos, and they are more likely than minorities to be lifted out of poverty by the taxpayer money that they get.[2] (This example indicates a relationship between race and receipt of government aid.)

In each of these examples, a relationship means that certain values of one variable tend to "go together" with certain values of the other variable. In Example 1, age "goes together" with immigration reform and being younger is associated with support of a path to citizenship. In Example 2, being white "goes together" with frequent use of government aid and being black or Latino goes with less frequent use of government aid.

In this chapter, we introduce one of the most common techniques used in the analysis of relationships between two variables: *cross-tabulation*. **Cross-tabulation** is a technique for analyzing the relationship between two variables that have been organized in a table. A cross-tabulation is a type of **bivariate analysis**, a method designed to detect and describe the relationship between two nominal or ordinal variables. We demonstrate not only how to detect whether two variables are associated but also how to determine the strength of the association and, when appropriate, its direction. We have already applied bivariate analysis in Chapter 8, using *t* tests and *Z* tests to determine the difference between two means or proportions.[3]

Cross-tabulation   A technique for analyzing the relationship between two nominal or ordinal variables that have been organized in a table.

Bivariate analysis   A statistical method designed to detect and describe the relationship between two nominal or ordinal variables.

▣ INDEPENDENT AND DEPENDENT VARIABLES

In the social sciences, an important aspect in research design and statistics is the distinction between the *independent variable* and the *dependent variable*. These terms, first introduced in Chapter 1, are used throughout this chapter as well as in the following chapters, and therefore, it is important that you understand the distinction between them.

In each of the illustrations given, there are two variables: an independent and a dependent variable. In Example 1, the purpose of the research is to explain support for immigration reform laws. One of the variables hypothesized as being connected to support for immigration reform is age. Therefore, support for immigration reform is the dependent variable, and age is the independent variable. In Example 2, the object of the investigation is to examine the common stereotype that people of color use government aid more than white Americans. The investigator is trying to explain differences in the use of government aid using race as an explanatory variable. Therefore, the use of government aid is the dependent variable, and race is the independent variable.

The statistical techniques discussed in this and the remaining chapters help the researcher decide the strength of the relationship between the independent and dependent variables.

✓ *Learning Check*

For some variables, whether they are the independent or dependent variables depends on the research question. If you are still having trouble distinguishing between an independent and a dependent variable, go back to Chapter 1 for a detailed discussion.

▣ HOW TO CONSTRUCT A BIVARIATE TABLE: RACE AND HOME OWNERSHIP

A **bivariate table** displays the distribution of one variable across the categories of another variable. It is obtained by classifying cases based on their joint scores on two nominal or ordinal variables. It can be thought of as a series of frequency distributions joined to make one table. The data in Table 9.1 represent a sample of General Social Survey (GSS) respondents by race and whether they own or rent their homes (in this case, both variables are nominal-level measurements).

To make sense of these data, we must first construct the table in which these individual scores will be classified. In Table 9.2, the 17 respondents have been classified according to joint scores on race and home ownership.

The table has the following features typical of most bivariate tables:

1. The table's title is descriptive, identifying its content in terms of the two variables.

Table 9.1   Race and Home Ownership for 17 GSS Respondents

Respondent	Race	Home Ownership
1	Black	Own
2	Black	Own
3	White	Rent
4	White	Rent
5	White	Own
6	White	Own
7	White	Own
8	Black	Rent
9	Black	Rent
10	Black	Rent
11	White	Own
12	White	Own
13	White	Rent
14	White	Own
15	Black	Rent
16	White	Own
17	Black	Rent

Table 9.2 Home Ownership by Race (absolute
frequencies), GSS

| Home Ownership | Race | | Row marginals (row total) |
	Black	White	
Own	2	7	9
Rent	5	3	8
	7	10	17 Total cases (*N*)

Column marginals (column total)

2. It has two dimensions, one for race and one for home ownership. The variable *home owner-ship* is represented in the rows of the table, with one row for owners and another for renters. The variable *race* makes up the columns of the table, with one column for each racial group. A table may have more columns and more rows, depending on how many categories the variables represent. For example, had we included a group of Latinos, there would have been three columns (not including the row total column). Usually, the independent variable is the **column variable** and the dependent variable is the **row variable**.

3. The intersection of a row and a column is called a **cell**. For example, the two individuals represented in the upper left cell are blacks who are also home owners.

4. The column and row totals are the frequency distribution for each variable, respectively. The column total is the frequency distribution for *race,* the row total for *home ownership.* Row and column totals are sometimes called **marginals**. The total number of cases (*N*) is the number reported at the intersection of the row and column totals. (These elements are all labeled in the table.)

5. The table is a 2 × 2 table because it has two rows and two columns (not counting the marginals). We usually refer to this as an *r* × *c* table, in which *r* represents the number of rows and *c* the number of columns. Thus, a table in which the row variable has three categories and the column variable has two categories would be designated as a 3 × 2 table.

6. The source of the data should also be clearly noted in a source note to the table. This is consistent with what we reviewed in Chapter 2.

Bivariate table   A table that displays the distribution of one variable across the categories of another variable.

Column variable   A variable whose categories are the columns of a bivariate table.

Row variable   A variable whose categories are the rows of a bivariate table.

Cell   The intersection of a row and a column in a bivariate table.

Marginals   The row and column totals in a bivariate table.

✓ *Learning Check*

Examine Table 9.2. Make sure you can identify all the parts just described and that you understand how the numbers were obtained. Can you identify the independent and dependent variables in the table? You will need to know this to convert the frequencies to percentages.

▣ HOW TO COMPUTE PERCENTAGES IN A BIVARIATE TABLE

To compare home ownership status for blacks and whites, we need to convert the raw frequencies to percentages because the column totals are not equal. Recall from Chapter 2 that percentages are especially useful for comparing two or more groups that differ in size. There are two basic rules for computing and analyzing percentages in a bivariate table:

1. Calculate percentages within each category of the independent variable.

2. Interpret the table by comparing the percentage point difference for different categories of the independent variable.

Calculating Percentages Within Each Category of the Independent Variable

The first rule means that we have to calculate percentages within each category of the variable that the investigator defines as the independent variable. When the independent variable is arrayed in the *columns*, we compute percentages within each column separately. The frequencies within each cell and the row marginals are divided by the total of the column in which they are located, and the column totals should sum to 100%. When the independent variable is arrayed in the *rows*, we compute percentages within each row separately. The frequencies within each cell and the column marginals are divided by the total of the row in which they are located, and the row totals should sum to 100%.

In our example, we are interested in *race* as the independent variable and in its relationship with *home ownership*. Therefore, we are going to calculate percentages by using the column total of each racial group as the base of the percentage. The percentage of black respondents who own their homes is obtained by dividing the number of black home owners by the total number of blacks in the sample.

Table 9.3   Home Ownership by Race (in percentages)

Home Ownership	Race		
	Black	*White*	*Total*
Own	28.6%	70.0%	52.9%
Rent	71.4%	30.0%	47.1%
Total	100%	100%	100%
(*N*)	(7)	(10)	(17)

Table 9.3 presents percentages based on the data in Table 9.2. Notice that the percentages in each column add up to 100%, including the total column percentages. Always show the *N*s that are used to compute the percentages—in this case, the column totals.

Comparing the Percentages Across Different Categories of the Independent Variable

The second rule tells us to compare how home ownership varies between blacks and whites. Comparisons are made by examining differences between percentage points across different categories of the independent variable. Some researchers limit their comparisons to categories with at least a 10 percentage point difference. In our comparison, we can see that there is a 41.4 percentage point difference between the percentages of white home owners (70%) and black home owners (28.6%). In other words, in this group, whites are more likely to be home owners than blacks.[4] Therefore, we can conclude that one's race appears to be associated with the likelihood of being a home owner.

Note that the same conclusion would be drawn had we compared the percentage of black and white renters. However, since the percentages of home owners and renters within each racial group sum to 100%, we need to make only one comparison. In fact, for any 2 × 2 table, only one comparison needs to be made to interpret the table. For a larger table, more than one comparison can be made and used in interpretation.

Practice constructing a bivariate table. Use Table 9.1 to create a percentage bivariate table. Compare your table with Table 9.3. Did you remember all the parts? Are your calculations correct? If not, go back and review this section. Remember, you must correctly identify the independent variable so that you know whether to percentage across the rows or down the columns.

✓ *Learning Check*

▣ HOW TO DEAL WITH AMBIGUOUS RELATIONSHIPS BETWEEN VARIABLES

Sometimes it isn't apparent which variable is independent or dependent; sometimes the data can be viewed either way. In this case, you might compute both row and column percentages. For example, Table 9.4 presents three sets of figures for the variables SPANKING and FEFAM for a sample of 127 GSS respondents: (a) the absolute frequencies, (b) the column percentages, and (c) the row percentages. SPANKING is measured with the survey question "Do you favor spanking to discipline a child?" The variable FEFAM measures whether the respondent agrees or disagrees with the statement "a man should work and a woman should stay at home." Table 9.4b shows that respondents who strongly disagree with spanking a child are less likely to agree with the FEFAM statement than those who strongly agree with spanking (10% compared with 50%). Table 9.4c shows that individuals who strongly agree that a man should work and a woman should stay at home are more likely to agree with spanking than those who disagree with the statement on men's and women's roles (94% compared with 63%).

Thus, percentaging within each *column* (Table 9.4b) allows us to examine the hypothesis that spanking (the independent variable) is associated with agreement with the FEFAM statement (the dependent variable). When we percentage within each *row* (Table 9.4c), the hypothesis is that agreement or disagreement with the FEFAM statement (the independent variable) may be related to SPANKING (the dependent variable).[5]

Table 9.4   The Different Ways Percentages Can Be Computed: SPANKING by FEFAM

FEFAM	SPANKING		Row Total
	Strongly Agree	Strongly Disagree	
a. Absolute frequencies			
Strongly agree	48	3	51
Strongly disagree	48	28	76
Column total	96	31	127
b. Column percentages (column totals as base)			
Strongly agree	50%	10%	40%
Strongly disagree	50%	90%	60%
Column total	100% (96)	100% (31)	100% (127)
c. Row percentages (row totals as base)			
Strongly agree	94%	6%	100% (51)
Strongly disagree	63%	37%	100% (76)
Column total	76%	24%	100% (127)

Finally, it is important to understand that ultimately what guides the construction and interpretation of bivariate tables is the theoretical question posed by the researcher. Although the particular example in Table 9.4 makes sense if interpreted using row or column percentages, not all data can be interpreted this way. For example, a table comparing women's and men's attitudes toward sexual harassment in the workplace could provide a sensible explanation in only one direction. Gender might influence a person's attitude toward sexual harassment; however, a person's attitude toward sexual harassment certainly couldn't influence her or his gender. Therefore, either row or column percentages are appropriate, depending on the way the variables are arrayed, but not both.

▣ READING THE RESEARCH LITERATURE: PLACE OF DEATH IN AMERICA

The guidelines for constructing and interpreting bivariate tables discussed in this chapter are not always strictly followed. Most bivariate tables presented in the professional literature are a good deal more complex than those we have just been describing. Let's conclude this section with a typical example of how bivariate tables are presented in social science literature. The following example is drawn from a 2007 study by Andrea Gruneir, Vincent Mor, Sherry Weitzen, Rachael Truchil, Joan Teno, and Jason Roy on understanding variations on the sites of death in America.

According to the researchers, driven in part by the increasing aging of the U.S. population, there has been an increase in the level of public and professional concern about the quality of end-of-life care. Though public surveys confirm that most would prefer to die at home, the majority of Americans die in an institutional setting, such as a hospital or care facility. Acute care hospitals are still the number one site of death for people with chronic illnesses. In this study, Gruneir and her colleagues explored the likelihood of home versus hospital or nursing home death, explaining that

> where individuals spend their last days of life is influenced by individual demographic and clinical characteristics as well as to the degree which their community has an interest in and sufficient wealth to invest in service resources such as hospitals, nursing homes, or home and hospice care services. (p. 359)

The study examines differences in the place of death by gender, age, marital status, race/ethnicity, education, and cause of death. Researchers relied on data from the 1997 National Vital Statistics System. The data set includes death certificates for 1,402,167 deaths, identifying place of death as either acute care hospital, nursing care facility, home, or other.

Table 9.5 shows the results of the survey. Follow these steps in examining it:

1. Identify the dependent variable and the type of unit of analysis it describes (such as individual, city, or child). Here the dependent variable is *place of death in 1997*. The categories for this variable are "hospital," "home," or "nursing home." The type of unit used in this table is individual.

Table 9.5   Place of Death, 1997, by Gender, Age, Marital Status, Race/Ethnicity, and Education (percentages reported)

	Place of Death, 1997		
	Hospital (N = 740,405)	*Home* (N = 330,447)	*Nursing Home* (N = 331,315)
Gender			
Male	57.3	25.5	17.2
Female	48.7	21.8	29.5
Age (years)			
<65	64.4	29.2	6.4
65–74	59.3	28.1	12.6
75–84	52.7	22.7	24.6
85–94	40.8	16.9	42.3
95+	28.0	14.5	57.6
Marital status			
Never married	55.9	21.5	22.7
Married	59.0	26.9	14.1
Widowed	45.2	19.9	34.9
Divorced	54.6	26.1	19.2
Not stated	57.1	24.9	18.0
Race/ethnicity			
White	49.7	24.2	26.1
Black	66.4	20.2	13.5
Hispanic	65.2	22.7	12.1
Other/unknown	63.4	21.7	14.9
Education (years)			
<9	50.8	20.3	29.0
9–11	54.7	23.0	22.3
12	53.5	23.8	22.7
13–15	51.9	26.4	21.7
16+	50.8	27.7	21.6
Unknown	56.8	18.8	24.4

Source: Andrea Gruneir, Vincent Mor, Sherry Weitzen, Rachael Truchil, Joan Teno, and Jason Roy, "Where People Die: A Multilevel Approach to Understanding Influences on Site of Death in America," *Medical Care Research Review* 64 (2007): 351–378.

2. Identify the independent variables included in the table and the categories of each. There are five independent variables: gender, age, marital status, race/ethnicity, and education. For the first variable, gender, the categories are "male" and "female." Review the table to determine the categories for the remaining variables.

3. Clarify the structure of the table. Note that the independent variables are arrayed in the rows of the table and the dependent variable, *place of death,* is arrayed in the columns. The table is divided into five panels, one for each independent variable. There are actually five bivariate tables here—one for each independent variable.

   Since the independent variables are arrayed in the rows, percentages are calculated within each row separately, with the row totals (not shown) serving as the bases for the percentages. From the table, we know that the largest number of deaths in 1997 occurred in hospitals—740,405 died in hospitals compared with 330,447 who died at home and 331,315 who died in nursing homes. Combining categories for hospital and nursing home, as Gruneir and colleagues explained, the majority of Americans died in an institutional setting rather than at home. Though not calculated, the row percentages should total 100%.

4. Using Table 9.5, we can make a number of comparisons, depending on which independent variable we are examining. For example, to determine the relationship between gender and place of death, compare the percentages between men and women. For example, we know that the largest percentages for both men and women are in the category "died in hospital." Yet looking at each place-of-death category, we see that the percentages are about the same for those who died at home (25.5 male, 21.8 female), but greater differences exist between the percentage of those who died in nursing homes (17.2 male, 29.5 female) and in hospitals (57.3 male, 48.7 female). A higher percentage of women than men were reported dying in nursing homes, while a higher percentage of men than women were reported dying in hospitals.

   You can make similar comparisons to determine the association between age, marital status, race/ethnicity, and education.

5. Finally, what conclusions can you draw about variations in place of death? The researchers offer this interpretation of the findings presented in the table.

   The frequency of nursing home death increased with age and among the oldest adults, nursing homes were the most common site of death. A greater percentage of women than men died in the nursing home (29.5% vs. 17.2%) but the converse was seen in other sites of death. Married and divorced decedents showed the greatest frequency of home death (26.9% and 26.1%, respectively) while widowed decedents showed the greatest frequency of nursing home death (34.9%). Approximately half of all white decedents died in hospital but well over 60% of each other racial/ethnic group died in hospital. Of those who died outside the hospital, white decedents were equivalently split between home and nursing home while other groups more frequently died at home than in nursing home. (p. 363)

✓ *Learning*
Check

Use Table 9.5 to verify each of the following conclusions drawn by the researchers about the place of death: (1) Among the oldest adults, nursing homes were the most common site of death. (2) A greater percentage of women died in nursing homes than men. (3) Marital status is related to home death. (4) Approximately half of all white respondents died in hospitals, while more than 60% of all other racial/ethnic groups died in hospitals. Can you explain these patterns? What other questions do these patterns raise about place of death?

▣ THE PROPERTIES OF A BIVARIATE RELATIONSHIP

In this section, we present some detailed observations that we may want to make about the "properties" of a bivariate association. These properties can be expressed as three questions to ask when examining a bivariate relationship:[6]

1. Does there appear to be a relationship?

2. How strong is it?

3. What is the direction of the relationship?

The Existence of the Relationship

Based on Table 9.6, we want to examine whether the frequency of church attendance by respondents had an effect on their support for abortion. Support for abortion was measured with the following question: "Please tell me whether or not you think it should be possible for a pregnant woman to obtain a legal abortion if the woman wants it for any reason." Frequency of church attendance was determined by asking respondents to indicate how often they attend religious services.[7]

Let's hypothesize that those who attend church frequently are more likely to be pro-life. We are not suggesting that church attendance necessarily "causes" pro-life attitudes, but that perhaps there is an indirect connection between the two. For example, perhaps those who attend church less frequently are more likely to want decisions about the body to be made on an individual basis through the right to choose an abortion.

In this formulation, church attendance is said to "influence" attitudes toward abortion, so it is the independent variable; therefore, percentages are calculated within each category of church attendance (church attendance is the column variable).

A relationship is said to exist between two variables in a bivariate table if the percentage distributions vary across the different categories of the independent variable, in this case church attendance. We can easily see that the percentage who support abortion changes across the different levels of church attendance. Of those who never attend church, 55% are pro-choice; of those who infrequently attend church, 50% are pro-choice; and of those who frequently attend church, 26% are pro-choice. Table 9.6 indicates that church attendance and support for abortion are associated as hypothesized.

If church attendance were unrelated to attitudes toward abortion among GSS respondents, then we would expect to find equal percentages of respondents who are pro-choice (or anti-choice)

Table 9.6 Support for Abortion by Church Attendance

Abortion	Church Attendance			Total
	Never	*Infrequently*	*Frequently*	*Total*
Yes	55%	50%	26%	43%
No	45%	50%	74%	57%
Total	100%	100%	100%	100%
(*N*)	(111)	(212)	(157)	(480)

regardless of the level of church attendance. Table 9.7 is a fictional representation of a strictly hypothetical pattern of no association between abortion attitudes and church attendance. The percentage of respondents who are pro-choice in each category of church attendance is equal to the overall percentage of respondents in the sample who are pro-choice (43%).

The Strength of the Relationship

In the preceding section, we saw how to establish whether an association exists in a bivariate table. If it does, how do we determine the strength of the association between the two variables? A quick method is to examine the percentage difference across the different categories of the independent variable. The larger the percentage difference across the categories, the stronger the association.

In the hypothetical example of no relationship between church attendance and attitude toward abortion (Table 9.7), there is a 0% difference between the columns. At the other extreme, if all respondents who never attended church were pro-choice and none of the respondents who frequently attended church were pro-choice, a perfect relationship would be manifested in a 100% difference. Most relationships, however, will be somewhere in between these two extremes. In fact, we rarely see a situation with either a 0% or a 100% difference. Going back to the observed percentages in Table 9.6, we find the largest percentage difference between respondents who

Table 9.7 Support for Abortion by Church Attendance (a hypothetical illustration of no relationship)

Abortion	Church Attendance			Total
	Never	*Infrequently*	*Frequently*	*Total*
Yes	43%	43%	43%	43%
No	57%	57%	57%	57%
Total	100%	100%	100%	100%
(*N*)	(111)	(212)	(157)	(480)

never attend church and respondents who frequently attend church (55% − 26% = 29%). The difference between respondents who infrequently attend church and respondents who frequently attend church (50% − 26% = 24%), though not as large, is nonetheless substantial, indicating a moderate relationship between church attendance and attitudes toward abortion.

Percentage differences are a rough indicator of the strength of a relationship between two variables. Later in this chapter, we discuss measures of association that provide a more standardized indicator of the strength of an association.

The Direction of the Relationship

When both the independent and dependent variables in a bivariate table are measured at the ordinal level or the interval-ratio level, we can talk about the relationship between the variables as being either positive or negative. A **positive** bivariate relationship exists when the variables vary in the same direction. Higher values of one variable "go together" with higher values of the other variable. In a **negative** bivariate relationship, the variables vary in opposite directions: higher values of one variable "go together" with lower values of the other variable (and the lower values of one go together with the higher values of the other).

Positive relationship   A bivariate relationship between two variables measured at the ordinal level or higher in which the variables vary in the same direction.

Negative relationship   A bivariate relationship between two variables measured at the ordinal level or higher in which the variables vary in opposite directions.

Table 9.8, from the International Social Survey Programme, displays a positive relationship between willingness to pay higher taxes and willingness to pay higher prices. Examine each category separately. For respondents who are unwilling to pay higher prices, an unwillingness to pay higher taxes is most typical (91.5%). For respondents who are indifferent to paying higher prices, the most common response is to be indifferent to paying higher taxes (55.1%); and finally,

Table 9.8   Willingness to Pay Higher Taxes by Willingness to Pay Higher Prices: A Positive Relationship

Willingness to Pay Higher Taxes	Willingness to Pay Higher Prices		
	Unwilling	Indifferent	Willing
Unwilling	91.5%	36.4%	23.6%
Indifferent	5.1%	55.1%	18.6%
Willing	3.4%	8.5%	57.8%
Total	100%	100%	100%
(N)	(529)	(352)	(532)

Table 9.9 Support for Attendance of Religious Services by Educational Level: A Negative Relationship

Attendance of Religious Services	Educational Level		
	None	*Secondary Degree*	*University Degree*
Never	5.2%	32.5%	37.3%
Infrequently	28.6%	35.0%	34.9%
2 to 3 times per month or more	66.2%	32.5%	27.8%
Total	100%	100%	100%
(*N*)	(77)	(237)	(126)

for respondents who are willing to pay higher prices, a willingness to pay higher taxes is most typical (57.8%). This is a positive relationship, with a willingness to pay higher prices associated with a willingness to pay higher taxes and an unwillingness to pay higher prices associated with an unwillingness to pay higher taxes.

Table 9.9, also from the International Social Survey Programme, shows a negative association between educational level and attendance of religious services for a sample of about 400 international respondents.[8] Individuals with no education typically attended religious services two to three times per month or more (66.2%). Individuals with secondary degrees (i.e., roughly, the U.S. equivalent to high school) typically attended religious services infrequently, ranging from monthly to several times a year (35.0%); and for individuals who had completed work at a university, the most common category was "never," meaning that they never attended religious services (37.3%). The relationship is a negative one because as educational level increases, the frequency of attendance of religious services decreases.

The examination of a possible relationship between two variables, however, is only a first step in data analysis. Having established through bivariate analysis that the independent and dependent variables are associated, we seek to further interpret and understand the nature of this relationship. **Elaboration** is a process designed to further explore a bivariate relationship, involving the introduction of additional variables, called **control variables**. By adding a control variable to our analysis, we are considering or "controlling" for the variable's effect on the bivariate relationship. Each potential control variable represents an alternative explanation for the bivariate relationship under consideration. The details of elaboration will not be considered in this text.

Elaboration   A process designed to further explore a bivariate relationship; it involves the introduction of control variables.

Control variable   An additional variable considered in a bivariate relationship. The variable is controlled for when we take into account its effect on the variables in the bivariate relationship.

▣ THE CHI-SQUARE TEST AND MEASURES OF ASSOCIATION

We extend our examination of the educational experience by focusing on first-generation college students—that is, students whose parents never completed a postsecondary education. Most first-generation students begin college at 2-year programs or at community colleges. According to W. Elliot Inman and Larry Mayes (1999), since first-generation college students represent a large segment of the community college population, they bring with them a set of distinct goals and constraints. Understanding their experiences and their demographic backgrounds may allow more intentional recruiting, retention, and graduation efforts. Inman and Mayes set out to examine first-generation college students' experiences, but they began first by determining who was most likely to be a first-generation college student.

Data from Inman and Mayes's study are presented in Table 9.10, a bivariate table, which includes *gender* and *first-generation college status*. From the table, we know that a higher percentage of women than men reported being first-generation college students, 46.6% versus 35.4%.

The percentage differences between men and women in first-generation college status, shown in Table 9.10, suggest that there is a relationship. If gender and first-generation college status were not associated, we would expect the same percentage of men and women to be first-generation college students. Similarly, we would expect to see the same percentage of men and women who are nonfirsts. These percentages should be equal to the percentage of "firsts" and "nonfirsts" in the sample as a whole (categories used by Inman and Mayes). The last column of Table 9.10—the row marginals—displays these percentages: 41.9% of all respondents were first-generation students, whereas 58.1% were nonfirsts. Therefore, if there were no association between gender and first-generation college status, we would expect to see 41.9% of the men and 41.9% of the women in the sample as first-generation students. Similarly, 58.1% of the men and 58.1% of the women would not be.

Table 9.10   Percentages of Men and Women Who Are First-Generation College Students

First Generation	Men	Women	Total
Firsts	35.4%	46.6%	41.9%
	(691)	(1,245)	(1,936)
Nonfirsts	64.6%	53.4%	58.1%
	(1,259)	(1,425)	(2,684)
Total (N)	100.0%	100.0%	100.0%
	(1,950)	(2,670)	(4,620)

Source: Adapted from W. Elliot Inman and Larry Mayes, "The Importance of Being First: Unique Characteristics of First Generation Community College Students," *Community College Review* 26, no. 3 (1999): 8. Copyright © North Carolina State University. Published by SAGE Publications.

Table 9.11 shows these hypothetical expected percentages. Because the percentage distributions of the variable *first-generation college status* are identical for men and women, we can say that Table 9.11 demonstrates a perfect model of "no association" between the variable *first-generation college status* and the variable *gender.*

If there is an association between gender and first-generation college status, then at least some of the observed percentages in Table 9.10 should differ from the hypothetical expected percentages shown in Table 9.11. Conversely, if gender and first-generation college status are not associated, the observed percentages should approximate the expected percentages shown in Table 9.11. In a cell-by-cell comparison of Tables 9.10 and 9.11, you can see that there is quite a disparity between the observed percentages and the hypothetical percentages. For example, in Table 9.10, 35.4% of the men reported that they were first-generation college students, whereas the corresponding cell for Table 9.11 shows that 41.9% of the men reported the same. The remaining three cells reveal similar discrepancies.

Are the disparities between the observed and expected percentages large enough to convince us that there is a genuine pattern in the population? The *chi-square* statistic helps us answer this question. It is obtained by comparing the actual observed frequencies in a bivariate table with the frequencies that are generated under an assumption that the two variables in the cross-tabulation are not associated with each other. If the observed and expected values are very close, the chi-square statistic will be small. If the disparities between the observed and expected values are large, the chi-square statistic will be large.

▣ THE CONCEPT OF CHI-SQUARE AS A STATISTICAL TEST

The **chi-square test** (pronounced kai-square and written χ^2) is an inferential statistical technique designed to test for significant relationships between two variables organized in a bivariate table. The test has a variety of research applications and is one of the most widely used tests in the social sciences. The chi-square test requires no assumptions about the shape of the population distribution from which a sample is drawn. It can be applied to nominally or ordinally measured variables (including grouped interval-level data).

Table 9.11 Percentages of Men and Women Who Are First-Generation College Students: Hypothetical Data Showing No Association

First Generation	Men	Women	Total
Firsts	41.9%	41.9%	41.9%
			(1,936)
Nonfirsts	58.1%	58.1%	58.1%
			(2,684)
Total (*N*)	100.0%	100.0%	100.0%
	(1,950)	(2,670)	(4,620)

Chi-square test   An inferential statistical technique designed to test for a significant relationship between two nominal or ordinal variables organized in a bivariate table.

The chi-square test can also be applied to the distribution of scores for a single variable. Also referred to as the goodness-of-fit test, the chi-square test can compare the actual distribution of a variable with a set of expected frequencies. This application is not presented in this chapter.

▣ THE CONCEPT OF STATISTICAL INDEPENDENCE

When two variables are not associated (as in Table 9.11), one can say that they are **statistically independent**. That is, an individual's score on one variable is independent of his or her score on the second variable. We identify statistical independence in a bivariate table by comparing the distribution of the dependent variable in each category of the independent variable. When two variables are statistically independent, the percentage distributions of the dependent variable within each category of the independent variable are identical. The hypothetical data presented in Table 9.11 illustrate the notion of statistical independence. Based on Table 9.11, we would say that first-generation college status is independent of one's gender.[9]

Statistical independence   The absence of association between two cross-tabulated variables. The percentage distributions of the dependent variable within each category of the independent variable are identical.

✓ *Learning Check*

The data we will use to practice calculating chi-square are also from Inman and Mayes's research. We will examine the relationship between age (independent variable) and first-generation college status (the dependent variable), as shown in the following bivariate table:

Age and First-Generation College Status

First-Generation Status	Age 19 Years or Younger	20 Years or Older	Total
Firsts	916 (33.7%)	1,018 (53.6%)	1,934 (41.9%)
Nonfirsts	1,802 (66.3%)	881 (46.4%)	2,683 (58.1%)
Total (N)	2,718 (100.0%)	1,899 (100.0%)	4,617 (100.0%)

Source: Adapted from W. Elliot Inman and Larry Mayes, "The Importance of Being First: Unique Characteristics of First Generation Community College Students," *Community College Review* 26, no. 3 (1999): 8.

Construct a bivariate table (in percentages) showing no association between age and first-generation college status.

▣ THE STRUCTURE OF HYPOTHESIS TESTING WITH CHI-SQUARE

The chi-square test follows the same five basic steps as the statistical tests presented in Chapter 8: (1) making assumptions, (2) stating the research and null hypotheses and selecting alpha, (3) selecting the sampling distribution and specifying the test statistic, (4) computing the test statistic, and (5) making a decision and interpreting the results. Before we apply the five-step model to a specific example, let's discuss some of the elements that are specific to the chi-square test.

The Assumptions

The chi-square test requires no assumptions about the shape of the population distribution from which the sample was drawn. However, like all inferential techniques, it assumes random sampling. It can be applied to variables measured at a nominal and/or an ordinal level of measurement.

Stating the Research and the Null Hypotheses

The research hypothesis (H_1) proposes that the two variables are related in the population.

H_1: The two variables are related in the population. (Gender and first-generation college status are statistically dependent.)

Like all other tests of statistical significance, the chi-square is a test of the null hypothesis. The null hypothesis (H_0) states that no association exists between two cross-tabulated variables in the population, and therefore, the variables are statistically independent.

H_0: There is no association between the two variables in the population. (Gender and first-generation college status are statistically independent.)

✓ *Learning Check*

Refer to the data in the previous Learning Check. Are the variables age and first-generation college status statistically independent? Write out the research and the null hypotheses for your practice data.

The Concept of Expected Frequencies

Assuming that the null hypothesis is true, we compute the cell frequencies that we would expect to find if the variables are statistically independent. These frequencies are called **expected frequencies** (and are symbolized as f_e). The chi-square test is based on cell-by-cell comparisons between the expected frequencies (f_e) and the frequencies actually observed (**observed frequencies** are symbolized as f_o).

Expected frequencies (f_e)   The cell frequencies that would be expected in a bivariate table if the two variables were statistically independent.

Observed frequencies (f_o)   The cell frequencies actually observed in a bivariate table.

Calculating the Expected Frequencies

The difference between f_o and f_e will determine the likelihood that the null hypothesis is true and that the variables are, in fact, statistically independent. When there is a large difference between f_o and f_e, it is unlikely that the two variables are independent, and we will probably reject the null hypothesis. On the other hand, if there is little difference between f_o and f_e, the variables are probably independent of each other, as stated by the null hypothesis (and therefore, we will not reject the null hypothesis).

The most important element in using chi-square to test for the statistical significance of cross-tabulated data is the determination of the expected frequencies. Because chi-square is computed on actual frequencies instead of on percentages, we need to calculate the expected frequencies based on the null hypothesis.

In practice, the expected frequencies are more easily computed directly from the row and column frequencies than from the percentages. We can calculate the expected frequencies using this formula:

$$f_e = \frac{(Column\ marginal)(Row\ marginal)}{N} \tag{9.1}$$

To obtain the expected frequencies for any cell in any cross-tabulation in which the two variables are assumed independent, multiply the row and column totals for that cell and divide the product by the total number of cases in the table.

Let's use this formula to recalculate the expected frequencies for our data on gender and first-generation college status as displayed in Table 9.10. Consider the men who were first-generation college students (the upper left cell). The expected frequency for this cell is the product of the column total (1,950) and the row total (1,936) divided by all the cases in the table (4,620):

$$f_e = \frac{(1,936)(1,950)}{4,620} = 817.14$$

For men who are nonfirsts (the lower left cell), the expected frequency is

$$f_e = \frac{(2,684)(1,950)}{4,620} = 1,132.86$$

Next, let's compute the expected frequencies for women who are first-generation college students (the upper right cell):

$$f_e = \frac{(1,936)(2,670)}{4,620} = 1,118.86$$

Finally, the expected frequency for women who are nonfirsts (the lower right cell) is

$$f_e = \frac{(2,684)(2,670)}{4,620} = 1,551.14$$

These expected frequencies are displayed in Table 9.12.

Note that the table of expected frequencies contains identical row and column marginals as the original table (Table 9.10). Although the expected frequencies usually differ from the observed frequencies (depending on the degree of relationship between the variables), the row and column marginals must always be identical with the marginals in the original table.

Table 9.12 Expected Frequencies of Men and Women and First-Generation College Status

First Generation	Men	Women	Total
Firsts	817.14	1,118.86	1,936
Nonfirsts	1,132.86	1,551.14	2,684
Total (N)	1,950	2,670	4,620

✓ **Learning Check**

Refer to the data in the Learning Check on page 228. Calculate the expected frequencies for age and first-generation college status and construct a bivariate table. Are your column and row marginals the same as in the original table?

Calculating the Obtained Chi-Square

The next step in calculating chi-square is to compare the differences between the expected and observed frequencies across all cells in the table. In Table 9.13, the expected frequencies are shown next to the corresponding observed frequencies. Note that the difference between the observed and expected frequencies in each cell is quite large. Is it large enough to be significant? The way we decide is by calculating the **obtained chi-square** statistic:

$$\chi^2 = \sum \frac{(f_o - f_e)^2}{f_e} \tag{9.2}$$

where

f_o = observed frequency

f_e = expected frequency

Chi-square (obtained)   The test statistic that summarizes the differences between the observed (f_o) and the expected (f_e) frequencies in a bivariate table.

Table 9.13   Observed and Expected Frequencies of Men and Women Who Are First-Generation College Students

First Generation	Men f_o	Men f_e	Women f_o	Women f_e	Total
Firsts	691	817.14	1,245	1,118.86	1,936
Nonfirsts	1,259	1,132.86	1,425	1,551.14	2,684
Total (N)	1,950		2,670		4,620

According to this formula, for each cell, subtract the expected frequency from the observed frequency, square the difference, and divide by the expected frequency. After performing this operation for every cell, sum the results to obtain the chi-square statistic.

Let's follow these procedures using the observed and expected frequencies from Table 9.13. Our calculations are displayed in Table 9.14. The obtained chi-square statistic, 57.99, summarizes the differences between the observed frequencies and the frequencies that we would expect to see if the null hypothesis were true and the variables—gender and first-generation college status—were not associated. Next, we need to interpret our obtained chi-square statistic and decide whether it is large enough to allow us to reject the null hypothesis.

✓ *Learning Check*

Using the format of Table 9.14, construct a table to calculate chi-square for age and first-generation college status.

The Sampling Distribution of Chi-Square

In Chapter 8, we learned that test statistics such as Z and t have characteristic sampling distributions that tell us the probability of obtaining a statistic, assuming that the null hypothesis is true. In the same way, the sampling distribution of chi-square tells the probability of getting values of chi-square, assuming no relationship exists in the population.

Table 9.14 Calculating Chi-Square

Gender and First-Generation College Status	f_o	f_e	$f_o - f_e$	$(f_o - f_e)^2$	$\dfrac{(f_o - f_e)^2}{f_e}$
Men/firsts	691	817.14	−126.14	15,911.2996	19.47
Men/nonfirsts	1,259	1,132.86	126.14	15,911.2996	14.04
Women/firsts	1,245	1,118.86	126.14	15,911.2996	14.22
Women/nonfirsts	1,425	1,551.14	−126.14	15,911.2996	10.26

$$\chi^2 = \Sigma \frac{(f_o - f_e)^2}{f_e} = 57.99$$

Like other sampling distributions, the chi-square sampling distributions depend on the degrees of freedom. In fact, the chi-square sampling distribution is not one distribution, but—like the *t* distribution—is a family of distributions. The shape of a particular chi-square distribution depends on the number of degrees of freedom. This is illustrated in Figure 9.1, which shows chi-square distributions for 1, 5, and 9 degrees of freedom. Here are some of the main properties of the chi-square distributions that can be observed in this figure:

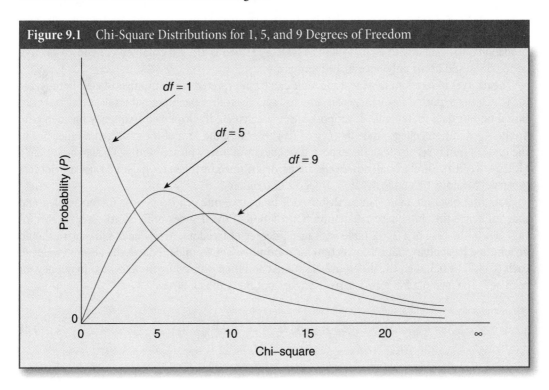

Figure 9.1 Chi-Square Distributions for 1, 5, and 9 Degrees of Freedom

- The distributions are positively skewed. The research hypothesis for the chi-square statistic is always a one-tailed test.
- Chi-square values are always positive. The minimum possible value is zero, with no upper limit to its maximum value. A chi-square of zero means that the variables are completely independent and the observed frequencies in every cell are equal to the corresponding expected frequencies.
- As the number of degrees of freedom increases, the chi-square distribution becomes more symmetrical and, with degrees of freedom greater than 30, begins to resemble the normal curve.

Determining the Degrees of Freedom

In Chapter 8, we defined degrees of freedom (*df*) as the number of values that are free to vary. With cross-tabulation data, we find the degrees of freedom by using the following formula:

$$df = (r - 1)(c - 1) \tag{9.3}$$

where

r = the number of rows

c = the number of columns

Thus, Table 9.10 with 2 rows and 2 columns has $(2 - 1)(2 - 1)$ or 1 degree of freedom. If the table had 3 rows and 2 columns, it would have $(3 - 1)(2 - 1)$ or 2 degrees of freedom.

Appendix C shows values of the chi-square distribution for various degrees of freedom. Notice how the table is arranged with the degrees of freedom listed down the first column and the level of significance (or *P* values) arrayed across the top. For example, with 5 degrees of freedom, the probability associated with a chi-square as large as 15.086 is .01. An obtained chi-square as large as 15.086 would occur only once in 100 samples.

The degrees of freedom in a bivariate table can be interpreted as the number of cells in the table for which the expected frequencies are free to vary, given that the marginal totals are already set. Based on our data in Table 9.12, suppose we first calculate the expected frequency for men who are first-generation college students (f_e = 817.14). Because the sum of the expected frequencies in the first column is set at 1,950, the expected frequency of men who are nonfirsts has to be 1,132.86 (1,950 − 817.14). Similarly, all other cells are predetermined by the marginal totals and are not free to vary. Therefore, this table has only 1 degree of freedom.

Data in a bivariate table can be distorted if by chance one cell is over- or undersampled and may influence the chi-square calculation. Calculation of the degrees of freedom compensates for this, but in the case of a 2 × 2 table with just 1 degree of freedom, the value of chi-square should be adjusted by applying Yates's correction for continuity. The formula reduces the absolute value of each ($f_o - f_e$) by 0.5; then the difference is squared and then divided by the expected frequency for each cell. The formula for Yates's correction for continuity is as follows:

$$\chi_c^2 = \sum \frac{\left(\left| f_o - f_e \right| - 0.5 \right)^2}{f_e} \tag{9.4}$$

✓ *Learning Check*

Based on Appendix C, identify the probability for each chi-square value (df in parentheses):

- *12.307 (15)*
- *20.337 (21)*
- *54.052 (24)*

Making a Final Decision

With Yates's correction, the corrected chi-square is 57.54. Refer to Table 9.15 for calculations. We can see that 57.54 does not appear on the first row ($df = 1$); in fact, it exceeds the largest chi-square value of 10.827 ($P = .001$). We can establish that the probability of obtaining a chi-square of 57.54 is less than .001 if the null hypothesis were true. If our alpha was preset at .05, the probability of 10.827 would be well below this. Therefore, we can reject the null hypothesis that gender and first-generation college status are not associated in the population from which our sample was drawn. Remember, the larger the chi-square statistic, the smaller the P value, providing us with more evidence to reject the null hypothesis. We can be very confident of our conclusion that there is a relationship between gender and first-generation college status in the population because the probability of this result occurring owing to sampling error is less than .001, a very rare occurrence.

Table 9.15 Calculating Yates's Correction

Gender and First-Generation College Status	$\|f_o - f_e\|$	$(\|f_o - f_e\| - .50)^2$	f_e	$\dfrac{(\|f_o - f_e\| - .5)^2}{f_e}$
Male firsts	126.14	$(125.64)^2 = 15{,}785.41$	817.14	19.32
Male nonfirsts	126.14	$(125.64)^2 = 15{,}785.41$	1,132.86	13.93
Female firsts	126.14	$(125.64)^2 = 15{,}785.41$	1,118.86	14.11
Female nonfirsts	126.14	$(125.64)^2 = 15{,}785.41$	1,551.14	10.18
Total				57.54

✓ *Learning Check*

What decision can you make about the association between age and first-generation college status? Should you reject the null hypothesis at the .05 alpha level or at the .01 level?

▣ SAMPLE SIZE AND STATISTICAL SIGNIFICANCE FOR CHI-SQUARE

Although we found the relationship between gender and first-generation college status to be statistically significant, this in itself does not give us much information about the *strength* of the relationship or its *substantive significance* in the population. Statistical significance helps us evaluate only whether the argument (the null hypothesis) that the observed relationship occurred by chance is reasonable. It does not tell us anything about the relationship's theoretical importance or even if it is worth further investigation.

The distinction between statistical and substantive significance is important in applying any of the statistical tests discussed in Chapter 8. However, this distinction is of particular relevance for the chi-square test because of its sensitivity to sample size. The size of the calculated chi-square is directly proportional to the size of the sample, independent of the strength of the relationship between the variables.

For instance, suppose that we cut the observed frequencies for every cell in Table 9.10 exactly in half—which is equivalent to reducing the sample size by one half. This change will not affect the percentage distribution of firsts among men and women; therefore, the size of the percentage difference and the strength of the association between gender and first-generation college status will remain the same. However, reducing the observed frequencies by half will cut down our calculated chi-square by exactly half, from 57.54 to 28.77. (Can you verify this calculation?) Conversely, had we doubled the frequencies in each cell, the size of the calculated chi-square would have doubled, thereby making it easier to reject the null hypothesis.

This sensitivity of the chi-square test to the size of the sample means that a relatively strong association between the variables may not be significant when the sample size is small. Similarly, even when the association between variables is very weak, a large sample may result in a statistically significant relationship. However, just because the calculated chi-square is large and we are able to reject the null hypothesis by a large margin does not imply that the relationship between the variables is strong and substantively important.

Another limitation of the chi-square test is that it is sensitive to small expected frequencies in one or more of the cells in the table. Generally, when the expected frequency in one or more of the cells is below 5, the chi-square statistic may be unstable and lead to erroneous conclusions. There is no hard-and-fast rule regarding the size of the expected frequencies. Most researchers limit the use of chi-square to tables that either have no f_e values below 5 or have no more than 20% of the f_e values below 5.

Testing the statistical significance of a bivariate relationship is only a small step, although an important one, in examining a relationship between two variables. A significant chi-square suggests that a relationship, weak or strong, probably exists in the population and is not due to sampling fluctuation. However, to establish the strength of the association, we need to employ measures of association such as gamma, lambda (both covered later in this chapter), or Pearson's *r* (refer to Chapter 11). Used in conjunction, statistical tests of significance and measures of association can help determine the importance of the relationship and whether it is worth additional investigation.

◉ FOCUS ON INTERPRETATION: EDUCATION AND HEALTH ASSESSMENT

For the GSS, individuals were asked to identify their highest educational degrees and their levels of health (poor, moderate, good, and excellent). These data are shown in Table 9.16. The bivariate table shows a clear pattern of positive association between education (the independent variable) and health assessment (the dependent variable). For instance, whereas 35% of individuals with some college or more reported excellent health, 12.5% of those with less than a high school degree reported the same. Similarly, whereas only 2.1% of respondents with some college or more reported poor health, 12.5% of respondents with less than a high school degree fell into that category.

Table 9.16   Health by Educational Level, GSS

	Educational Level			
Health	*Less Than High School*	*High School Degree*	*Some College or More*	*Total*
Poor	16	26	6	48
	(12.5%)	(6.0%)	(2.1%)	(5.7%)
Moderate	44	79	39	162
	(34.4%)	(18.1%)	(13.9%)	(19.2%)
Good	52	213	137	402
	(40.6%)	(48.9%)	(48.9%)	(47.6%)
Excellent	16	118	98	232
	(12.5%)	(27.1%)	(35.0%)	(27.5%)
Total	128	436	280	844
	(100%)	(100.1%)	(99.9%)	(100%)

The differences in the levels of health among the three educational groups seem sizable. However, it is not clear whether these differences are due to chance or to sampling fluctuations, or whether they reflect a real pattern of association in the population. In the following discussion, we will not review our calculations (though they are presented in Table 9.17). Rather, our focus will be on the five-step model and drawing conclusions about the relationship between health and educational degree.

Making Assumptions. Our assumptions are as follows:

1. A random sample of $N = 844$ is selected.
2. The level of measurement of the variable *education* is ordinal.
3. The level of measurement of the variable *health* is ordinal.

Stating the Research and Null Hypotheses and Selecting Alpha. Our hypotheses are as follows:

H_1: There is a relationship between education and health in the population. (Education and health are statistically dependent.)

H_0: There is no relationship between education and health in the population. (Education and health are statistically independent.)

For this test, we'll select an alpha of .01.

Selecting the Sampling Distribution and Specifying the Test Statistic. The sampling distribution is chi-square; the test statistic is also chi-square.

Computing the Test Statistic. Degrees of freedom for Table 9.16 is Calculation of the/as follows:

$$df = (r-1)(c-1) = (4-1)(3-1) = (3)(2) = 6$$

The chi-square obtained is 53.96. The detailed calculations are shown in Table 9.17.

Making a Decision and Interpreting the Results. To determine if the observed frequencies are significantly different from the expected frequencies, we compare our calculated chi-square with Appendix C. With 6 degrees of freedom, our chi-square of 53.96 exceeds the largest listed chi-square value of 22.457 ($P = .001$). We determine that the probability of observing our obtained chi-square of 53.96 is less than .001, and less than our alpha of .01. We can reject the null hypothesis that there are no differences in health among the different educational groups. Thus, we conclude that in the population from which our sample was drawn, health does vary by educational attainment. The positive relationship between the two variables is significant.

✓ *Learning*
Check

> *For the bivariate table with age and first-generation college status, the value of the obtained chi-square is 181.15 with 1 degree of freedom. Based on Appendix C, we determine that its probability is less than .001. This probability is less than our alpha level of .05. We reject the null hypothesis of no relationship between age and first-generation college status. If we reduce our sample size by half, the obtained chi-square is 90.58. Determine the P value for 90.58. What decision can you make about the null hypothesis?*

▣ PROPORTIONAL REDUCTION OF ERROR: A BRIEF INTRODUCTION

Earlier, we introduced cross-tabulation, whereby the relationship between two variables was analyzed by making a number of percentage comparisons. Using the chi-square statistic, we also

Table 9.17 Calculating Chi-Square for Education and Health

Education and Health	f_o	f_e	$f_o - f_e$	$(f_o - f_e)^2$	$\dfrac{(f_o - f_e)^2}{f_e}$
Less than high school/poor	16	7.3	8.7	75.69	10.37
Less than high school/moderate	44	24.6	19.4	376.36	15.30
Less than high school/good	52	61.0	−9.0	81.0	1.33
Less than high school/excellent	16	35.2	−19.2	368.64	10.47
High school/poor	26	24.8	1.2	1.44	0.06
High school/moderate	79	83.7	−4.7	22.09	0.26
High school/good	213	207.7	5.3	28.09	0.13
High school/excellent	118	119.8	−1.8	3.24	0.03
Some college or more/poor	6	15.9	−9.9	98.01	6.16
Some college or more/moderate	39	53.7	−14.7	216.09	4.02
Some college or more/good	137	133.4	3.6	12.96	0.10
Some college or more/excellent	98	77.0	21	441	5.73

$$\chi^2 = \Sigma \frac{(f_o - f_e)^2}{f_e} = 53.96$$

examined whether two variables are statistically related. Now we review special **measures of association** for nominal and ordinal variables. Unlike chi-square, measures of association reflect the strength of the relationship and, at times, its direction (whether it is positive or negative). They also indicate the usefulness of predicting the dependent variable from the independent variable.

In this section, we discuss four measures of association: lambda (measures of association for nominal variables), gamma and Kendall's tau-*b* (measures of association between ordinal variables), and Cramer's *V* (a chi-square-related measure of association). In Chapter 11, we introduce Pearson's correlation coefficient, which is used for measuring bivariate associations between interval-ratio variables.

Measure of association   A single summarizing number that reflects the strength of a relationship, indicates the usefulness of predicting the dependent variable from the independent variable, and often shows the direction of the relationship.

Except for Cramer's *V*, all the measures of association discussed here and in Chapter 11 are based on the concept of the **proportional reduction of error**, often abbreviated as **PRE**. According to the concept of PRE, two variables are associated when information about one can help us improve our prediction of the other.

Proportional reduction of error (PRE) The concept that underlies the definition and interpretation of several measures of association. PRE measures are derived by comparing the errors made in predicting the dependent variable while ignoring the independent with errors made when making predictions that use information about the independent variable.

Table 9.18 may help us grasp intuitively the general concept of PRE. Using GSS 2010 data, Table 9.18 shows a moderate relationship between the independent variable, *educational attainment*, and the dependent variable, *support for abortion if the woman is poor and can't afford any more children*. The table shows that 69.0% of the respondents who did not receive bachelor's degrees were antiabortion, compared with only 45.9% of the respondents who had bachelor's degrees or more.

The conceptual formula for all[10] PRE measures of association is

$$PRE = \frac{E_1 - E_2}{E_1} \tag{9.5}$$

where

E_1 = errors of prediction made when the independent variable is ignored (Prediction 1)

E_2 = errors of prediction made when the prediction is based on the independent variable (Prediction 2)

Table 9.18 Support for Abortion by Degree

	Degree		
Support for Abortion	*Less Than Bachelor's*	*Bachelor's or More*	*Total*
No	462	124	586
	69.0%	45.9%	62.3%
Yes	208	146	354
	31.0%	54.1%	37.7%
Total	670	270	940
	100.0%	100.0%	100.0%

Source: GSS, 2010.

All PRE measures are based on comparing predictive error levels that result from each of the two methods of prediction. Let's say that we want to predict a respondent's position on abortion, but we do not know anything about the degree he or she has. Based on the row totals in Table 9.18, we could predict that every respondent in the sample is antiabortion because this is the modal category of the variable *abortion position*. With this prediction, we would make 354 errors because in fact 586 respondents in this group are antiabortion but 354 respondents are pro-choice. Thus,

$$E_1 = 940 - 586 = 354$$

How can we improve this prediction by using the information we have on each respondent's educational attainment? For our new prediction, we will use the following rule: If a respondent has less than a bachelor's degree, we predict that he or she will be antiabortion; if a respondent has a bachelor's degree or more, we predict that he or she is pro-choice. It makes sense to use this rule because we know, based on Table 9.18, that respondents with lower educational attainment are more likely to be antiabortion, while respondents who have bachelor's degrees or more are more likely to be pro-choice. Using this prediction rule, we will make 332 errors (instead of 354) because 124 of the respondents who have bachelor's degrees or more are actually antiabortion, whereas 208 of the respondents who have less than a bachelor's degree are pro-choice (124 + 208 = 332). Thus,

$$E_2 = 124 + 208 = 332$$

Our first prediction method, ignoring the independent variable (educational attainment), resulted in 354 errors. Our second prediction method, using information we have about the independent variable (educational attainment), resulted in 332 errors. If the variables are *associated*, the second method will result in fewer errors of prediction than the first method. The stronger the relationship is between the variables, the larger will be the reduction in the number of errors of prediction.

Let's calculate the proportional reduction of error for Table 9.18 using Formula 9.4. The proportional reduction of error resulting from using educational attainment to predict position on abortion is

$$PRE = \frac{354 - 332}{354} = 0.06$$

PRE measures of association can range from 0.0 to ±1.0. A PRE of zero indicates that the two variables are not associated; information about the independent variable will not improve predictions about the dependent variable. A PRE of ±1.0 indicates a perfect positive or negative association between the variables; we can predict the dependent variable without error using information about the independent variable. Intermediate values of PRE will reflect the strength of the association between the two variables and therefore the utility of using one to predict the other. The more the measure of association departs from 0.00 in either direction, the stronger the association. PRE measures of association can be multiplied by 100 to indicate the percentage improvement in prediction.

▣ WHAT IS STRONG? WHAT IS WEAK? A GUIDE TO INTERPRETATION

The more you work with various measures of association, the better the feel you will have for what particular values mean. Until you develop this skill, here are some guidelines regarding what is generally considered a strong relationship and what is considered a weak relationship.

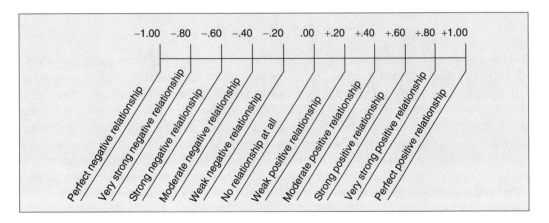

Keep in mind that these are only rough guidelines. Often, the interpretation for a measure of association will depend on the research context. A +0.30 in one research field will mean something a little different from a +0.30 in another research field. Zero, however, always means the same thing: no relationship.

A PRE of 0.06 indicates that there is a weak relationship between respondents' educational attainment and their positions on abortion. A PRE of 0.06 means that we have improved our prediction of respondents' position on abortion by just 6% ($0.06 \times 100 = 6.0\%$) by using information on their educational attainment.

▣ LAMBDA: A MEASURE OF ASSOCIATION FOR NOMINAL VARIABLES

In June 2013, Edward Snowden, a former National Security Agency employee, leaked confidential information to certain members of the American and international press, revealing that the federal government had collected personal data on approximately one third of American citizens after September 11, 2001, in a supposed effort to monitor potential terrorist attacks. Thanks to the passage of the USA PATRIOT Act in 2001, the federal government was more easily able to collect personal data (e.g., phone and Internet conversations) from U.S. civilians and noncivilians. Many citizens believed personal data surveillance and collection was reserved specifically for individuals suspected of engaging in terrorism. However, a significant amount of controversy emerged when Americans learned they had a one-in-three chance of being monitored by the federal government, including such things as their social media and e-mail accounts. Further controversy emerged when it was made known that the U.S. government had illegally tapped the European Union's offices, in addition to infiltrating its internal computer networks.[11]

The following is an examination of the relationship between party identification, the independent variable, and approval or disapproval of the federal government's collection of phone and Internet data for antiterrorism efforts, the dependent variable. Table 9.19 displays results from the Pew Research Center based on a nationally representative survey taken in June 2013. We can see 840 U.S. citizens classified by their party identification and their stances on collection of personal data for antiterrorism efforts. We will consider a respondent's party identification to be the independent variable and approval or disapproval of the federal government's collection of phone and Internet data for antiterrorism efforts to be the dependent variable.

Because party identification and approval or disapproval are nominal variables, we need to apply a measure of association suitable for calculating relationships between nominal variables. Such a measure will help us determine how strongly associated party identification is with one's position on the collection of phone and Internet data as part of antiterrorism efforts. **Lambda** is such a PRE measure.

Lambda   An asymmetrical measure of association, lambda is suitable for use with nominal variables and may range from 0.0 to 1.0. It provides us with an indication of the strength of an association between the independent and dependent variables.

A Method for Calculating Lambda

Take a look at Table 9.19 and examine the row totals, which show the distribution of the variable collection of data for antiterrorism efforts. If we had to predict whether Democrats would approve or disapprove of these data collection efforts, our best bet would be to guess the mode, which is that everyone approved of phone and Internet data collection. This prediction will result in the smallest possible error. The number of wrong predictions we make using this method is actually 383, since only 457 (the mode) out of 840 indicated that they approved of the collection of phone and Internet data for antiterrorism efforts (840 − 457 = 383).

Table 9.19   Position on Government Collection of Phone and Internet Data by Party Identification

Collection of Data for Antiterrorism Efforts	Party Identification		
	Republican	Democrat	Row Total
Approve	175	282	457
Disapprove	198	185	383
Column total	373	467	840

Source: Pew Research Center, "Public Split Over Impact of NSA Leak, but Most Want Snowden Prosecuted," June 17, 2013.

Now take another look at Table 9.19, but this time let's consider party identification when we predict approval or disapproval of phone and Internet data collection. Again, we can use the mode, but this time we apply it separately for Republicans and Democrats. The mode for Republicans is "disapprove" (198); therefore, we can predict that all Republicans disapprove of the collection of phone and Internet data for antiterrorism efforts. With this method of prediction, we make 175 errors, since 175 out of 373 Republicans approve of phone and Internet data collection (373 − 198 = 175). Next, we look at the group of Democrats. The mode for this group is "approve"; this will be our prediction for this group. This method of prediction results in 185 errors (467 − 282 = 185). The total number of errors is thus 175 + 185, or 360 errors.

Let's now put it all together and state the procedure for calculating lambda in more general terms.

1. Find E_1, the errors of prediction made when the independent variable is ignored. To find E_1, find the mode of the dependent variable and subtract its frequency from N. For Table 9.19,

$$E_1 = N - \text{Modal frequency}$$

$$E_1 = 840 - 457 = 383$$

2. Find E_2, the errors made when the prediction is based on the independent variable. To find E_2, find the modal frequency for each category of the independent variable, subtract it from the category total to find the number of errors, and then add up all the errors. For Table 9.19,

$$\text{Republicans} = 373 - 198 = 175$$

$$\text{Democrats} = 467 - 282 = 185$$

$$E_2 = 175 + 185 = 360$$

3. Calculate lambda using Formula 9.5:

$$\text{Lambda} = \frac{E_1 - E_2}{E_1} = \frac{383 - 360}{383} = 0.06$$

Lambda may range in value from 0.0 to 1.0. Zero indicates that there is nothing to be gained by using the independent variable to predict the dependent variable. A lambda of 1.0 indicates that by using the independent variable as a predictor, we are able to predict the dependent variable without any error. In our case, a lambda of 0.06 is less than one quarter of the distance between 0.0 and 1.0, indicating that for this sample of respondents, party identification and approval or disapproval of the government's collection of phone and Internet data for antiterrorism efforts are only slightly associated.

The proportional reduction of error indicated by lambda, when multiplied by 100, can be interpreted as follows: By using information on respondents' party identification to predict one's position on the collection of phone and Internet data, we have reduced our error of prediction by 6% (0.06 × 100 = 6%). In other words, if we rely on respondents' party identification to predict whether they approve or disapprove of the government's collection of phone and Internet data for antiterrorism efforts, we would reduce our error of prediction by 6 out of 100, or 6% (0.06 × 100).

Some Guidelines for Interpreting Lambda

Lambda is an **asymmetrical measure of association**. This means that lambda will vary depending on which variable is considered the independent variable and which the dependent variable. In our example, we considered one's approval or disapproval of the government's collection of phone and Internet data on behalf of antiterrorism efforts as the dependent variable and party identification as the independent variable, not vice versa. Had we considered, instead, party identification as the dependent variable and one's approval or disapproval of the government's collection of phone and Internet data as the independent variable, we would have obtained a slightly different lambda value.

Asymmetrical measure of association   A measure whose value may vary depending on which variable is considered the independent variable and which the dependent variable.

The method of calculation follows the same guidelines even when the variables are switched. However, exercise caution in calculating lambda, especially when the independent variable is arrayed in the rows rather than in the columns. To avoid confusion, it is safer to switch the variables and follow the convention of arraying the independent variable in the columns; then follow the exact guidelines suggested for calculating lambda. Remember, however, that although lambda can be calculated either way, ultimately what guides the decision of which variables to consider as independent or dependent is the theoretical question posed by the researcher.

Lambda is always zero in situations in which the mode for each category of the independent variable falls into the same category of the dependent variable. A problem with interpreting lambda arises in situations in which lambda is zero, but other measures of association indicate that the variables are associated. To avoid this potential problem, examine the percentage differences in the table whenever lambda is exactly equal to zero. If the percentage differences are very small (usually 5% or less), lambda is an appropriate measure of association for the table. However, if the percentage differences are larger, indicating that the two variables may be associated, lambda will be a poor choice as a measure of association. In such cases, we may want to discuss the association in terms of the percentage differences or select an alternative measure of association.

▣ CRAMER'S *V*: A CHI-SQUARE-RELATED MEASURE OF ASSOCIATION FOR NOMINAL VARIABLES

Cramer's *V* is an alternative measure of association that can be used for nominal variables. It is based on the value of chi-square (discussed earlier in this chapter) and ranges between 0 and 1, with 0 indicating no association and 1 indicating a perfect association. Because it cannot take negative values, it is considered a nondirectional measure. Unfortunately, Cramer's *V* is somewhat limited because the results cannot be interpreted using the PRE framework. It is calculated using the following formula:

$$\text{Cramer's } V = \sqrt{\frac{\chi^2}{N \times m}} \tag{9.6}$$

where $m =$ smaller of $(r-1)$ or $(c-1)$.

Earlier, we tested the hypothesis that education and health assessment are related in the population (Tables 9.16 and 9.17). The analysis yielded a chi-square value of 53.96, leading us to reject the null hypothesis that there are no differences in health among different educational groups. We concluded that in the population from which our sample was drawn, health does vary by educational attainment.

We can use Cramer's V to measure the relative strength of the association between health assessment and level of education using Formula 9.5.

$$\text{Cramer's } V = \sqrt{\frac{\chi^2}{N \times m}} = \sqrt{\frac{53.96}{844 \times 2}} = \sqrt{0.032} = 0.18$$

A Cramer's V of 0.18 tells us that there is a weak association between health assessment and level of education.

🔲 FOCUS ON INTERPRETATION: GAMMA AND KENDALL'S TAU-*b*

In this section, we discuss a way to measure and interpret an association between two *ordinal* variables. If there is an association between the two variables, knowledge of one variable will enable us to make better predictions of the other variable.

Let's look at a research example in which the association between two ordinal variables is considered. We want to examine the hypothesis that the higher one's educational level, the more satisfied he or she is with his or her financial situation. To examine this hypothesis, we selected two variables from the 2010 GSS: the variable education (EDUC), with those indicating that they had 11 or fewer years of education in one category and those reporting that they had 12 or more years of education in a second category. The variable satisfaction with financial situation (SATFIN) has two categories: "satisfied" and "unsatisfied."

Table 9.20 displays the cross-tabulation of these two variables, with education as the independent variable and satisfaction with financial situation as the dependent variable. We find that 80.7% of those with less than a high school degree are unsatisfied with their financial situation, as compared with 74.2% of those with at least a high school degree. The percentage difference (80.7% − 74.2% = 6.5%) suggests that the variables are related. We can examine the percentage difference across those who are satisfied with their financial situation (25.8% − 19.3% = 6.5%), and reach the same conclusion.

Table 9.20   Financial Satisfaction by Education

| | Education (X) | | |
Financial Satisfaction (Y)	High School or More	Less Than High School	Total
Satisfied	25.8%	19.3%	24.9%
	(332)	(40)	(372)
Unsatisfied	74.2%	80.7%	75.1%
	(957)	(167)	(1,124)
Total	100%	100%	100%
(N)	(1,289)	(207)	(1,496)

Source: GSS, 2010.

The next step in analyzing the relationship between education and financial satisfaction is to select a measure that will enable us to assess the strength and the direction (sign) of that relationship. We selected gamma because our variables (education and financial satisfaction) are ordinal variables. We will focus on interpreting gamma rather than calculating it.

The SPSS output showing the value of gamma is presented below. It reports that the gamma for the table is .183, indicating a weak positive relationship between education and financial satisfaction. The positive sign of gamma indicates that as education increases, so does the level of financial satisfaction. A gamma of .183 indicates that by using education to predict financial satisfaction, we've reduced our prediction error by 18.3%.

Symmetric Measures

		Value	Asymp. Std. Error[a]	Approx. T[b]	Approx. Sig.
Ordinal by Ordinal	Kendall's tau-b	.051	.024	2.128	.033
	Gamma	.183	.090	2.128	.033
N of Valid Cases		1496			

a. Not assuming the null hypothesis.

b. Using the asymptotic standard error assuming the null hypothesis.

Gamma and **Kendall's tau-*b*** are **symmetrical measures of association** suitable for use with ordinal variables or with dichotomous nominal variables. This means that their values will be the same regardless of which variable is the independent variable or the dependent variable. Thus, if we had wanted to predict education from financial satisfaction rather than the opposite, we would have obtained the same gamma. Both gamma and Kendall's tau-*b* can vary from 0.0 to ±1.0 and provide us with an indication of the strength and direction of the association between the variables. Gamma and Kendall's tau-*b* can be positive or negative. A gamma or Kendall's tau-*b* of 1.0 indicates that the relationship between the variables is positive and that the dependent variable can be predicted without any errors based on the independent variable. A gamma of −1.0 indicates a perfect, negative association between the variables. A gamma or a Kendall's tau-*b* of zero reflects no association between the two variables; hence there is nothing to be gained by using the independent variable to predict the dependent variable.

Gamma   A symmetrical measure of association suitable for use with ordinal variables or with dichotomous nominal variables. It can vary from 0.0 to ±1.0 and provides us with an indication of the strength and direction of the association between the variables.

*Kendall's tau-*b   A symmetrical measure of association suitable for use with ordinal variables. It can vary from 0.0 to ±1.0. It provides an indication of the strength and direction of the association between the variables. Kendall's tau-*b* will always be lower than gamma.

Symmetrical measure of association   A measure whose value will be the same when either variable is considered the independent variable or the dependent variable.

▣ USING ORDINAL MEASURES WITH DICHOTOMOUS VARIABLES

Measures of association for ordinal data are not influenced by the modal category as is lambda. Consequently, an ordinal measure of association might be preferable for tables when an association cannot be detected by lambda. We can use an ordinal measure for some tables where one or both variables would appear to be measured on a nominal scale. Dichotomous variables (those with only two categories) can be treated as ordinal variables for most purposes. In this chapter, we calculated lambda to examine the association between abortion attitudes and educational attainment (Table 9.18). Although both variables might be considered as nominal variables—because both are dichotomized (yes/no; high school or less/more than high school)—they could also be treated as ordinal variables. Thus, the association might also be examined using gamma, an ordinal measure of association.[12]

▣ FOCUS ON INTERPRETATION: THE GENDER GAP IN GUN CONTROL

Is there a gender gap in attitude toward gun control? Surveys have shown that women are generally more in favor of stricter gun laws than men.

Let's examine data from the GSS 2010 comparing men's and women's attitudes toward requiring gun permits. The table below shows that more women (83.2%) than men (67.0%) are in favor of gun permits. These findings confirm the results obtained by most public opinion polls.

			RESPONDENTS SEX		Total
			MALE	FEMALE	
FAVOR OR OPPOSE GUN PERMITS	FAVOR	Count	272	469	741
		% within RESPONDENTS SEX	67.0%	83.2%	76.4%
	OPPOSE	Count	134	95	229
		% within RESPONDENTS SEX	33.0%	16.8%	23.6%
Total		Count	406	564	970
		% within RESPONDENTS SEX	100.0%	100.0%	100.0%

The observed percentage difference between men and women is a useful indicator that there is a gender gap in opinions about gun control. However, we are seeking a measure that will enable us to assess the strength of that relationship. We can select gamma for our purposes because the independent and dependent variables are dichotomous.

The SPSS output showing the value of gamma is presented below. It reports that the gamma for the table is −0.417,[13] indicating a moderate relationship between sex and opinions about requiring gun permits. By using sex to predict men's and women's opinions about requiring gun permits, we've reduced our prediction error by almost half (41.7%).

Symmetric Measures					
		Value	Asymp. Std. Error[a]	Approx. T[b]	Approx. Sig.
Ordinal by Ordinal	Gamma	-.417	.064	-5.728	.000
N of Valid Cases		970			
a. Not assuming the null hypothesis.					
b. Using the asymptotic standard error assuming the null hypothesis.					

MAIN POINTS

- Bivariate analysis is a statistical technique designed to detect and describe the relationship between two variables. A relationship is said to exist when certain values of one variable tend to "go together" with certain values of the other variable.

- A bivariate table displays the distribution of one variable across the categories of another variable. It is obtained by classifying cases based on their joint scores for two variables.

- Percentaging bivariate tables are used to examine the relationship between two variables that have been organized in a bivariate table. The percentages are always calculated within each category of the independent variable.

- Bivariate tables are interpreted by comparing percentages across different categories of the independent variable. A relationship is said to exist if the percentage distributions vary across the categories of the independent variable.

- Variables measured at the ordinal or interval-ratio levels may be positively or negatively associated. With a positive association, higher values of one variable correspond to higher values of the other variable. When there is a negative association between variables, higher values of one variable correspond to lower values of the other variable.

- The chi-square test is an inferential statistical technique designed to test for a significant relationship between nominal or ordinal variables organized in a bivariate table. The test is conducted by testing the null hypothesis that no association exists between two cross-tabulated variables in the population, and therefore, the variables are statistically independent.

- The obtained chi-square (χ^2) statistic summarizes the differences between the observed frequencies (f_o) and the expected frequencies (f_e)—the frequencies we would have expected to see if the null hypothesis were true and the variables were not associated. Yates's correction for continuity is applied to all 2×2 tables.

- The sampling distribution of chi-square tells the probability of getting values of chi-square, assuming no relationship exists in the population. The shape of a particular chi-square sampling distribution depends on the number of degrees of freedom.

- Measures of association are single summarizing numbers that reflect the strength of the relationship between variables, indicate the usefulness of predicting the dependent from the independent variable, and often show the direction of the relationship.

- Proportional reduction of error (PRE) underlies the definition and interpretation of several measures of association. PRE measures are derived by comparing the errors made in predicting the dependent variable while ignoring the independent variable with errors made when making predictions that use information about the independent variable.

• Measures of association may be symmetrical or asymmetrical. When the measure is symmetrical, its value will be the same regardless of which of the two variables is considered the independent or dependent variable. In contrast, the value of asymmetrical measures of association may vary depending on which variable is considered the independent variable and which the dependent variable.

• Lambda is an asymmetrical measure of association suitable for use with nominal variables. It can range from 0.0 to 1.0 and gives an indication of the strength of an association between the independent and the dependent variables.

• Gamma is a symmetrical measure of association suitable for ordinal variables or for dichotomous nominal variables. It can vary from 0.0 to ±1.0 and reflects both the strength and direction of the association between two variables.

• Kendall's tau-*b* is a symmetrical measure of association suitable for use with ordinal variables. Unlike gamma, it accounts for pairs tied on the independent and dependent variable. It can vary from 0.0 to ±1.0. It provides an indication of the strength and direction of the association between two variables.

• Cramer's *V* is a measure of association for nominal variables. It is based on the value of chi-square and ranges between 0.0 and 1.0. Because it cannot take negative values, it is considered a nondirectional measure.

KEY TERMS

asymmetrical measure of association	expected frequencies (f_e)	proportional reduction of error (PRE)
bivariate analysis	gamma	row variable
bivariate table	Kendall's tau-*b*	statistical independence
cell	lambda	symmetrical measure of association
chi-square (obtained)	marginals	
chi-square test	measure of association	
column variable	negative relationship	
control variable	observed frequencies (f_o)	
cross-tabulation	positive relationship	
elaboration		

ⓈSAGE edge™

Sharpen your skills with SAGE edge at **edge.sagepub.com/ssdsess2e**. **SAGE edge for students** provides a personalized approach to help you accomplish your coursework goals in an easy-to-use learning environment.

CHAPTER EXERCISES

1. Use the following GSS data on fear, race, and home ownership for this exercise. Variables measure respondent's race, whether the respondent fears walking alone at night, and his or her home ownership.

Respondent	Race	Fear of Walking Alone	Rent/Own
1	W	N	R
2	B	N	R
3	W	Y	R
4	B	N	R
5	W	N	R
6	B	Y	O
7	W	Y	R
8	W	Y	R
9	W	N	O
10	W	N	O
11	W	Y	R
12	W	N	R
13	B	Y	O
14	W	N	R
15	B	N	O
16	B	N	R
17	W	N	O
18	W	N	O
19	B	N	R
20	W	N	O
21	B	Y	R

Notes: Race: B = black, W = white; fear: Y = yes, N = no; rent/own: R = rent, O = own.

a. Construct a bivariate table of frequencies for race and fear of walking alone at night. Which is the independent variable?

b. Calculate percentages for the table based on the independent variable. Describe the relationship between race and fear of walking alone using the table. What sampling issues are involved here?

c. Use the data to construct a bivariate table to compare fear of walking alone at night between people who own their homes and those who rent. Use percentages to show whether there is a difference between home owners and renters in fear of walking alone.

2. Do women and men have different opinions about affirmative action? Based on data from the GSS 2010, the output in Figure 9.2 shows respondents' sex (SEX) and attitudes toward affirmative action (DISCAFF: Are whites hurt by affirmative action?).

Figure 9.2   DISCAFF by SEX Cross-Tabulation

discaff WHITES HURT BY AFF. ACTION * sex RESPONDENTS SEX Crosstabulation

Count

		sex RESPONDENTS SEX		
		1 MALE	2 FEMALE	Total
discaff WHITES HURT BY AFF. ACTION	1 VERY LIKELY	56	103	159
	2 SOMEWHAT LIKELY	171	255	426
	3 NOT VERY LIKELY	168	193	361
Total		395	551	946

a. Which is the independent variable?
b. What are the differences in attitudes between men and women?
c. Test whether SEX and DISCAFF are independent (alpha = .05). What do you conclude?

3. Advocates of gay rights often argue that homosexuality is not a "preference" or a choice but rather an "orientation" that cannot be changed. One of your classmates argues that attitudes about homosexuality often influence political views. Those who think homosexual relations are wrong tend to be more conservative compared with those who do not think that homosexual relations are wrong. Use the following table based on the GSS 2010 to answer the questions.

	Homosexual Relations		
Political Views	Always Wrong	Not Wrong at All	Total
Liberal	82	155	237
	18.2%	45.4%	29.9%
Moderate	140	120	260
	31%	35.2%	32.8%
Conservative	229	66	295
	50.8%	19.4%	37.2%
Total	451	341	792
	100.0%	100.0%	99.9%

Note: Original GSS categories have been recoded for illustration purposes.

Exercises

    a. Based on your classmate's argument, what is the dependent variable? The independent variable?

    b. What percentage of those polled think that homosexual relations are always wrong?

    c. Using the percentages in the table, describe the relationship between views about homosexual relations and political orientation?

    d. Calculate the chi-square statistic for this table. Based on an alpha of .01, do you reject the null hypothesis? Explain.

4. We continue our examination of attitudes regarding homosexuality. Suppose that a classmate of yours suggests that views about homosexual relations can be explained by the frequency of church attendance. Your classmate shows you the following table taken from the 2010 GSS sample. (Frequencies are shown below.)

Homosexual Relations	Church Attendance			
	Never	Several Times a Year	Every Week	Total
Always wrong	50	52	136	238
Not wrong at all	111	36	38	185
Total	161	88	174	423

    a. Which is the dependent variable in this table? Which is the independent variable?

    b. Calculate the percentages using church attendance as the independent variable for each cell in the table. Is there a relationship between church attendance and views about homosexual relations? If so, how strong is it?

    c. Suppose that you respond to your classmate by stating that it is not church attendance that explains views about homosexual relations; rather, it is one's opinion about the nature of right and wrong (i.e., morality) that explains attitudes about homosexual relations. Why might there be a potential problem with your argument? Think in terms of assigning variables to the independent and dependent categories.

5. Youth were asked in the Monitoring the Future (MTF) 2011 survey to report how drunk they get when they consume alcohol. Responses for 361 male youth are reported by race.

Alchhowdrunk	Race			
	Black	White	Hispanic	Total
Not at all	10	71	14	95
A little	4	64	16	84
Moderate	10	108	19	137
Very	5	35	5	45
Total	29	278	54	361

Calculate the percentages using race as the independent variable. Is there a relationship between student race and level of drunkenness?

6. The educational level of Americans increased throughout the 20th century. The following U.S. census data show the level of education attained by American adults over the age of 25 years at several points in time.

	Educational Level	
Year	High School Graduate or More (%)	College Graduate or More (%)
1980	66.5	16.2
1990	77.6	21.3
1995	81.7	23.0
2000	84.1	25.6
2005	85.2	27.7
2010	87.1	30.3

Source: U.S. Census Bureau, *Statistical Abstract of the United State: 2012,* Table 229.

a. What is the direction of the relationship between each year and level of education?
b. Use percentage differences to describe the relationship. Why don't the percentages add to 100% by year? Is there a problem in analyzing the table?
c. Do these data support the idea that Americans were more educated in 2010 than in previous years?

7. In 2004, high school seniors were surveyed about their postsecondary expectations and plans. U.S. Department of Education data are presented for male and female students. Which group of students has higher educational expectations? Refer to the data to support your answer. (The row total for males will not equal to 100% due to rounding.)

	Students' Educational Expectations (percentages reported)				
Sex	Do Not Know Yet	High School or Less	Some College	Bachelor's Degree	Graduate/Advanced Degree
Male	9.4	6.9	20.5	34.4	28.9
Female	7.4	3.1	15.6	32.6	41.3

Source: Xianglei Chen, Joanna Wu, Shayna Tasoff, and Thomas Weko, *Postsecondary Expectations and Plans of the High School Senior Class of 2003–2004,* U.S. Department of Education NCES 2010-070 rev, 2010.

8. Illegal immigration in the United States is a complex matter, and people have diverse and conflicting ideas on how best to address it. The 2010 GSS contains several questions on this topic. For this exercise, we present the SPSS analysis of political party identification and the variable UNDOCK-ID, which measures support of the statement that children of illegal immigrants should qualify for citizenship (see Figure 9.3).

a. What percentage of Democrats indicate that children of immigrants should qualify for citizenship? What percentage of Republicans?

b. Based on an alpha of .01, what can you conclude about the relationship between political party identification and UNDOCKID?

Figure 9.3   Cross-Tabulation and Chi-Square for NPARTYID and UNDOCKID

npartyid Recoded party id * undockid US CITIZENSHIP FOR CHILDREN OF ILLEGAL IMMIGRANTS Crosstabulation

Count

		undockid US CITIZENSHIP FOR CHILDREN OF ILLEGAL IMMIGRANTS		
		1 YES, QUALIFY	2 NO, NOT QUALIFY	Total
npartyid Recoded party id	1.00 democrat	172	53	225
	2.00 independent	42	30	72
	3.00 republican	80	75	155
Total		294	158	452

Chi-Square Tests

	Value	df	Asymp. Sig. (2-sided)
Pearson Chi-Square	26.586[a]	2	.000
Likelihood Ratio	26.869	2	.000
Linear-by-Linear Association	25.670	1	.000
N of Valid Cases	452		

a. 0 cells (0.0%) have expected count less than 5. The minimum expected count is 25.17.

9. Does access to marijuana vary by the size of the community a teenager lives in? The MTF (2012) survey asked teens how easy it was to obtain marijuana (GWEED). Their responses are shown by residence size—small town, medium-sized city, and large city. Refer to Figure 9.4. Based on an alpha of .05, what would you conclude? Does marijuana access vary by residential city size?

10. Teens were asked in the MTF 2012 survey to report their levels of happiness (how are things these days?). The bivariate table includes responses organized by race. Based on an alpha of .05, test whether race and happiness are independent.

	Race of Respondent			
Response	Black	White	Hispanic	Total
Not happy	33	93	30	156
Pretty happy	116	488	119	723
Very happy	32	214	40	286
Total	181	795	189	1,165

Exercises

Figure 9.4   Cross-Tabulation and Chi-Square for NGREWUP and GWEED

Ngrewup Recoded grew up * gweed how easy is it to obtain marijuana? Crosstabulation

			gweed How easy is it do obtain marijuana?					
			1 PROB IMP: (1)	2 VRY DIFF: (2)	3 FRLY DIF: (3)	4 FRLY EAS: (4)	5 VRY EASY: (5)	Total
Ngrewup Recoded grew up	1.00 small town	Count	35	22	35	131	320	543
		% within Ngrewup Recoded grew up	6.4%	4.1%	6.4%	24.1%	58.9%	100.0%
	2.00 medium city	Count	14	8	18	87	183	310
		% within Ngrewup Recoded grew up	4.5%	2.6%	5.8%	28.1%	59.0%	100.0%
	3.00 large cityy	Count	17	14	24	99	232	386
		% within Ngrewup Recoded grew up	4.4%	3.6%	6.2%	25.6%	60.1%	100.0%
Total		Count	66	44	77	317	735	1239
		% within Ngrewup Recoded grew up	5.3%	3.6%	6.2%	25.6%	59.3%	100.0%

Chi-Square Tests

	Value	df	Asymp.Sig. (2-sided)
Pearson Chi-Squre	4.872[a]	8	.771
Likelihood Ratio	4.909	8	.767
Linear-by-Linear Association	1.492	1	.222
N of Valid Cases	1239		

a. 0 cells (0.0%) have expected count less than 5. The minimum expected count is 11.01.

11. In the following table, data for the sex of offenders and the sex of victims are reported (U.S. Department of Justice, Expanded Homicide Data Table 6, 2011).

	Sex of Offender	
Sex of Victim	*Male*	*Female*
Male	3,760	450
Female	1,590	140

a. Treating sex of offender as the independent variable, how many errors of prediction will be made if the independent variable is ignored?
b. How many fewer errors will be made if the independent variable is taken into account?
c. Combine your answers in (a) and (b) to calculate lambda. Discuss the relationship between these two variables.

Exercises

12. Let's continue our analysis of offenders and victims of violent crime. Is there a relationship between the races of violent offenders and their victims? Data from the U.S. Department of Justice (Expanded Homicide Data Table 6, 2011) are presented below.

	Race of Offender		
Race of Victim	White	Black	Other
White	2,630	448	33
Black	193	2,447	9
Other	180	45	99

a. Let's treat race of offender as the independent variable and race of victims as the dependent variable. If we first ignore the independent variable and try to predict race of victim, how many errors will we make?

b. If we now take into account the independent variable, how many errors of prediction will we make for those offenders who are white? Black offenders? Other offenders?

c. Combine the answers in (a) and (b) to calculate the proportional reduction in error for this table based on the independent variable. How does this statistic improve our understanding of the relationship between the two variables?

13. The GSS asked respondents to report their opinion on spanking as a method to discipline a child (SPANKING). Examine how respondents' attitudes toward spanking a child are associated with SEX, CLASS, and MARITAL (marital status). Using the SPSS output below, interpret the measures of association for each of the variable pairs.

Figure 9.5 Crosstabulation and Measures of Association for Spanking and Sex

Crosstab

Count

		RESPONDENTS SEX		
		MALE	FEMALE	Total
FAVOR SPANKING TO DISCIPLINE CHILD	STRONGLY AGREE	120	121	241
	AGREE	217	260	477
	DISAGREE	104	126	230
	STRONGLY DISAGREE	15	44	59
Total		456	551	1007

Symmetric Measures

		Value	Asymp. Std. Error	Approx. T	Approx. Sig.
Ordinal by Ordinal	Kendall's tau-b	.067	.029	2.305	.021
	Gamma	.118	.051	2.305	.021
N of Valid Cases		1007			

Figure 9.6 Crosstabulation and Measures of Association for Spanking and Class

Count

		SUBJECTIVE CLASS IDENTIFICATION (RECODED)				
		Upper Class	Middle Class	Working Class	Lower Class	Total
FAVOR SPANKING TO DISCIPLINE CHILD	STRONGLY AGREE	3	85	128	23	239
	AGREE	15	195	224	39	473
	DISAGREE	4	121	84	19	228
	STRONGLY DISAGREE	2	33	19	5	59
Total		24	434	455	86	999

Symmetric Measures

		Value	Asymp. Std. Error	Approx. T	Approx. Sig.
Ordinal by Ordinal	Kendall's tau-b	-.113	.028	-4.014	.000
	Gamma	-.178	.044	-4.014	.000
N of Valid Cases		999			

Figure 9.7 Crosstabulation and Measures of Association for Spanking and Marital Status

Crosstab

Count

		MARITAL STATUS					
		MARRIED	WIDOWED	DIVORCED	SEPARATED	NEVER MARRIED	Total
FAVOR SPANKING TO DISCIPLINE CHILD	STRONGLY AGREE	120	20	29	8	64	241
	AGREE	245	33	70	12	117	477
	DISAGREE	99	23	43	9	56	230
	STRONGLY DISAGREE	31	6	6	2	14	59
Total		495	82	148	31	251	1007

Directional Measures

			Value	Asymp. Std. Error	Approx. T	Approx. Sig.
Nominal by Nominal	Lambda	Symmetric	.000	.000	.	.
			.000	.000	.	.
			.000	.000	.	.

Exercises

14. Does the belief that women are not suited for politics vary by gender and/or educational attainment? GSS 2010 respondents were asked if they believed that women are not suited for politics (FEPOL). Examine how this variable is associated with respondent's sex (SEX). Examine and interpret the output below. Would lambda be appropriate as a measure of association? Why or why not? Would gamma and Kendall's tau-*b* be appropriate measures? Why or why not?

Figure 9.8 Crosstabulation and Measures of Association for Sex and Fepol

WOMEN NOT SUITED FOR POLITICS * RESPONDENTS SEX Crosstabulation

Count

| | | RESPONDENTS SEX | | |
		MALE	FEMALE	Total
WOMEN NOT SUITED FOR POLITICS	AGREE	83	112	195
	DISAGREE	349	421	770
Total		432	533	965

Directional Measures

			Value	Asymp. Std. Error	Approx. T	Approx. Sig.
Nominal by Nominal	Lambda	Symmetric	.000	.000	.	.
			.000	.000	.	.
			.000	.000	.	.

Symmetric Measures

		Value	Asymp. Std. Error	Approx. T	Approx. Sig.
Ordinal by Ordinal	Kendall's tau-b	-.022	.032	-.695	.487
	Gamma	-.056	.081	-.695	.487
N of Valid Cases		965			

Chapter 10

Analysis of Variance

Chapter Learning Objectives

❖ Understanding the application of an analysis of variance (ANOVA) model
❖ Assessing significance and interpretation of the F statistic

Many research questions require us to look at multiple samples or groups, at least more than two at a time. We may be interested in studying the influence of ethnic identity (white, African American, Asian American, Latino/a) on church attendance, the influence of social class (lower, working, middle, and upper) on President Barack Obama's job-approval ratings, or the effect of educational attainment (less than high school, high school graduate, some college, and college graduate) on household income. Note that each of these examples requires a comparison between multiple demographic or ethnic groups, more than the two-group comparisons that we reviewed in Chapter 8. While it would be easy to confine our analyses between two groups, our social world is much more complex and diverse.

Let's say that we're interested in examining educational attainment—on average, how many years of education do Americans achieve? For 2010, the U.S. census found that 87% of adults (25 years and older) had completed at least a high school degree, and 30% of all adults had attained at least a bachelor's degree.[1] During his first term of office, President Obama pledged that the United States would have the world's highest proportion of college graduates by 2020. Special attention has been paid to the educational achievement of Latino students. Data from the U.S. census, as well as from the U.S. Department of Education, confirm that Latino students continue to have lower levels of educational achievement than other racial or ethnic groups.

In Chapter 8, Testing Hypotheses, we introduced statistical techniques to assess the difference between two sample means or proportions. For our example in Table 8.2, we compared the difference in educational attainment for blacks and whites. But what if we wanted to examine separate groups of men and women by their race or ethnicity? Is there a significant variation in educational attainment among black women, Hispanic women, black men, and Hispanic men?

Table 10.1   Educational Attainment (measured in years) for Four GSS Groups

Black Men $n_1 = 6$	Hispanic Men $n_2 = 4$	Black Women $n_3 = 6$	Hispanic Women $n_4 = 5$
16	14	16	14
12	12	18	12
14	11	16	12
12	11	14	13
12		16	14
12		12	

We've taken a random sample of 21 men and women from the General Social Survey (GSS), grouped them into four demographic categories, and included their educational attainment in Table 10.1. With the *t*-test statistic we covered in Chapter 8, we could analyze only two samples at a time. We would have to analyze the mean educational attainment of black women versus Hispanic women, black women versus black men, black women versus Hispanic men, and so on. (Confirm that we would have to analyze six different pairs.) In the end, we would have a tedious series of *t*-test statistic calculations, and we still wouldn't be able to answer our original question: Is there a difference in educational attainment among all *four* demographic groups?

There is a statistical technique that will allow us to examine all four groups or samples simultaneously. This technique is called **analysis of variance (ANOVA)**. ANOVA follows the same five-step model of hypothesis testing that we used with the *t* test and *Z* test for proportions (in Chapter 8) and chi-square (in Chapter 9). In this chapter, we review the calculations for ANOVA and discuss two applications of ANOVA from the research literature.

Analysis of variance (ANOVA)   An inferential statistics technique designed to test for a significant relationship between two variables in two or more groups or samples.

🔲 UNDERSTANDING ANALYSIS OF VARIANCE

Recall that the *t* test examines the difference between two means, $\overline{Y}_1 - \overline{Y}_2$, while the null hypothesis assumes that there is no difference between them: $\mu_1 = \mu_2$. Rejecting the null hypothesis means that there is a significant difference between the two mean scores (or the populations from which the samples were drawn). In our Chapter 8 example, we analyzed the difference between mean years of education for blacks and whites. Based on our *t*-test statistic, we rejected the null hypothesis, concluding that white men and women, on average, have significantly more years of education than black men and women do.

The logic of ANOVA is the same but extending to two or more groups. For the data presented in Table 10.1, ANOVA will allow us to examine the variation among four means $\left(\overline{Y}_1, \overline{Y}_2, \overline{Y}_3, \text{and } \overline{Y}_4 \right)$, and the null hypothesis can be stated as follows: $\mu_1 = \mu_2 = \mu_3 = \mu_4$. Rejecting the null hypothesis for ANOVA indicates that there is significant variation among the four samples (or the four populations from which the samples were drawn) and that at least one of the sample means is significantly different from the others. In our example, it suggests that years of education (the dependent variable) do vary by group membership (the independent variable). When ANOVA procedures are applied to data with one dependent and one independent variable, it is called a **one-way ANOVA**.

One-way ANOVA   Analysis of variance application with one dependent variable and one independent variable.

The means, standard deviations, and variances for the samples have been calculated and are shown in Table 10.2. Note that the four mean educational years are not identical, with black women having the highest educational attainment. Also, based on the standard deviations, we can tell that the samples are relatively homogeneous, with deviations within 1.00 to 2.07 years of the mean. We already know that there is a difference between the samples, but the question remains: Is this difference significant? Do the samples reflect a relationship between demographic group membership and educational attainment in the general population?

✓ *Learning*
Check

We've calculated the mean and standard deviation scores for each group in Table 10.2. Compute each mean (Chapter 3) and standard deviation (Chapter 4) and confirm that our statistics are correct.

Table 10.2   Means, Variances, and Standard Deviations for Four GSS 2006 Groups

Black Men $n_1 = 6$	*Hispanic Men* $n_2 = 4$	*Black Women* $n_3 = 6$	*Hispanic Women* $n_4 = 5$
16	14	16	14
12	12	18	12
14	11	16	12
12	11	14	13
12		16	14
12		12	

(Continued)

Table 10.2   (Continued)

| Black Men | Hispanic Men | Black Women | Hispanic Women |
$n_1 = 6$	$n_2 = 4$	$n_3 = 6$	$n_4 = 5$
$\bar{Y}_1 = 13.00$	$\bar{Y}_2 = 12.00$	$\bar{Y}_3 = 15.33$	$\bar{Y}_4 = 13.00$
$S_1 = 1.67$	$S_2 = 1.41$	$S_3 = 2.07$	$S_4 = 1.00$
$S_1^2 = 2.79$	$S_2^2 = 1.99$	$S_3^2 = 4.28$	$S_4^2 = 1.00$
$\bar{Y} = 13.48$			

To determine whether the differences are significant, ANOVA examines the differences *between* our four samples, as well as the difference *within* a single sample. The differences can also be referred to as variance or variation, which is why ANOVA is the analysis of *variance*. What is the difference between one sample's mean score and the overall mean? What is the variation of individual scores within one sample? Are all the scores alike (no variation), or is there a broad variation in scores? ANOVA allows us to determine whether the variance between samples is larger than the variance within the samples. If the variance is larger between samples than the variance within samples, we know that educational attainment varies significantly across the samples. It would support the notion that group membership explains the variation in educational attainment.

▣ THE STRUCTURE OF HYPOTHESIS TESTING WITH ANOVA

The Assumptions

ANOVA requires several assumptions regarding the method of sampling, the level of measurement, the shape of the population distribution, and the homogeneity of variance.

1. Independent random samples are used. Our choice of sample members from one population has no effect on the choice of sample members from the second, third, or fourth population. For example, the selection of Hispanic men has no effect on the selection of any other sample.

2. The dependent variable, years of education, is an interval-ratio level of measurement. Some researchers also apply ANOVA to ordinal-level measurements.

3. The population is normally distributed. Although we cannot confirm whether the populations are normal, given that our N is so small, we must assume that the population is normally distributed to proceed with our analysis.

4. The population variances are equal. Based on our calculations in Table 10.2, we see that the sample variances, although not identical, are relatively homogeneous.[2]

Stating the Research and the Null Hypotheses and Setting Alpha

The research hypothesis (H_1) proposes that at least one of the means is different. We do not identify which one(s) will be different, or larger or smaller, we only predict that a difference does exist.

H_1: At least one mean is different from the others.

ANOVA is a test of the null hypothesis of no difference between any of the means. Since we're working with four samples, we include four μs in our null hypothesis.

H_0: $\mu_1 = \mu_2 = \mu_3 = \mu_4$

As we did in other models of hypothesis testing, we'll have to set our alpha. Alpha is the level of probability at which we'll reject our null hypothesis. For this example, we'll set alpha at .05.

The Concepts of Between and Within Total Variance

A word of caution before we proceed: Since we're working with four different samples and a total of 21 respondents, we'll have a lot of calculations. It's important to be consistent with your notations (don't mix up numbers for the different samples) and to be careful with your calculations.

Our primary set of calculations has to do with the two types of variance: between-group variance and within-group variance. The estimate of each variance has two parts, the sum of squares and degrees of freedom (*df*).

The **between-group sum of squares** or *SSB* measures the difference in average years of education between our four groups. Sum of squares is the short form for "sum of squared deviations." For *SSB*, what we're measuring is the sum of squared deviations between each sample mean and the overall mean score. The formula for the *SSB* can be presented as follows:

$$SSB = \sum n_k \left(\overline{Y}_k - \overline{Y} \right)^2 \tag{10.1}$$

where

n_k = the number of cases in a sample (*k* represents the number of different samples)

$\overline{Y}_k$ = the mean of a sample

$\overline{Y}$ = the overall mean

SSB can also be understood as the amount of variation in the dependent variable (years of education) that can be attributed to or explained by the independent variable (the four demographic groups).

Between-group sum of squares (SSB)   The sum of squared deviations between each sample mean and the overall mean score.

Within-group sum of squares or *SSW* measures the variation of scores within a single sample or, as in our example, the variation in years of education within one group. *SSW* is also referred to as the amount of unexplained variance, since this is what remains after we consider the effect of the specified independent variable. The formula for *SSW* measures the sum of squared deviations within each group, between each individual score with its sample mean.

$$SSW = \Sigma\left(Y_i - \overline{Y}_k\right)^2 \tag{10.2}$$

where

Y_i = each individual score in a sample

$\overline{Y}_k$ = the mean of a sample

Even with our small sample size, if we were to use Formula 10.2, we'd have a tedious and cumbersome set of calculations. Instead, we suggest using the following computational formula for within-group variation or *SSW*:

$$SSW = \Sigma Y_i^2 - \Sigma\frac{\left(\Sigma Y_k\right)^2}{n_k} \tag{10.3}$$

where

Y_i^2 = the squared scores from each sample

ΣY_k = the sum of the scores of each sample

n_k = the number of cases in a sample

Within-group sum of squares (SSW)   Sum of squared deviations within each group, calculated between each individual score and the sample mean.

Together, the explained (*SSB*) and unexplained (*SSW*) variances compose the amount of total variation in scores. The **total sum of squares** or *SST* can be represented by

$$SST = \Sigma\left(Y_i - \overline{Y}\right)^2 = SSB + SSW \tag{10.4}$$

where

Y_i = each individual score

$\overline{Y}$ = the overall mean

Total sum of squares (SST)   The total variation in scores, calculated by adding *SSB* and *SSW*.

The second part of estimating the between-group and within-group variances is calculating the degrees of freedom. Degrees of freedom are also discussed in Chapters 8 and 9. For ANOVA, we have to calculate two degrees of freedom. For SSB, the degrees of freedom are determined by

$$df_b = k - 1 \tag{10.5}$$

where k is the number of samples.

For SSW, the degrees of freedom are determined by

$$df_w = N - k \tag{10.6}$$

where

N = total number of cases

k = number of samples

Finally, we can estimate the between-group variance by calculating **mean square between.** Simply stated, mean squares are averages computed by dividing each sum of squares by its corresponding degrees of freedom. Mean square between can be represented by

$$\text{Mean square between} = SSB/df_b \tag{10.7}$$

and the within-group variance, or **mean square within** can be represented by

$$\text{Mean square within} = SSW/df_w \tag{10.8}$$

Mean square between   Sum of squares between divided by its corresponding degrees of freedom.

Mean square within   Sum of squares within divided by its corresponding degrees of freedom.

The *F* Statistic

Together the mean square between (Formula 10.7) and mean square within (Formula 10.8) compose the *F* **ratio** or *F* **statistic.** Developed by R. A. Fisher, the *F* statistic is the ratio of between-group variance to within-group variance and is determined by Formula 10.9:

$$F = \frac{\text{Mean square between}}{\text{Mean square within}} = \frac{SSB/df_b}{SSW/df_w} \tag{10.9}$$

We know that a larger F obtained statistic means that there is more between-group variance than within-group variance, increasing the chances of rejecting our null hypothesis. In Table 10.3, we present additional calculations to compute F.

Let's calculate between-group sum of squares and degrees of freedom based on Formulas 10.1 and 10.5. The calculation for SSB is

$$\sum n_k \left(\overline{Y}_k - \overline{Y} \right)^2 = 6(13.00 - 13.48)^2 + 4(12.00 - 13.48)^2 + 6(15.33 - 13.48)^2 + 5(13.00 - 13.48)^2$$
$$= 31.83$$

The degrees of freedom for SSB is $k - 1$ or $4 - 1 = 3$. Based on Formula 10.7, the mean square between is

$$\text{Mean square between} = \frac{31.83}{3} = 10.61$$

Table 10.3 Computational Worksheet for ANOVA

Black Men $n_1 = 6$	*Hispanic Men* $n_2 = 4$	*Black Women* $n_3 = 6$	*Hispanic Women* $n_4 = 5$
16	14	16	14
12	12	18	12
14	11	16	12
12	11	14	13
12		16	14
12		12	
$\overline{Y}_1 = 13.00$	$\overline{Y}_2 = 12.00$	$\overline{Y}_3 = 15.33$	$\overline{Y}_4 = 13.00$
$S_1 = 1.67$	$S_2 = 1.41$	$S_3 = 2.07$	$S_4 = 1.00$
$S_1^2 = 2.79$	$S_2^2 = 1.99$	$S_3^2 = 4.28$	$S_4^2 = 1.00$
$\sum Y_1 = 78$	$\sum Y_2 = 48$	$\sum Y_3 = 92$	$\sum Y_4 = 65$
$\sum Y_1^2 = 1028$	$\sum Y_2^2 = 582$	$\sum Y_3^2 = 1432$	$\sum Y_4^2 = 849$

$$\overline{Y} = 13.48$$

The within-group sum of squares and degrees of freedom are based on Formulas 10.3 and 10.6. The calculation for *SSW* is

$$SSW = \sum Y_i^2 - \sum \frac{(\sum Y_k)^2}{n_k} = (1,028 + 582 + 1,432 + 849) - \left(\frac{78^2}{6} + \frac{48^2}{4} + \frac{92^2}{6} + \frac{65^2}{5} \right)$$

$$= 3,891 - 3,845.67$$

$$= 45.33$$

The degrees of freedom for *SSW* is $N - k = 21 - 4 = 17$. Based on Formula 10.8, the mean square within is

$$\text{Mean square within} = \frac{45.33}{17} = 2.67$$

Finally, our calculation of *F* is based on Formula 10.9:

$$F = \frac{10.61}{2.67} = 3.97$$

F *ratio or* F *statistic* The test statistic for ANOVA, calculated as the ratio of mean square between to mean square within.

Making a Decision

To determine the probability of calculating an *F* statistic of 3.97, we rely on Appendix D, the distribution of the *F* statistic. Appendix D lists the corresponding values of the *F* distribution for various degrees of freedom and two levels of significance, .05 and .01.

Since we set alpha at .05, we'll refer to the table marked "$P = .05$." Note that Appendix D includes two *dfs*. These refer to our degrees of freedom, $df_1 = df_b$ and $df_2 = df_w$.

Because of the two degrees of freedom, we'll have to determine the probability of our *F* obtained differently than we did with the *t* test or chi-square. For this ANOVA example, we'll have to determine the corresponding *F*, also called the *F* critical, when $df_b = 3$ and $df_w = 17$, and $\alpha = .05$.

Based on Appendix D, the *F* critical is 3.20, while our *F* obtained (the one that we calculated) is 3.97. Since our *F* obtained is greater than the *F* critical ($3.97 > 3.20$), we know that its probability is <.05, extending into the shaded area. (If our *F* obtained were <3.20, we could determine that its probability was greater than our alpha of .05, in the unshaded area of the *F* distribution curve. Refer to Figure 10.1.) We can reject the null hypothesis of no difference and conclude that there is a significant difference in educational attainment between the four groups.

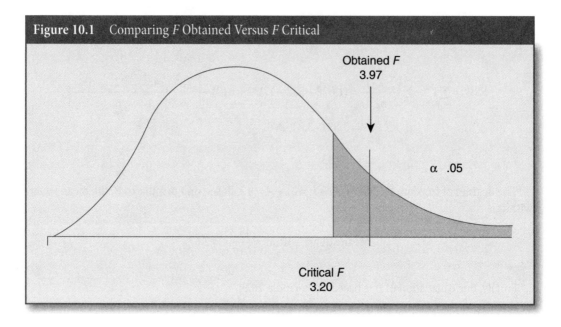

Figure 10.1 Comparing *F* Obtained Versus *F* Critical

▣ THE FIVE STEPS IN HYPOTHESIS TESTING: A SUMMARY

To summarize, we've calculated an analysis of variance test examining the difference between four demographic groups and their average years of education.

Making Assumptions.

1. Independent random samples are used.

2. The dependent variable, years of education, is an interval-ratio level of measurement.

3. The population is normally distributed.

4. The population variances are equal.

Stating the Research and Null Hypotheses and Selecting Alpha.

H_1: At least one mean is different from the others.

H_0: $\mu_1 = \mu_2 = \mu_3 = \mu_4$

$\alpha = .05$

Selecting the Sampling Distribution and Specifying the Test Statistic. The *F* distribution and *F* statistic are used to test the significance of the difference between the four sample means.

Computing the Test Statistic. We need to calculate the between-group and within-group variation (sum of squares and degrees of freedom). We estimate $SSB = 10.61$ ($df_b = 3$) and $SSW = 2.67$ ($df_w = 17$). Based on Formula 10.9,

$$F = \frac{10.61}{2.67} = 3.97$$

Making a Decision and Interpreting the Results. We reject the null hypothesis of no difference and conclude that the groups are different in their educational attainment. Our *F* **obtained** of 3.97 is greater than the *F* **critical** of 3.20. The probability of 3.97 is <.05. *F* doesn't advise us about which groups are different, only that educational attainment does differ significantly by demographic group members. However, based on the sample data, we know that the only group to achieve a college education average was black women (15.33 years). If we were to rank the remaining means, second highest educational attainment was among black men (13.00) and Hispanic women (13.00), and finally, the lowest educational attainment was for Hispanic men (12.00). The mean education years for all the three groups were at least 2 years lower than the mean score for black women. Educational attainment does differ significantly by race and gender group membership.

F *obtained*   The *F*-test statistic that is calculated.

F *critical*   The *F*-test statistic that corresponds to the alpha level, df_w, and df_b (as in Appendix D).

▣ FOCUS ON INTERPRETATION: ARE IMMIGRANTS GOOD FOR AMERICA'S ECONOMY?

We rely on SPSS to examine the relationship between respondents' political views (three ordinal categories: liberal, moderate, and conservative) and their levels of agreement with this statement: Immigrants are generally good for America's economy. The variable IMMAMECO has five ordinal categories: 1 = agree strongly; 2 = agree; 3 = neither; 4 = disagree; and 5 = disagree strongly. We will set alpha at .05 to assess our results. SPSS output is presented in Figure 10.2.

The ANOVA output includes two tables, Descriptives and ANOVA. In the Descriptives table, the *N*, mean, and standard deviation are reported for each group and the entire sample, along with the 95% confidence interval for each mean.

The *F* obtained is reported in the ANOVA table, along with its level of significance (or probability). For these data, *F* obtained is 3.201 with a significance of .043. Since the level of significance is less than our alpha (.043 < .05), we reject the null hypothesis and conclude that at least one of the means is significantly different. Notice that the lowest IMMAMECO score is 2.42, for those in the liberal group. Moderates and conservatives are tied at 2.76 on the five-point scale.

Figure 10.2 SPSS ANOVA Output: Political Views and IMMAMECO

Descriptives

Immameco IMMIGRANTS GOOD FOR AMERICA

	N	Mean	Std. Deviation	Std. Error	95% Confidence interval for Mean		Minimum	Maximum
					Lower Bound	Upper Bound		
1.00 liberal	76	2.42	.868	.100	2.22	2.62	1	4
2.00 moderate	75	2.76	.998	.115	2.53	2.99	1	5
3.00 conservative	75	2.76	.984	.114	2.53	2.99	1	5
Total	226	2.65	.961	.064	2.52	2.77	1	5

ANOVA

Immameco IMMIGRANTS GOOD FOR AMERICA

	Sum of Squares	df	Mean Square	F	Sig.
Between Groups	5.795	2	2.898	3.201	.043
Within Groups	201.886	223	.905		
Total	207.681	225			

▣ READING THE RESEARCH LITERATURE: STRESSES AND STRAINS AMONG GRANDMOTHER CAREGIVERS

Musil and colleagues (2009) examined the family life stresses and strains affecting grandmothers involved in caregiving for grandchildren.[3] Previous studies suggested that grandmother caregivers have more depressive symptoms than their noncaregiving peers. The sample comprises grandmothers, divided into three caregiving groups: primary, multigenerational, or noncaregiver. The groups were defined as follows:

> Primary caregiver grandmothers had responsibility for raising their grandchildren without parents living in the home. Multigenerational grandmothers lived in a home with one or more grandchildren and the grandchild(ren)'s parent(s). Noncaregiver grandmothers did not live with or provide regular babysitting for grandchildren but lived within 1 hour or 50 miles of grandchildren and had an ongoing relationship with them. (p. 395)[4]

The researchers measured family life stresses and strains based on several existing scales.

- The level of strain and stresses (conflict, difficulty) was measured for general intrafamily strain (conflict among children, difficulty in managing children), family life stresses, financial (increasing financial debts), transitions (a member lost or quit a job, moved into a new home), family legal (incidents of physical abuse or aggression), family loss (child died), family care (child became seriously ill or injured), and pregnancy (teenager became pregnant).
- Social support was assessed based on the Duke Social Support Index, measuring both subjective and instrumental dimensions of support. Instrumental support items measured the extent to which friends and family offered assistance or help in specific situations. A higher score indicates high instrumental support. Subjective support was measured by items about

feelings of support and involvement with friends and family. A higher score indicates a high level of subjective support.

- Resourcefulness was measured by the Self-Control Schedule. A higher score indicates greater resourcefulness.
- Depressive symptoms were evaluated based on a 20-item Center for Epidemiological Studies–Depression Scale. Higher scores indicate an increased clinical depression.

A portion of Musil and her colleagues' findings are presented in Table 10.4. In the table, they report the mean score and standard deviation for each stress and strain area, social support, resourcefulness, and depressive symptoms, using analysis of variance to compare the results for each grandmother group. They highlight the significant differences in the following paragraph.

There were significant differences between groups in intrafamily strain: Primary caregivers reported the most strain [Table 10.4, this chapter]; there were no differences in the family life

Table 10.4 Means, Standard Deviations, and ANOVA Results by Caregiver Group

Variables	Primary (n = 183)		Multigenerational (n = 136)		Noncaregivers (n = 167)		F Test
	M	SD	M	SD	M	SD	
Intrafamily strain	4.4	2.8	3.9	2.9	2.7	2.3	18.4***
Family life stresses (aggregate)	5.2	3.2	5.4	3.1	4.6	2.9	4.1
Financial	1.2	0.8	1.1	0.8	0.9	0.8	5.8*
Transitions	1.7	1.5	2.3	1.7	1.9	1.5	4.7**
Family legal	0.7	0.9	0.4	0.7	0.4	0.7	9.9***
Family loss	0.7	0.7	0.7	0.7	0.6	0.7	0.2
Family care	0.8	1.1	0.8	1.0	0.7	1.0	0.2
Pregnancy	0.1	0.2	0.1	0.2	0.0	0.2	2.0
Support—instrumental	7.7	3.4	9.7	2.2	8.6	2.8	19.2***
Support—subjective	11.1	3.1	11.8	2.4	12.2	2.6	7.9***
Resourcefulness	3.2	0.6	3.2	0.6	3.3	0.6	2.2
Depressive symptoms	15.8	11.3	12.4	10.4	11.5	10.6	8.0***

Note: *$P < .05$, **$P < .01$, ***$P < .001$.

stresses summary score. There were significant differences between grandmother caregiver groups on specific family life stresses.... Post hoc tests showed that noncaregivers reported fewer financial strains than primary and multigenerational grandmothers, and primary caregivers reported significantly more family legal problems. Multigenerational grandmothers reported more transitions than primary caregivers. There were no significant between-group differences on family-care strains, loss, or pregnancy strain. There were significant between-group differences in support, but not resourcefulness. Noncaregivers reported more subjective support than primary caregivers. Grandmothers in multigenerational homes reported the most instrumental support and primary caregivers reported the least. Primary caregivers reported higher depressive symptoms than grandmothers in the other two groups. (p. 399)[5]

Notice how F-test results are not reported in their summary. However, we know from Table 10.4 which model was significant. For example, the ANOVA model for intrafamily strain produced an F obtained of 18.4 (significant at the .001 level). As we review the mean scores for each group, the highest level of intrafamily strain was reported by primary caregivers (a mean score of 4.4), followed by multigenerational caregivers (3.9). Musil and colleagues conclude that these results reflect the "more complex family situations in these homes." Apart from the need to coordinate the schedules of grandchildren and the adults in the household, they identify the additional relationship strains of lack of privacy, less discretionary time, and conflict with birth parents as sources of intrafamily strain.

✓ *Learning*
Check

For the ANOVA model for intrafamily strain, what is the F *critical? What information do you need to determine the* F *critical? Assume alpha = .05.*

MAIN POINTS

• Analysis of variance (ANOVA) procedures allow us to examine the variation in means in more than two samples. To determine whether the difference in mean scores is significant, ANOVA examines the differences between multiple samples, as well as the differences within a single sample.

• One-way ANOVA is a procedure using one dependent variable and one independent variable. The five-step hypothesis testing model is applied to one-way ANOVA.

• The test statistic for ANOVA is F. The F statistic is the ratio of between-group variance to within-group variance.

KEY TERMS

analysis of variance (ANOVA)
between-group sum of squares (SSB)
F critical

F obtained
F ratio or F statistic
mean square between
mean square within
one-way ANOVA

total sum of squares (SST)
within-group sum of squares (SSW)

$SAGE edge™

Sharpen your skills with SAGE edge at **edge.sagepub.com/ssdsess2e**. **SAGE edge for students** provides a personalized approach to help you accomplish your coursework goals in an easy-to-use learning environment.

CHAPTER EXERCISES

1. In Chapter 9 we analyzed the relationship between social class and health assessment. We continue the analysis here for a random sample of 32 GSS cases. Health is measured according to a four-point scale: 1 = excellent, 2 = good, 3 = fair, and 4 = poor. Four social classes are reported here: lower, working, middle, and upper. Present the five-step model for these data, using alpha = .05.

Lower Class	Working Class	Middle Class	Upper Class
3	2	2	2
2	1	3	1
2	3	1	1
2	2	1	2
3	2	2	1
3	2	3	1
4	3	3	1
4	3	1	2

2. Data for a subsample of 19 Monitoring the Future (MTF) respondents are presented below, examining their race and their responses to FRSMOKE ("How many of your friends smoke cigarettes?"). Responses are measured 1= none, 2 = a few, 3 = some, 4 = most, and 5 = all.

Hispanic	White	Black
4	5	3
3	5	2
3	5	1
4	4	2
5	4	3
3	3	2
1		

a. What is the independent variable?
b. Complete the five-step model for these data, using alpha = .05.

3. In this exercise, let's examine the relationship between educational degree and church attendance. We selected a sample of 30 International Social Science Programme respondents, noting their educational status (no degree, secondary degree, and university degree) and their level of church attendance (0 = never, 1 = infrequently, and 2 = two to three times per month or more).

a. Complete the five-step model for these data, using alpha = .05.

b. If alpha were changed to .01, would your Step 5 decision change? Explain.

No Degree	Secondary Degree	University Degree
2	2	0
1	2	0
1	2	0
2	1	0
2	1	1
2	0	1
0	2	0
2	1	1
2	1	2
2	2	1

4. Based on data from the GSS10SDSS, we examine the relationship between highest educational degree and agreement with the statement "Financial dependence on others is one of my greatest fears about old age." The variable FINDEPND is measured on an ordinal scale: 1 = strongly agree, 2 = agree, 3 = neither, 4 = disagree, and 5 = strongly disagree. Is there a relationship between educational attainment and level of agreement with the statement?

Set alpha at .01 and test the null hypothesis of equal means.

Figure 10.3   ANOVA Output for FINDEPND and DEGREE

FINDEPND WORRY ABOUT DEPENDENCE

	N	Mean	Std. Deviation	Std. Error	95% Confidence interval for Mean Lower Bound	Upper Bound	Minimum	Maximum
0 LT HIGH SCHOOL	32	2.53	1.218	.215	2.09	2.97	1	5
1 HIGH SCHOOL	156	2.63	1.214	.097	2.44	2.82	1	5
2 JUNIOR COLLEGE	26	2.88	1.243	.244	2.38	3.39	1	5
3 BACHELOR	50	2.88	1.304	.184	2.51	3.25	1	5
3 GRADUATE	45	3.49	1.199	.179	3.13	3.85	1	5
Total	309	2.81	1.259	.072	2.66	2.95	1	5

Anova

FINDEPND WORRY ABOUT DEPENDENCE

	Sum of Squares	df	Mean Square	F	Sig.
Between Groups	28.767	4	7.192	4.757	.001
Within Groups	459.583	304	1.512		
Total	488.350	308			

Exercises

5. GSS 2010 respondents were asked about their agreement with the following statement: "Parents ought to provide financial help to their adult children when the children are having financial difficulty." Their responses to HELPKIDS were measured on the same agreement scale as FINDEPND in Question 4. Set alpha at .05 and test the null hypothesis of equal group means.

Figure 10.4   ANOVA Output for HELPKIDS and DEGREE

Descriptives

HELPKIDS PARENTS ADULT CHILDREN FINANCIALLY

	N	Mean	Std. Deviation	Std. Error	95% Confidence interval for Mean		Minimum	Maximum
					Lower Bound	Upper Bound		
0 LT HIGH SCHOOL	32	2.53	1.270	.224	2.07	2.99	1	5
1 HIGH SCHOOL	155	2.59	.811	.065	2.46	2.72	1	5
2 JUNIOR COLLEGE	26	2.65	1.056	.207	2.23	3.08	1	5
3 BACHELOR	50	2.60	1.030	.146	2.31	2.89	1	5
3 GRADUATE	45	2.62	.984	.147	2.33	2.92	1	5
Total	308	2.60	.945	.054	2.49	2.70	1	5

Anova

HELPKIDS PARENTS ADULT CHILDREN FINANCIALLY

	Sum of Squares	df	Mean Square	F	Sig.
Between Groups	.253	4	.063	.070	.991
Within Groups	273.825	303	.904		
Total	274.078	307			

6. We examine the relationship between educational attainment and agreement with the statement "Immigrants take jobs away from people who were born in America" as measured in the GSS 2010. Responses to IMMJOBS are measured according to the same five-point scale as FINDEPND in Question 4.
   a. Set alpha at .05 and test the null hypothesis of equal means.
   b. If alpha were set at .01, would your decision remain the same?

Figure 10.5   ANOVA Output for IMMJOBS and DEGREE

Descriptives

Immjobs IMMIGRANTS TAKE JOBS AWAY

	N	Mean	Std. Deviation	Std. Error	95% Confidence interval for Mean		Minimum	Maximum
					Lower Bound	Upper Bound		
0 LT HIGH SCHOOL	53	2.96	1.315	.181	2.60	3.32	1	5
1 HIGH SCHOOL	189	2.63	1.081	.079	2.48	2.79	1	5
2 JUNIOR COLLEGE	35	2.97	1.071	.181	2.60	3.34	1	5
3 BACHELOR	70	3.27	1.020	.122	3.03	3.51	1	5
3 GRADUATE	36	3.42	1.079	.180	3.05	3.78	1	5
Total	383	2.90	1.137	.058	2.79	3.02	1	5

ANOVA

Immjobs IMMIGRANTS TAKE JOBS AWAY

	Sum of Squares	df	Mean Square	F	Sig.
Between Groups	32.931	4	8.233	6.746	.000
Within Groups	461.298	378	1.220		
Total	494.230	382			

7. Based on a sample of 21 MTF respondents, we present their racial/ethnic backgrounds and the numbers of school days missed in the past 4 weeks.
   a. Complete the five-step model for these data, setting alpha at .05.
   b. If alpha were set at .01, would your decision change? Explain.

White	Black	Hispanic
4	1	4
5	2	3
3	2	5
4	1	1
4	3	5
4	4	2
6	3	2

8. We selected a sample of 14 MTF respondents. We present their numbers of moving (traffic) violations in the past 12 months along with their residential areas (residential area is the independent variable). Complete the five-step model for these data, using alpha = .05.

Small Town	Medium-Sized City	Large City
0	2	3
0	3	4
1	1	4
2	1	3
1		2

9. Nan Sook Park and her colleagues (2012) investigated racial/ethnic differences in predictors of self-rated health and the use of sociocultural resources. Their data are based on the Survey of Older Floridians, a statewide sample of white, African American, Cuban, and non-Cuban Hispanic seniors.

We present ANOVA results for two sociocultural resources. Social support was measured with the question "In times of trouble, can you count on at least some of your family and friends?" (1 = hardly ever, 2 = some of the time, and 3 = most of the time). Religious attendance was measured according to the following scale: 1 = never or almost never to 5 = more than once a week. Mean scores are presented for each racial/ethnic group, along with the standard deviations in parentheses.

| | Racial/Ethnic Group Mean (Standard Deviation) | | | | |
| | | | | | |
Variable	*Whites* (n = 503)	*African Americans* (n = 360)	*Cubans* (n = 328)	*Non-Cuban Hispanics* (n = 241)	**F**
Social support	2.85 (0.47)	2.75 (0.57)	2.73 (0.60)	2.58 (0.70)	12.17***
Religious attendance	2.79 (1.57)	3.94 (1.21)	2.74 (1.49)	3.37 (1.43)	56.43***

Source: Nan Sook Park, Yuri Jan, Beom Lee and David Chiriboga. "Racial/Ethnic Differences in Predictors of Self-Rated Health: Findings From the Survey of Older Floridians." *Research on Aging* 35, no. 1 (2012): 207.

*** $P < .001$

Review each measure of sociocultural resource and determine whether the null hypothesis would be rejected. Set alpha at .05 for each.

10. Is there a significant difference in Internet use per week between different educational groups? Using data from the GSS10SSDS, we ran an ANOVA model using DEGREE (educational attainment) as the independent variable and WWHR (WWW hours per week) as the dependent variable. Based on an alpha of .05, what can you conclude?

Figure 10.6 ANOVA Output for WWWHR and DEGREE

Descriptives

wwwhr WWW HOURS PER WEEK

| | N | Mean | Std. Deviation | Std. Error | 95% Confidence Interval for Mean | | Minimum | Maximum |
					Lower Bound	Upper Bound		
0 LT HIGH SCHOOL	18	3.56	5.772	1.361	.69	6.43	0	25
1 HIGH SCHOOL	165	8.90	12.166	.947	7.03	10.77	0	80
2 JUNIOR COLLEGE	22	7.14	8.817	1.880	3.23	11.05	0	40
3 BACHELOR	84	11.18	10.404	1.135	8.92	13.44	0	50
4 GRADUATE	34	11.32	14.954	2.565	6.11	16.54	0	66
Total	323	9.33	11.689	.650	8.05	10.61	0	80

ANOVA

wwwhr WWW HOURS PER WEEK

	Sum of Squares	df	Mean Square	F	Sig.
Between Groups	1158.308	4	289.577	2.150	.075
Within Groups	42835.246	318	134.702		
Total	43993.554	322			

11. Based on data from MTF 2011, we present ANOVA results for grade point average (GPA) by student race. Set alpha at .05 and test the null hypothesis of equal means.

Figure 10.7   ANOVA Output for GPA and RACE

Descriptives

High School GPA (recoded from "grade")

	N	Mean	Std. Deviation	Std. Error	95% Confidence Interval for Mean		Minimum	Maximum
					Lower Bound	Upper Bound		
BLACK:(1)	166	3.0572	.63440	.04924	2.9600	3.1544	1.00	4.00
WHITE:(2)	758	3.2323	.63649	.02312	3.1869	3.2777	1.00	4.00
HISPANIC:(3)	179	3.0905	.66169	.04946	2.9929	3.1881	1.00	4.00
Total	1103	3.1830	.64399	.01939	3.1449	3.2210	1.00	4.00

ANOVA

High School GPA (recoded from "grade")

	Sum of Squares	df	Mean Square	F	Sig.
Between Groups	6.001	2	3.001	7.318	.001
Within Groups	451.018	1100	.410		
Total	457.020	1102			

Chapter 11

Regression and Correlation

Chapter Learning Objectives

- ❖ Understanding linear relations and prediction rules
- ❖ Constructing and interpreting straight-line graphs and finding the best-fitting line
- ❖ Calculating and interpreting a and b
- ❖ Understanding the meaning of prediction errors
- ❖ Calculating and interpreting the coefficient of determination (r^2) and Pearson's correlation coefficient (r)

Many research questions require the analysis of relationships between interval-ratio variables. Social scientists, for instance, frequently measure variables such as educational attainment, family size, and household income. Bivariate regression analysis provides us with the tools to express a relationship between two interval-ratio variables in a concise way.[1]

The U.S. Census Bureau collects and reports an array of information about the United States and its residents. Annual household income and educational attainment are two of the many characteristics regularly monitored. Table 11.1 displays the percentages of state residents with bachelor's degrees in 2009 for the 10 most populated states. Also presented are the mean, variance, and range for these data.

In examining Table 11.1 and the descriptive statistics, notice the variability in educational attainment. The percentage of state residents with bachelor's degrees ranges from a low of 24.10% for the state of Ohio to a high of 34.50% for the state of New Jersey.

One possible explanation for the differences in educational attainment is the economic conditions of these states. We would expect states with higher household incomes to also have larger percentages of their residents attain college degrees. Table 11.2 displays the median household incomes in 2009 for 10 of the most populated states. Note that the median household income ranges widely from a low of $44,736 for the state of Florida to a high of $68,342 for the state of New Jersey.

Table 11.1   Percentages of State Residents With Bachelor's Degrees, 2009

State	Percentage of State Residents With Bachelor's Degrees
California	29.90
Texas	25.50
New York	32.40
Florida	25.30
Illinois	30.60
Pennsylvania	26.40
Ohio	24.10
Michigan	24.60
Georgia	27.50
New Jersey	34.50

$$\text{Mean } X = \overline{X} = \frac{\sum X}{N} = \frac{280.8}{10} = 28.08$$

$$\text{Variance } X = S_X^2 = \frac{\sum \left(X - \overline{X}\right)^2}{N-1} = \frac{115.04}{9} = 12.78$$

$$\text{Range } X = 34.5 - 24.1 = 10.4$$

Source: U.S. Census Bureau, *Statistical Abstract of the United States*, 2012, Table 233.

▣ THE SCATTER DIAGRAM

Let's examine the possible relationship between the interval-ratio variables *percentage of state residents with bachelor's degrees* and *median household income*. One quick visual method used to display such a relationship between two interval-ratio variables is the **scatter diagram** (or **scatterplot**). Often used as a first exploratory step in regression analysis, a scatter diagram can suggest whether two variables are associated.

Scatter diagram (scatterplot)   A visual method used to display a relationship between two interval-ratio variables.

Table 11.2 Median Household Income ($)

State	Median Household Income ($)
California	58,931
Texas	48,259
New York	54,659
Florida	44,736
Illinois	53,966
Pennsylvania	49,520
Ohio	45,395
Michigan	45,255
Georgia	47,590
New Jersey	68,342

$$\text{Mean } X = \overline{X} = \frac{\Sigma X}{N} = \frac{516,653}{10} = \$51,665.30$$

$$\text{Variance } X = S_X^2 = \frac{\Sigma(X-\overline{X})^2}{N-1} = \frac{506,395,248}{9} = \$56,266,139$$

$$\text{Range } X = \$68,342 - \$44,736 = \$23,606$$

Source: U.S. Census Bureau, *American Community Survey,* 2009.

The scatter diagram showing the relationship between educational attainment and household income for the 10 states is shown in Figure 11.1. In a scatter diagram, the scales for the two variables form the vertical and horizontal axes of a graph. Usually, the independent variable, X, is arrayed along the horizontal axis and the dependent variable, Y, along the vertical axis. Because differences in the median household income are hypothesized to account for differences in the percentage of state residents with bachelor's degrees, household income is assumed as the independent variable and is arrayed along the horizontal axis. Educational attainment, the dependent variable, is arrayed along the vertical axis. In Figure 11.1, each dot represents a state; its location lies at the exact intersection of that state's percentage of residents with bachelor's degrees and its median household income.

Note that there is an apparent tendency for states with lower median household incomes (e.g., Ohio and Texas) to also have lower percentages of residents with bachelor's degrees, whereas in

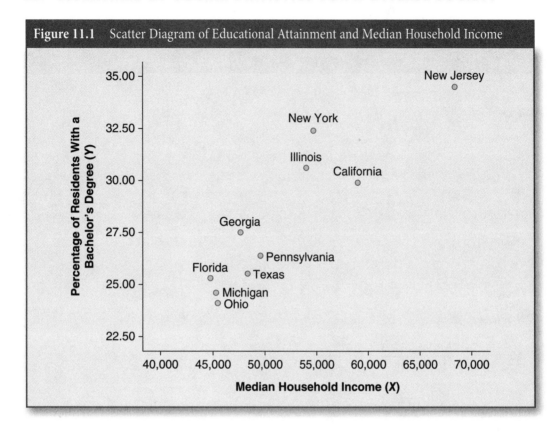

Figure 11.1   Scatter Diagram of Educational Attainment and Median Household Income

states with higher median household incomes (e.g., New Jersey and California), there are higher percentages of residents with bachelor's degrees. In other words, we can say that median household income and educational attainment are positively associated.

Scatter diagrams can also illustrate a negative association between two variables. For example, Figure 11.2 displays the association between median household income and larceny/theft crime rates for the 10 most populated states using data from the *Statistical Abstract of the United States*. Figure 11.2 suggests that a low median household income is associated with a higher rate of larceny/theft. Conversely, high median household income seems to be associated with a lower larceny/theft rate (see Table 11.5 for data).

▣ LINEAR RELATIONS AND PREDICTION RULES

Scatter diagrams provide a useful but only preliminary step in exploring a relationship between two interval-ratio variables. We need a more systematic way to express this relationship. Let's examine Figures 11.1 and 11.2 again. They allow us to understand how household income is related to educational attainment (Figure 11.1) and larceny (Figure 11.2). The relationships displayed are by no

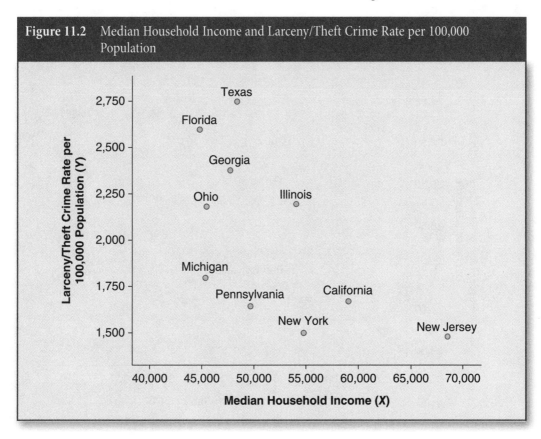

Figure 11.2 Median Household Income and Larceny/Theft Crime Rate per 100,000 Population

means perfect, but the trends are apparent. In the first case (Figure 11.1), as state income increases, so does the percentage of residents with bachelor's degrees. In the second case (Figure 11.2), as state income increases, larceny/theft crime rate decreases.

One way to evaluate these relationships is by expressing them as *linear relationships*. A **linear relationship** allows us to approximate the observations displayed in a scatter diagram with a straight line. In a perfectly linear relationship, all the observations (the dots) fall along a straight line (a **perfect** relationship is sometimes called a **deterministic relationship**), and the line itself provides a predicted value of Y (the vertical axis) for any value of X (the horizontal axis). For example, in Figure 11.3, we have superimposed a straight line on the scatterplot originally displayed in Figure 11.1. Using this line, we can obtain a predicted value of the percentage of state residents with bachelor's degrees for any value of household income, by reading up to the line from the income axis and then over to the percentage with a bachelor's degree axis (indicated by the dotted lines). For example, the predicted value of the percentage of state residents with bachelor's degrees in a state with a \$50,000 median household income is approximately 27%. Similarly, for a state with a \$55,000 median household income, we would predict that approximately 29% of its residents would have bachelor's degrees.

Figure 11.3 A Straight-Line Graph for Median Household Income and Percentage of Residents With Bachelor's Degrees

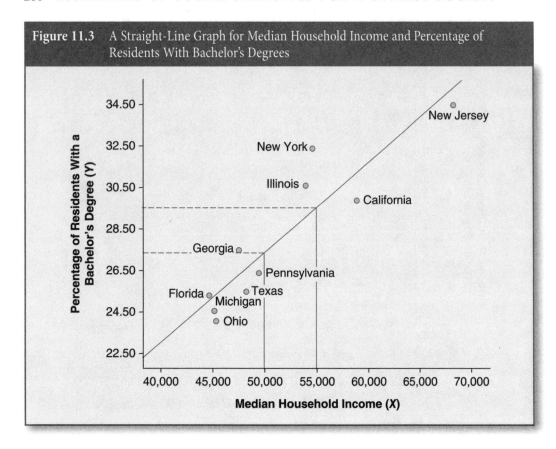

Linear relationship A relationship between two interval-ratio variables in which the observations displayed in a scatter diagram can be approximated with a straight line.

Deterministic (perfect) linear relationship A relationship between two interval-ratio variables in which all the observations (the dots) fall along a straight line. The line provides a predicted value of *Y* (the vertical axis) for any value of *X* (the horizontal axis).

As indicated in Figure 11.3, for the 10 states surveyed, the actual relationship between income and the percentage of state residents with bachelor's degrees is not perfectly linear. Although some of the states lie very close to the line, none fall exactly on the line, and some deviate from it considerably. Are there other lines that provide a better description of the relationship between income and the percentage of state residents with bachelor's degrees?

In Figure 11.4, we have drawn two additional lines that approximate the pattern of relationship shown by the scatter diagram. In each case, notice that even though some of the states lie close to

Figure 11.4   Alternative Straight-Line Graph for Median Household Income and Percentage of Residents With Bachelor's Degrees

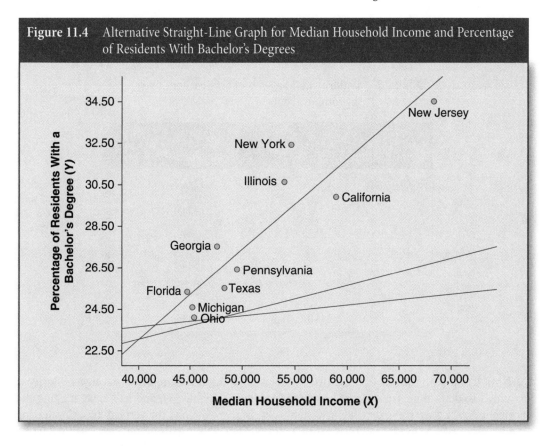

the line, all fall considerably short of perfect linearity. Is there one line that provides the best linear description of the relationship between median household income and the percentage of residents with bachelor's degrees? How do we choose such a line? What are its characteristics? Before we describe a technique for finding the straight line that most accurately describes the relationship between two variables, we first need to review some basic concepts about how straight-line graphs are constructed.

✓ *Learning Check*

Use Figure 11.3 to predict the percentage of residents with bachelor's degrees in a state with a median household income of $47,500 and one with a median household income of $50,000.

Constructing Straight-Line Graphs

To illustrate the fundamentals of straight-line graphs, let's take a simple example. Suppose that in a local school system, teachers' salaries are completely determined by seniority. New teachers begin with an annual salary of $12,000, and for each year of seniority, their salaries increase by $2,000. The seniority and annual salaries of six hypothetical teachers are presented in Table 11.3.

Table 11.3   Seniority and Salaries of Six Teachers (hypothetical data)

X Seniority (in years)	Y Salary (in dollars)
0	12,000
1	14,000
2	16,000
3	18,000
4	20,000
5	22,000

Now, let's plot the values of these two variables on a graph (Figure 11.5). Because seniority is assumed to determine salary, let it be our independent variable (X), and let's array it along the horizontal axis. Salary, the dependent variable (Y), is arrayed along the vertical axis. Connecting the six observations in Figure 11.5 gives us a straight-line graph. This graph allows us to obtain a predicted salary value for any value of seniority level simply by reading from the specific seniority level up to the line and then over to the salary axis. For instance, we have marked the lines going up from a seniority level of 7 years and then over to the salary axis. We can see that a teacher with 7 years of seniority makes $26,000.

The relationship between salary and seniority, as depicted in Table 11.3 and Figure 11.5, can also be described with the following algebraic equation:

$$Y = 12,000 + 2,000(X)$$

where

X = seniority (in years)

Y = salary (in dollars)

This equation allows us to correctly predict salary (Y) for any value for seniority (X) that we plug into the equation. For example, the salary of a teacher with 5 years of seniority is

$$Y = 12,000 + 2,000(5) = 12,000 + 10,000 = \$22,000$$

Figure 11.5 A Perfect Linear Relationship Between Seniority (in years) and Annual Salary (in thousand dollars) of Six Teachers (hypothetical)

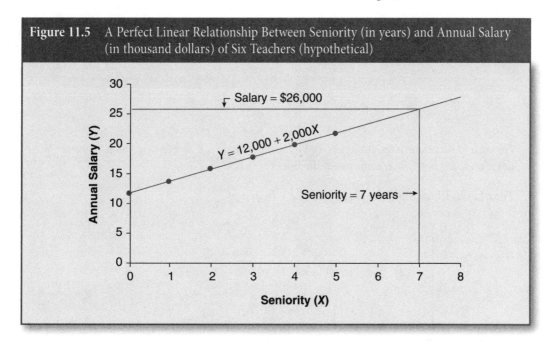

Note that we can also plug in values of X that are not shown in Table 11.3. For example, the salary of a teacher with 10 years of seniority is

$$Y = 12,000 + 2,000(10) = 12,000 + 20,000 = \$32,000$$

The equation describing the relationship between seniority and salary is an equation for a straight line. The equations for all straight-line graphs have the same general form:

$$Y = a + b(X) \qquad (11.1)$$

where

Y = the predicted score on the dependent variable

X = the score on the independent variable

a = the **Y-intercept**, or the point where the line crosses the Y-axis; therefore, a is the value of Y when X is 0

b = the slope of the regression line, or the change in Y with a unit change in X. In our example, $a = 12,000$ and $b = 2,000$. That is, a teacher will make \$12,000 with 0 years of seniority but then her or his salary will go up by \$2,000 with each year of seniority.

Y-intercept (a) *The point where the regression line crosses the Y-axis, or the value of* Y *when* X *is 0.*

Learning Check

Use the linear equation describing the relationship between seniority and salary of teachers to obtain the predicted salary of a teacher with 12 years of seniority.

✓ *Learning Check*

For each of these four lines, as X goes up by 1 unit, what does Y do? Be sure you can answer this question using both the equation and the line.

Four Lines: Illustrating the Slope and the Y-Intercept

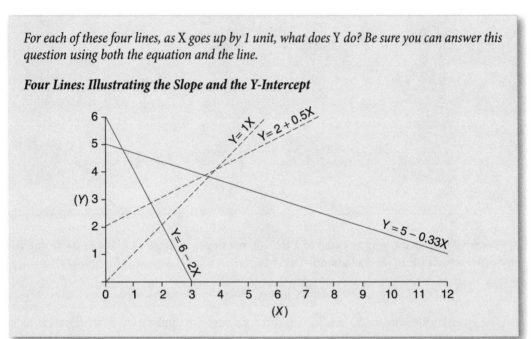

Finding the Best-Fitting Line

The straight line displayed in Figure 11.5 and the linear equation representing it ($Y = 12,000 + 2,000[X]$) provide a very simple depiction of the relationship between seniority and salary because salary (the Y variable) is completely determined by seniority (the X variable). When each value of Y is completely determined by X, all the points (observations) lie on the line, and the relationship between the two variables is a deterministic, or perfectly linear, relationship.

However, most relationships we study in the social sciences are not deterministic, and we are not able to come up with a linear equation that allows us to predict Y from X with perfect accuracy. We are much more likely to find relationships approximating linearity, but in which numerous cases don't follow this trend perfectly. For instance, in reality, teachers' salaries are not completely determined by seniority, and therefore, knowing years of seniority will not provide us with a perfect prediction of salary.

When the dependent variable (Y) is not completely determined by the independent variable (X), not all (sometimes none) of the observations will lie exactly on the line. Look back at Figure 11.4, our example of the percentage of state residents with bachelor's degrees in relation to the state's median household income. Though each line represents a linear equation showing us how the percentage of state residents with bachelor's degrees rises with a state's median household income, we do not have a perfect prediction in any of the lines. Although all three lines approximate the linear trend suggested by the scatter diagram, very few of the observations lie exactly on any of the lines, and some deviate from them considerably.

Given that none of the lines is perfect, our task is to choose one line—the *best-fitting line*. But which is the best-fitting line?

Defining Error

The best-fitting line is the one that generates the least amount of error, also referred to as the *residual*. Look again at Figure 11.3. For each income level, the line (or the equation that this line represents) predicts a value of Y. Texas, for example, with a median household income of \$48,259, gives us a predicted value for Y of 26.7%. But the actual value for Texas is 25.5% (see also Table 13.1). Thus, we have two values for Y: (1) a predicted Y, which we symbolize as $\hat{Y}$ and which is generated by the prediction equation, also called the *linear regression equation* $Y = a + b(X)$, and (2) the observed Y, symbolized simply as Y. Thus, for Texas, $\hat{Y} = 26.7\%$, whereas $Y = 25.5\%$.

We can think of the residual as the difference between the observed Y (Y) and the predicted Y ($\hat{Y}$). If we symbolize the residual as e, then

$$e = Y - \hat{Y}$$

The residual for Texas is 25.5% – 26.7% = –1.2 percentage points.

The Residual Sum of Squares (Σe^2)

We want a line or a prediction equation that minimizes e for each individual observation. However, any line we choose will minimize the residual for some observations but may maximize it for others. We want to find a prediction equation that minimizes the residuals over all observations.

There are many mathematical ways of defining the residuals. Statisticians prefer to work with the third method—squaring and summing the residuals over all observations. The result is the *residual sum of squares*, or Σe^2. Symbolically, Σe^2 is expressed as

$$\Sigma e^2 = \Sigma (Y - \hat{Y})^2$$

The Least Squares Line

The best-fitting regression line is that line where the sum of the squared residuals, or Σe^2, is at a minimum. Such a line is called the **least squares line** (or **best-fitting line**), and the technique that produces this line is called the **least squares method**. The technique involves choosing a and b for the equation such that Σe^2 will have the smallest possible value. In the next section, we use the data from the 10 states to find the least squares equation. But before we continue, let's review where we are so far.

Least squares line (best-fitting line)   A line where the residual sum of squares, or Σe^2, is at a minimum.

Least squares method   The technique that produces the least squares line.

Computing *a* and *b* for the Prediction Equation

Through the use of calculus, it can be shown that to figure out the values of *a* and *b* in a way that minimizes Σe^2, we need to apply the following formulas:

$$b = \frac{S_{YX}}{S_X^2} \tag{11.2}$$

$$a = \bar{Y} - b(\bar{X}) \tag{11.3}$$

where

S_{YX} = the covariance of *X* and *Y*

S_X^2 = the variance of *X*

$\bar{Y}$ = the mean of *Y*

$\bar{X}$ = the mean of *X*

a = the *Y*-intercept

b = the slope of the line

These formulas assume that *X* is the independent variable and *Y* is the dependent variable.

Before we compute *a* and *b*, let's examine these formulas. The denominator for *b* is the variance of the variable *X*. It is defined as follows:

$$\text{Variance } (X) = S_X^2 = \frac{\Sigma(X - \bar{X})^2}{N - 1}$$

This formula should be familiar to you from Chapter 4. The numerator (S_{YX}), however, is a new term. It is the covariance of *X* and *Y* and is defined as

$$\text{Covariance } (X, Y) = S_{YX} = \frac{\Sigma(X - \bar{X})(Y - \bar{Y})}{N - 1} \tag{11.4}$$

The covariance is a measure of how X and Y vary together. Basically, the covariance tells us to what extent higher values of one variable "go together" with higher values on the second variable (in which case we have a positive covariation) or with lower values on the second variable (which is a negative covariation). Take a look at this formula. It tells us to subtract the mean of X from each X score and the mean of Y from each Y score, and then take the product of the two deviations. The results are then summed for all the cases and divided by $N - 1$.

In Table 11.4, we show the computations necessary to calculate the values of a and b for our 10 states. To calculate the covariance, we first subtract from each X score (Column 3) and from each Y score (Column 5) to obtain the mean deviations. We then multiply these deviations for every observation. The products of the mean deviations are shown in Column 7. For example, for the first observation, California, the mean deviation for median household income is 7,265.7 (58,931 − 51,665.3 = 7,265.7); for the percentage of residents with bachelor's degrees, it is 1.82 (29.90 − 28.08 = 1.82). The product of these deviations, 13,233.57 ($7,265.7 \times 1.82 = 13,223.57$), is shown in Column 7. The sum of these products, shown at the bottom of Column 7, is 220,301.66. Dividing it by 9 ($N - 1$), we get the covariance of 24,477.96.

The covariance is a measure of the linear relationship between two variables, and its value reflects both the strength and the direction of the relationship. The covariance will be close to zero when X and Y are unrelated; it will be larger than zero when the relationship is positive and smaller than zero when the relationship is negative.

Now let's substitute the values for the covariance and the variance from Table 11.4 to calculate b:

$$b = \frac{S_{YX}}{S_X^2} = \frac{24{,}477.96}{56{,}266{,}139} = 0.0004$$

Once b has been calculated, finding a, the intercept, is simple:

$$a = \bar{Y} - b(\bar{X}) = 28.08 - 0.0004(51{,}665.3) = 7.41$$

The prediction equation is therefore

$$\hat{Y} = 7.41 + 0.0004(X)$$

This equation can be used to obtain a predicted value for the percentage of state residents who have bachelor's degrees given a state's median household income. For example, for a state with a median household income of $48,000, the predicted percentage is

$$\hat{Y} = 7.41 + 0.0004(48{,}000) = 26.61$$

Similarly, for a state with a median household income of $56,000, the predicted value is

$$\hat{Y} = 7.41 + 0.0004(56{,}000) = 29.81$$

Table 11.4   Worksheet for Calculating *a* and *b* for the Regression Equation

	(1)	(2)	(3)	(4)	(5)	(6)	(7)
	Median Household Income	Percentage With Bachelor's Degrees					
State	X	Y	$(X-\bar{X})$	$(X-\bar{X})^2$	$(Y-\bar{Y})$	$(Y-\bar{Y})^2$	$(X-\bar{X})(Y-\bar{Y})$
California	58,931	29.9	7,265.7	52,790,396	1.82	3.31	13,223.57
Texas	48,259	25.5	−3,406.3	11,602,880	−2.58	6.66	8,788.25
New York	54,659	32.4	2,993.7	8,962,240	4.32	18.66	12,932.78
Florida	44,736	25.3	−6,929.3	48,015,198	−2.78	7.73	19,263.45
Illinois	53,966	30.6	2,300.7	5,293,220	2.52	6.35	5,797.76
Pennsylvania	49,520	26.4	−2,145.3	4,602,312	−1.68	2.82	3,604.10
Ohio	45,395	24.1	−6,270.3	39,316,662	−3.98	15.84	24,955.79
Michigan	45,255	24.6	−6,410.3	41,091,946	−3.48	12.11	22,307.84
Georgia	47,590	27.5	−4,075.3	16,608,070	−0.58	0.34	2,363.67
New Jersey	68,342	34.5	16,676.7	278,112,323	6.42	41.22	107,064.41
	ΣX=516,653	ΣY=280.8	0[a]	506,395,248	0[a]	115.0	220,301.66

$$\text{Mean } X = \bar{X} = \frac{\Sigma X}{N} = \frac{516,653}{10} = 51,665.3$$

$$\text{Mean } Y = \bar{Y} = \frac{\Sigma Y}{N} = \frac{280.8}{10} = 28.08$$

$$\text{Variance}(X) = S_X^2 = \frac{\Sigma(X-\bar{X})^2}{N-1} = \frac{506,395,248}{9} = 56,266,139$$

$$\text{Standard Deviation}(X) = S_X = \sqrt{56,266,139} = 7,501.08$$

$$\text{Variance}(Y) = S_Y^2 = \frac{\Sigma(Y-\bar{Y})^2}{N-1} = \frac{115.0}{9} = 12.78$$

$$\text{Standard Deviation}(Y) = S_Y = \sqrt{12.78} = 3.57$$

$$\text{Covariance}(X,Y) = S_{XY} = \frac{\Sigma(X-\bar{X})(Y-\bar{Y})}{N-1} = \frac{220,301.66}{9} = 24,477.96$$

a. Answers may differ because of rounding; however, the exact values of these column totals, properly calculated, will always be equal to zero.

Now we can plot the straight-line graph corresponding to the regression equation. To plot a straight line, we need only two points, where each point corresponds to an X, Y value predicted by the equation. We can use the two points we just obtained: (1) X = $48,000, $\hat{Y}$ = 26.61% and (2) X = $56,000, $\hat{Y}$ = 29.81% . In Figure 11.6, the regression line is plotted over the scatter diagram we first displayed in Figure 11.1.

> *Use the prediction equation to calculate the predicted values of Y for New York, Georgia, and Ohio. Verify that the regression line in Figure 11.6 passes through these points.*

✓ *Learning Check*

Interpreting *a* and *b*

Now let's interpret the coefficients *a* and *b* in our equation. The *b* coefficient is equal to 0.0004%. This tells us that the percentage of state residents with bachelor's degrees will increase by 0.0004% for every $1 increment in their state's median household income. Similarly, an increase of $10,000 in a state's median household income corresponds to a 4% increase in the percentage of state residents with bachelor's degrees.

Figure 11.6 The Best-Fitting Line for Median Household Income and Percentage of State Residents With Bachelor's Degrees

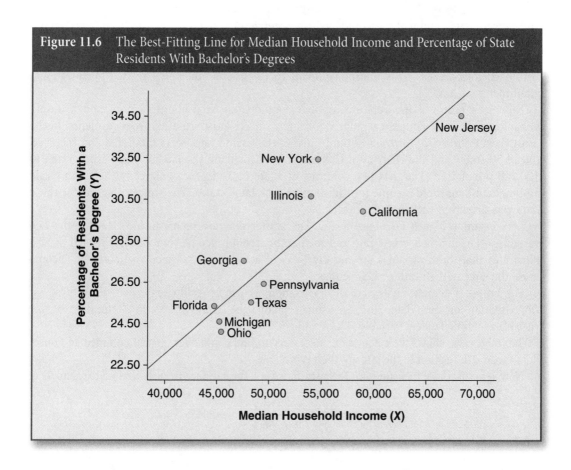

Note that because the relationships between variables in the social sciences are inexact, we don't expect our regression equation to make perfect predictions for every individual case. However, even though the pattern suggested by the regression equation may not hold for every individual state, it gives us a tool by which to make the best possible guess about how a state's median household income is associated, on average, with the percentage of state residents with bachelor's degrees. We can say that the slope of 0.0004% is the estimate of this underlying relationship.

The Y-intercept a is the predicted value of Y when $X = 0$. Thus, it is the point at which the regression line and the Y-axis intersect. With $a = 7.41$, we would predict that very few residents (7.41%) of a state with a median household income equal to zero would have obtained bachelor's degrees. Note, however, that no state has an income as low as zero. As a general rule, be cautious when making predictions for Y based on values of X that are outside the range of the data. Thus, when the lowest value for X is far above zero, the intercept may not have a clear substantive interpretation.

▣ STATISTICS IN PRACTICE: MEDIAN HOUSEHOLD INCOME AND CRIMINAL BEHAVIOR

In our ongoing example, we have looked at the association between median household income and educational attainment measured by the percentage of state residents with bachelor's degrees. The regression equation we have estimated from the data collected by the Census Bureau in 10 states shows that as a state's median household income rises, so does its percentage of residents with bachelor's degrees. This finding confirms that household income is related to educational attainment.

Now let's examine the relationship between median household income and criminal behavior. The Census Bureau regularly collects a variety of information from residents. In the most recent *Statistical Abstract of the United States*, crime rates were tabulated and reported for all 50 states and the District of Columbia. Let's focus on the larceny/theft crime rate per 100,000 population. The first two columns of Table 11.5 display the larceny/theft crime rate and median household income for the 10 most populated states. The scatter diagram for these data was displayed earlier, in Figure 11.2.

Let's examine Figure 11.2 once again. The scatter diagram seems to indicate that the two variables—larceny/theft crime rate and median household income—are linearly related. It also illustrates that these variables are negatively associated; that is, as median household income rises, the larceny/theft crime rate declines.

For a more systematic analysis of the association, we need to estimate the least squares regression equation for these data. Since we want to predict the larceny/theft crime rate, we treat this variable as our dependent variable (Y).

Table 11.5 also shows the calculations necessary to find a and b for our data on median household income in relation to the larceny/theft crime rate.

Now, let's substitute the values for the covariance and the variance from Table 11.5 to calculate b:

$$b = \frac{S_{XY}}{S_X^2} = \frac{-2,154,753}{56,266,139} = -0.04$$

Table 11.5   Median Household Income and the Larceny/Theft Crime Rate for 10 States

	(1) Median Household Income	(2) Larceny/ Theft Crime Rate	(3)	(4)	(5)	(6)	(7)
State	X	Y	$(X-\bar{X})$	$(X-\bar{X})^2$	$(Y-\bar{Y})$	$(Y-\bar{Y})^2$	$(X-\bar{X})(Y-\bar{Y})$
California	58,931	1,663	7,265.7	52,790,396	−348.7	121,592	−2,533,550
Texas	48,259	2,741	−3,406.3	11,602,880	729.3	531,878	−2,484,215
New York	54,659	1,494	2,993.7	8,962,240	−517.7	268,013	−1,549,838
Florida	44,736	2,589	−6,929.3	48,015,198	577.3	333,275	−4,000,285
Illinois	53,966	2,188	2,300.7	5,293,220	176.3	31,082	405,613
Pennsylvania	49,520	1,636	−2,145.3	4,602,312	−375.7	141,150	805,989
Ohio	45,395	2,173	−6,270.3	39,316,662	161.3	26,018	−1,011,399
Michigan	45,255	1,791	−6,410.3	41,091,946	−220.7	48,708	1,414,753
Georgia	47,590	2,369	−4,075.3	16,608,070	357.3	127,663	−1,456,105
New Jersey	68,342	1,473	16,676.7	278,112,323	−538.7	290,198	−8,983,738
	$\Sigma X = 516,653$	$\Sigma Y = 20,117$	0^a	506,395,248	0^a	1,919,578	−19,392,774

$$\text{Mean } X = \bar{X} = \frac{\Sigma X}{N} = \frac{516,653}{10} = 51,665.3$$

$$\text{Mean } Y = \bar{Y} = \frac{\Sigma Y}{N} = \frac{20,117}{10} = 2,011.7$$

$$\text{Variance}(X) = S_X^2 = \frac{\Sigma(X-\bar{X})^2}{N-1} = \frac{506,395,248}{9} = 56,266,139$$

$$\text{Standard Deviation}(X) = S_X = \sqrt{56,266,139} = 7,501.08$$

$$\text{Variance}(Y) = S_Y^2 = \frac{\Sigma(Y-\bar{Y})^2}{N-1} = \frac{1,919,578}{9} = 213,286$$

$$\text{Standard Deviation}(Y) = S_Y = \sqrt{213,286} = 462$$

$$\text{Covariance}(X,Y) = S_{XY} = \frac{\Sigma(X-\bar{X})(Y-\bar{Y})}{N-1} = \frac{-19,392,774}{9} = -2,154,753$$

a. Answers may differ because of rounding; however, the exact values of these column totals, properly calculated, will always be equal to zero.

Once b has been calculated, finding a, the intercept, is simple:

$$a = \overline{Y} - b(\overline{X}) = 2,011.7 - (-0.04)(51,665.3) = 4,078$$

The prediction equation is therefore

$$\hat{Y} = 4,078 + (-0.04)X$$

This equation can be used to obtain a predicted value for a state's larceny/theft crime rate given the state's median household income.

▣ METHODS FOR ASSESSING THE ACCURACY OF PREDICTIONS

So far, we have developed two regression equations that are helping us make state-level predictions about educational attainment and criminal behavior. But in both cases, our predictions are far from perfect. If we examine Figures 11.6 and 11.7, we can see that we fail to make accurate predictions in every case. Though some of the states lie pretty close to the regression line, hardly any lie directly on the line—an indication that some error of prediction was made.

We saw earlier that one way to judge the accuracy of the predictions is to "eyeball" the scatterplot. The closer the observations are to the regression line, the better the "fit" between the predictions and the actual observations. Still, we want a more systematic method for making such a judgment. We need a measure that tells us how accurate a prediction the regression model provides. The *coefficient of determination*, or r^2, is such a measure. It tells us how well the bivariate regression model fits the data. Both r^2 and r measure the strength of the association between two interval-ratio variables.

The **coefficient of determination (r^2)** reflects the proportion of the total variation in the dependent variable, Y, explained by the independent variable, X. An r^2 of 0.83 means that by using median household income and the linear prediction rule to predict Y—the percentage of state residents with bachelor's degrees—we have reduced the error of prediction by 83% (0.83×100). We can also say that the independent variable (median household income) explains about 83% of the variation in the dependent variable (the percentage of state residents with a bachelor's degree), as illustrated in Figure 11.8.

Coefficient of determination (r^2)   A PRE measure reflecting the proportional reduction of error that results from using the linear regression model. It reflects the proportion of the total variation in the dependent variable, Y, explained by the independent variable, X.

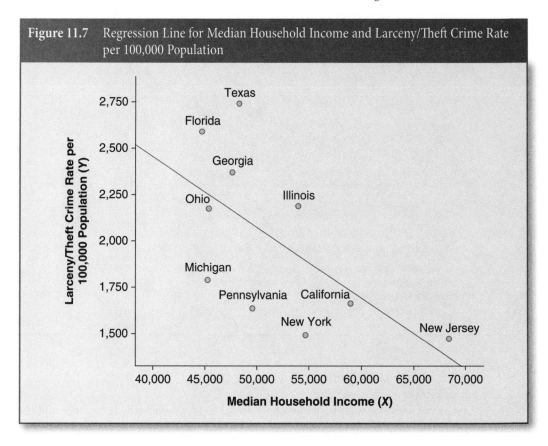

Figure 11.7   Regression Line for Median Household Income and Larceny/Theft Crime Rate per 100,000 Population

The coefficient of determination ranges from 0.0 to 1.0. An r^2 of 1.0 means that by using the linear regression model, we have reduced uncertainty by 100%. It also means that the independent variable accounts for 100% of the variation in the dependent variable. With an r^2 of 1.0, all the observations fall along the regression line, and the prediction error is equal to 0.0. An r^2 of 0.0 means that using the regression equation to predict Y does not improve the prediction of Y. Figure 11.9 shows r^2 values near 0.0 and near 1.0. In Figure 11.9a, where r^2 is approximately 1.0, the regression model provides a good fit. In contrast, a very poor fit is evident in Figure 11.9b, where r^2 is near zero. An r^2 near zero indicates either poor fit or a well-fitting line with a b of zero.

Calculating r^2

Another method for calculating r^2 uses the following equation:

$$r^2 = \frac{\left[Covariance(X,Y)^2 \right]}{\left[\text{Variance}(X) \right]\left[\text{Variance}(Y) \right]} = \frac{S_{YX}^2}{S_X^2 \, S_Y^2} \tag{11.5}$$

Figure 11.8 A Pie Graph Approach to r^2

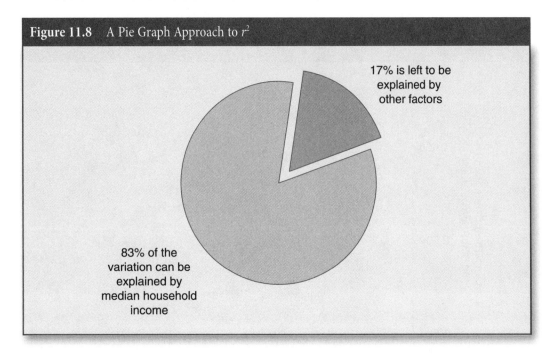

17% is left to be explained by other factors

83% of the variation can be explained by median household income

This formula tells us to divide the square of the covariance of X and Y by the product of the variance of X and the variance of Y.

To calculate r^2 for our example, we can go back to Table 11.4, where the covariance and the variances for the two variables have already been calculated:

$$S_{XY} = 24{,}477.96$$

$$S_X^2 = 56{,}266{,}139$$

$$S_Y^2 = 12.78$$

Therefore,

$$r^2 = \frac{(24{,}477.96)^2}{(56{,}266{,}139)(12.78)} = \frac{599{,}170{,}526}{719{,}081{,}256} = 0.83$$

Since we are working with actual values for median household income, its metric, or measurement, values are different from the metric values for the dependent variable, the percentage of state residents with bachelor's degrees. We must account for this measurement difference if we elect to use the variances and covariance to calculate r^2 (Formula 11.5). The remedy is actually quite simple.

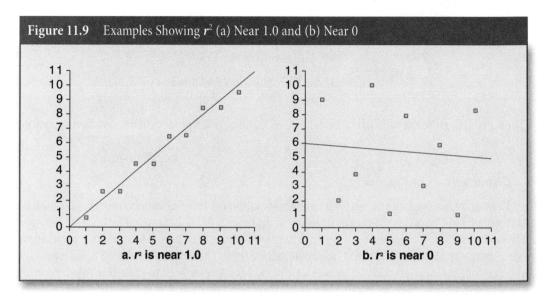

Figure 11.9 Examples Showing r^2 (a) Near 1.0 and (b) Near 0

a. r^2 is near 1.0

b. r^2 is near 0

All we have to do is multiply our obtained r^2, 0.83, by 100 to obtain 83. You might be wondering, "Why multiply the obtained r^2 by 100?" The answer is that we multiply by 100 because the dependent variable, percentage of residents with a bachelor's degree, is measured as a percentage ranging from 1 to 100.

We can multiply r^2 by 100 to obtain the percentage of variation in the dependent variable explained by the independent variable. An r^2 of 0.83 means that by using median household income and the linear prediction rule to predict Y, the percentage of state residents with bachelor's degrees, we have reduced uncertainty of prediction by 83% (0.83×100). We can also say that the independent variable (median household income) explains 83% of the variation in the dependent variable (the percentage of state residents with a bachelor's degree), as illustrated in Figure 11.8.

Pearson's Correlation Coefficient (*r*)

In the social sciences, it is the square root of r^2, or r—known as **Pearson's correlation coefficient**—that is most often used as a measure of association between two interval-ratio variables:

$$r = \sqrt{r^2}$$

Pearson's correlation coefficient (r)   The square root of r^2; it is a measure of association for interval-ratio variables, reflecting the strength of the linear association between two interval-ratio variables. It can be positive or negative in sign.

Pearson's r is usually computed directly by using the following definitional formula:

$$r = \frac{\left[Covariance(X,Y)\right]}{\left[Standard\,deviation(X)\right]\left[Standard\,deviation(Y)\right]} = \frac{S_{YX}}{S_X S_Y}$$

Thus, r is defined as the ratio of the covariance of X and Y to the product of the standard deviations of X and Y.

Characteristics of Pearson's r

Pearson's r is a measure of relationship or association for interval-ratio variables. Like gamma (introduced in Chapter 9), it ranges from 0.0 to ± 1.0, with 0.0 indicating no association between the two variables. An r of +1.0 means that the two variables have a perfect positive association; −1.0 indicates that it is a perfect negative association. The absolute value of r indicates the strength of the linear association between two variables. (Refer back to the guide to interpreting the strength of this relationship on page 242.) Thus, a correlation of −0.75 demonstrates a stronger association than a correlation of 0.50. Figure 11.10 illustrates a strong positive relationship, a strong negative relationship, a moderate positive relationship, and a weak negative relationship.

Unlike the b coefficient, r is a symmetrical measure. That is, the correlation between X and Y is identical to the correlation between Y and X. In contrast, b may be different when the variables are switched—for example, when we use Y as the independent variable rather than as the dependent variable.

To calculate r for our example of the relationship between median household income and the percentage of state residents with bachelor's degrees, let's return to Table 11.4, where the covariance and the standard deviations for X and Y have already been calculated:

$$r = \frac{S_{XY}}{S_X S_Y} = \frac{24,477.96}{(7,501.08)(3.57)} = \frac{24,477.96}{26,778.86} = 0.91$$

A correlation coefficient of 0.91 indicates that there is a strong positive linear relationship between median household income and the percentage of state residents with a bachelor's degree.

Note that we could have just taken the square root of r^2 to calculate r, because $r = \sqrt{r^2}$ or $\sqrt{0.83} = 0.91$. Similarly, if we first calculate r, we can obtain r^2 simply by squaring r (be careful not to lose the sign of r).

▣ STATISTICS IN PRACTICE: TEEN PREGNANCY AND SOCIAL INEQUALITY

The United States has by far the highest rate of teenage pregnancy of any industrialized nation. The pregnancy rate for U.S. teens aged between 15 and 19 years was 95.9 pregnancies per 1,000 women in 1990. Although in 2010, the teen pregnancy rate reached its lowest point in more than 70 years

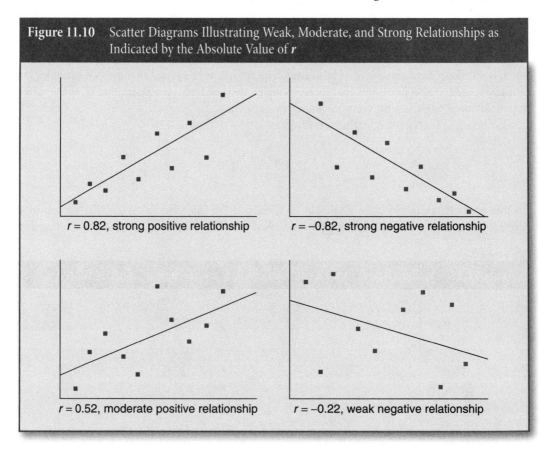

Figure 11.10   Scatter Diagrams Illustrating Weak, Moderate, and Strong Relationships as Indicated by the Absolute Value of *r*

$r = 0.82$, strong positive relationship

$r = -0.82$, strong negative relationship

$r = 0.52$, moderate positive relationship

$r = -0.22$, weak negative relationship

(34.2 per 1,000 women aged 15–19),[2] this rate is still higher than that of any other industrialized nation. These high rates have been attributed, among other factors, to the high rate of poverty and inequality in the United States.

The association between teen pregnancy and poverty and social inequality has been well documented both nationally and internationally. Teen pregnancy rates are higher among people living in poverty, and industrial societies that have done the most to reduce social inequality tend to have the lowest rates of teen pregnancy.[3] The noted sociologist William Wilson has claimed that the disappearance of hundreds of low-skilled jobs in the past 25 years and the resulting increase in unemployment, especially in the inner cities, has led to the increase in teenage pregnancy rates and to welfare dependency.[4] Teenagers living in areas of high unemployment, poverty, and lack of opportunities are six to seven times more likely to become unwed parents.[5]

To examine the degree to which economic factors influence teenage pregnancy rates, we analyze state-by-state data on unemployment rates in 2009 and teenage pregnancy rates in 2010. Using unemployment rate and pregnancy rate, both interval-ratio variables, we can examine the hypothesis that states with higher unemployment rates will tend to have higher teenage pregnancy rates.

Figure 11.11 shows the scatter diagram for unemployment rate and teenage pregnancy rate. Because we are assuming that the unemployment rate in 2009 can predict the teen pregnancy rate in 2010, we will treat unemployment rate as our independent variable, X. Teen pregnancy rate, then, is our dependent variable, Y. The scatter diagram seems to suggest that the two variables are linearly related. It also illustrates that these variables are positively associated; that is, as the state's unemployment rate rises, the teen pregnancy rate rises as well.

Our bivariate regression equation for 40 states[6] is

$$\hat{Y} = 17.312 + 1.567(X)$$

In this prediction equation, the **slope (b)** is 1.567 and the intercept (a) is 17.312. The positive slope, 1.567, confirms our earlier impression, based on the scatter diagram, that the relationship

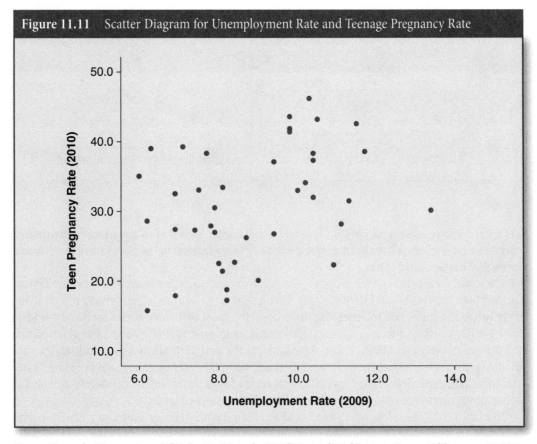

Figure 11.11   Scatter Diagram for Unemployment Rate and Teenage Pregnancy Rate

Sources: Center for Disease Control, "Births: Final Data for 2010," *National Vital Statistics Report*: 61(1), August 28, 2012. Bureau of Labor Statistics, "Unemployment Rates for States Annual Average Rankings Year: 2009"

between the unemployment rate and teenage pregnancy rate is positive. In other words, the higher the unemployment rate, the higher the pregnancy rate. A *b* equal to 1.567 means that for every 1 percentage point increase in the unemployment rate, the pregnancy rate for teens aged 15 to 19 years will increase by 1.567 pregnancies per 1,000 women. The intercept, *a*, of 17.312 indicates that with full employment (an unemployment rate of 0), the teen pregnancy rate will be 17.312 pregnancies per 1,000 women. The regression line corresponding to this linear regression equation is shown in Figure 11.12.

Slope (b)   The amount of change in a dependent variable per unit change in an independent variable.

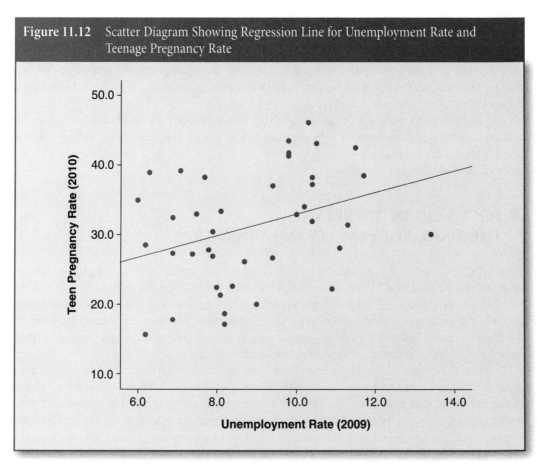

Figure 11.12   Scatter Diagram Showing Regression Line for Unemployment Rate and Teenage Pregnancy Rate

Sources: Center for Disease Control, "Births: Final Data for 2010," *National Vital Statistics Report:* 61(1), August 28, 2012. Bureau of Labor Statistics, "Unemployment Rates for States Annual Average Rankings Year: 2009."

Based on the linear regression equation, we could predict the teenage pregnancy rate for any state based on its unemployment rate in 2009. For example, with a 2009 unemployment rate of 9.8%, Alabama's predicted 2010 teen pregnancy rate is

$$\hat{Y} = 17.312 + 1.567(9.8) = 32.67$$

With a higher unemployment rate of 13.4%, the predicted 2010 teen pregnancy rate for Michigan is

$$\hat{Y} = 17.312 + 1.567(13.4) = 38.31$$

We also calculated r and r^2 for these data. We obtained an r of 0.342 and an r^2 of $0.342^2 = 0.117$. An r of 0.342 indicates that there is a moderate positive relationship between the unemployment rate in 2009 and the teen pregnancy rate in 2010.

The coefficient of determination, r^2, measures the proportional reduction of error that results from using the linear regression model to predict teen pregnancy rates. An r^2 of 0.117 means that by using the regression equation, our prediction of teen pregnancy rates is improved by 11.7% (0.117×100) over the prediction we would make using the mean pregnancy rate alone. We can also say that the independent variable (unemployment rate) explains 11.7% of the variation in the dependent variable (teen pregnancy rate).

This analysis deals with only one factor affecting teen pregnancy rates. Other important socio-economic indicators likely to affect teen pregnancy rates—such as poverty rates, welfare policies, and expenditures on education—would also need to be considered for a complete analysis of the determinants of teenage pregnancy.

回 FOCUS ON INTERPRETATION: THE MARRIAGE PENALTY IN EARNINGS

Among factors commonly associated with earnings are human capital variables (e.g., age, education, work experience, and health) and labor market variables (such as the unemployment rate and the structure of occupations). Individual characteristics, such as gender, race, and ethnicity, also explain disparities in earnings. In addition, marital status has been linked to differences in earnings. Although marriage was associated with higher earnings for men, for women it carried a penalty; married women tended to earn less at every educational level than single women.

The lower earnings of married women have been related to differences in labor force experience. Getting married and being the mothers of young children tended to limit women's choice of jobs to those that may offer flexible working hours but are generally low paying and offer fewer opportunities for promotion. Moreover, married women tended to be out of the labor market longer and to have fewer years on the job than single women. When they reentered the job market or began their careers after their children were grown, they competed with coworkers with considerably more work experience and on-the-job training. (Women who need to become financially independent after divorce or widowhood may share some of the same liabilities as married women.)

Past studies have shown that the returns for formal education were generally lower for married women and that single women earned more for each year of formal education than married women.

We explore this issue by analyzing the bivariate relationship between level of education and personal income among 307 single and married women (working full-time) included in the 2010 General Social Survey (GSS).[7] We are assuming that level of education (measured in years) can predict personal income, and therefore, we treat education as our independent variable, X. Personal income (measured in dollars), then, is the dependent variable, Y. Since both are interval-ratio variables, we can use bivariate regression analysis to examine the difference in returns for education.

We will rely on SPSS to calculate the bivariate regression equations for both single and married women working full-time. We will not present the calculations of a, b, and r^2, because we want to focus on interpreting the SPSS output. The output for single women includes two tables. The output for married women is not shown here.

The regression equation coefficients are listed in the Coefficients table, under the column labeled "B" (see Figure 11.13). The coefficient for EDUC, or b, is 2,637.41; the intercept term, or a, identified in the "(Constant)" row, is −11,810.19. Using these terms, we can express the bivariate relationship between education and income for single women working full-time as

$$\hat{Y}(\text{single}) = -11,810.19 + 2,637.41(X)$$

The regression equation tells us that for every one-year increase in education we can predict an increase of $2,637.41 in the annual income of single women in our sample who work full-time.

Figure 11.13   SPSS Regression Output for Single Women

Model Summary[a]

Model	R	R Square	Adjusted R Square	Std. Error of the Estimate
1	.436[b]	.190	.176	15391.651

a. MARITAL STATUS = NEVER MARRIED

b. Predictors: (Constant), HIGHEST YEAR OF SCHOOL COMPLETED

Coefficients[a,b]

Model		Unstandardized Coefficients		Standardized Coefficients	t	Sig.
		B	Std. Error	Beta		
1	(Constant)	-11810.194	10835.606		-1.090	.280
	HIGHEST YEAR OF SCHOOL COMPLETED	2637.410	714.091	.436	3.693	.000

a. MARITAL STATUS = NEVER MARRIED

b. Dependent Variable: RS INCOME IN CONSTANT $

The bivariate regression equation for married women working full-time is (output not shown here) is:

$$\hat{Y}(\text{married}) = -32,946.04 + 4,155.15(X)$$

This analysis shows that in fact it is married women who benefit more from the effect of education on their income. For single women one year of education is worth \$2,637.41; for married women it is worth \$4,155.15. These results are puzzling and contradict literature that has suggested that married women were placed at a disadvantage when compared with single women because of their prolonged absence from the labor market due to child-rearing responsibilities. However, with evolving family patterns, many more married women remain in the labor force even when their children are very young. This may account for the larger return for education for married women.

One should be extremely cautious when interpreting these data. In all likelihood, the difference in earnings between single and married women is due to numerous other factors, such as occupation, family size, ethnicity, and race. To test some of this idea, we would need to use more advanced statistical techniques not covered in this text.

SPSS also calculates r^2 for these data. These results are presented in the table titled "Model Summary." The coefficient of determination (r^2) is labeled "R square." For single women, r^2 is 0.190; for married women, it is 0.275 (not shown). These coefficients indicate that for both groups there is a weak to moderate relationship between education and earnings (the relationship is substantially stronger for married women).

Using the regression equation, our prediction of income for single women is improved by 19.0% (0.190×100) over the prediction we would make using the mean alone. For married women, there is more of an improvement in prediction, 27.5% (0.275×100).

We can use these regression equations to predict the difference in annual income between a single woman and a married woman, both with 16 years of education and working full-time:

$$\hat{Y}(\text{single}) = -11,810.19 + 2,637.41(16) = 30,388.37$$

$$\hat{Y}(\text{married}) = -32,946.04 + 4,155.15(16) = 33,536.36$$

This analysis deals with only one factor affecting earnings—the level of education. Other important factors associated with earnings—including occupation, seniority, race/ethnicity, and age—would need to be considered for a complete analysis of the differences in earnings between single and married women.

MAIN POINTS

- A scatter diagram (also called scatter-plot) is a quick visual method used to display relationships between two interval-ratio variables.

- Equations for all straight lines have the same general form:

$$\hat{Y} = a + b(X)$$

- The best-fitting regression line is that line where the residual sum of squares, or Σe^2, is at a minimum. Such a line is called the least squares line, and the technique that produces this line is called the least squares method.

- The coefficient of determination (r^2) and Pearson's correlation coefficient (r) measure how well the regression model fits the data. Pearson's r also measures the strength of the association between the two variables.

KEY TERMS

coefficient of
  determination (r^2)
deterministic
  (perfect)
  linear
  relationship

least squares line
  (best-fitting line)
least squares method
linear relationship
Pearson's correlation
  coefficient (r)

scatter diagram
  (scatterplot)
slope (b)
Y-intercept (a)

⑤SAGE edge™

Sharpen your skills with SAGE edge at **edge.sagepub.com/ssdsess2e**. **SAGE edge for students** provides a personalized approach to help you accomplish your coursework goals in an easy-to-use learning environment.

CHAPTER EXERCISES

1. Based on the following eight countries, examine the data to determine the extent of the relationship between simply being concerned about the environment and actually giving money to environmental groups.

Country	Percentage Concerned	Percentage Donating Money
Austria	35.5	27.8
Denmark	27.2	22.3
Netherlands	30.1	44.8
Philippines	50.1	6.8
Russia	29.0	1.6
Slovenia	50.3	10.7
Spain	35.9	7.4
United States	33.8	22.8

Source: International Social Survey Programme, 2000.

Exercises

    a. Construct a scatterplot of the two variables, placing percentage concerned about the environment on the horizontal or X-axis and the percentage donating money to environmental groups on the vertical or Y-axis.

    b. Does the relationship between the two variables seem linear? Describe the relationship.

    c. Find the value of the Pearson correlation coefficient that measures the association between the two variables and offer an interpretation.

2. There is often thought to be a relationship between a person's educational attainment and the number of children he or she has. The hypothesis is that as one's educational level increases, he or she has fewer children. Investigate this conjecture with the following scatterplot and regression output produced from the 2010 GSS and interpret the results (see Figures 11.14 and 11.15).

3. The condition and health of our environment is a growing concern. Let's examine the relationship between a country's gross national product (GNP) and the percentage of respondents willing to pay higher prices for goods to protect the environment. The following table displays information for five countries selected at random.

Country	GNP per Capita	Percentage Willing to Pay
United States	29.24	44.9
Ireland	18.71	53.3
Netherlands	24.78	61.2
Norway	34.31	40.7
Sweden	25.58	32.6

Source: International Social Survey Programme, 2000.

    a. Calculate the correlation coefficient between a country's GNP and the percentage of its residents willing to pay higher prices to protect the environment. What is its value?

    b. Provide an interpretation for the coefficient.

4. The SPSS output shown in Figure 11.16 displays the relationship between education (measured in years) and television viewing per day (measured in hours) based on 2010 GSS data. We can hypothesize that as educational attainment increases, hours of television viewing will decrease, indicating a negative relationship between the two variables. Discuss the significance of the overall model based on F and its P values. Is the relationship between education and television viewing significant?

5. Before calculating a correlation coefficient or a regression equation, it is always important to examine a scatter diagram between two variables to see how well a straight line fits the data. If a straight line does not appear to fit, other curves can be used to describe the relationship (this subject is not discussed in our text).

The SPSS scatterplot in Figure 11.17 and output shown in Figure 11.18 display the relationship between education (measured in years) and television viewing (measured in hours) based on 2010 GSS data. We can hypothesize that as educational attainment increases, hours of television viewing will decrease, indicating a negative relationship between the two variables.

Interpret the results of the scatterplot, as well as the "Coefficients" and "Model Summary" output.

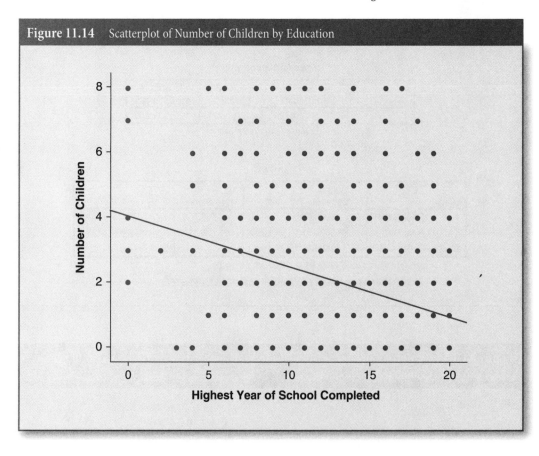

Figure 11.14　Scatterplot of Number of Children by Education

Figure 11.15　Linear Regression Output Specifying the Relationship Between Education and Number of Children

Coefficients[a]

Model		Unstandardized Coefficients		Standardized Coefficients	t	Sig.
		B	Std. Error	Beta		
1	(Constant)	4.060	.191		21.220	.000
	HIGHEST YEAR OF SCHOOL COMPLETED	-.155	.014	-.279	-11.203	.000

a. Dependent Variable: NUMBER OF CHILDREN

Model Summary

Model	R	R Square	Adjusted R Square	Std. Error of the Estimate
1	.279[a]	.078	.077	1.667

a. Predictors: (Constant), HIGHEST YEAR OF SCHOOL COMPLETED

Figure 11.16 ANOVA Output for Education and Television Viewing

Model Summary

Model	R	R Square	Adjusted R Square	Std. Error of the Estimate
1	.268[a]	.072	.071	2.553

a. Predictors: (Constant), HIGHEST YEAR OF SCHOOL COMPLETED

ANOVA[a]

Model		Sum of Squares	df	Mean Square	F	Sig.
1	Regression	506.785	1	506.785	77.757	.000[b]
	Residual	6563.151	1007	6.518		
	Total	7069.937	1008			

a. Dependent Variable: HOURS PER DAY WATCHING TV

b. Predictors: (Constant), HIGHEST YEAR OF SCHOOL COMPLETED

Figure 11.17 Scatterplot of Hours of Television Viewing per Day by Highest Year of School Completed

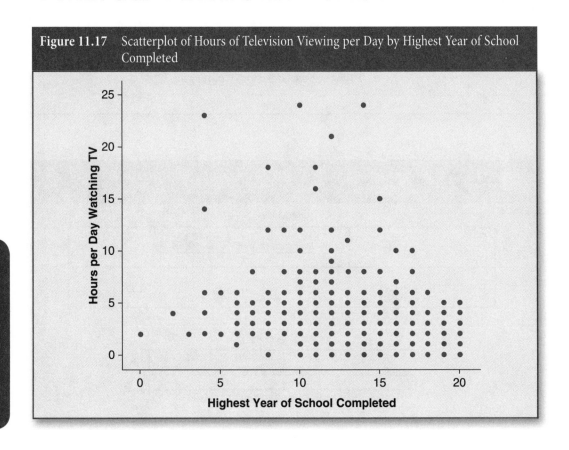

Figure 11.18 Linear Regression Output Specifying the Relationship Between Education and Hours Spent per Day Watching Television

Coefficients[a]

Model		Unstandardized Coefficients		Standardized Coefficients	t	Sig.
		B	Std. Error	Beta		
1	(Constant)	6.130	.363		16.884	.000
	HIGHEST YEAR OF SCHOOL COMPLETED	-.231	.026	-.268	-8.818	.000

a. Dependent Variable: HOURS PER DAY WATCHING TV

Model Summary

Model	R	R Square	Adjusted R Square	Std. Error of the Estimate
1	.268[a]	.072	.071	2.553

a. Predictors: (Constant), HIGHEST YEAR OF SCHOOL COMPLETED

6. Social scientists have long been interested in the aspirations and achievements of people in the United States. Research on social mobility, status, and educational attainment has provided convincing evidence on the relationship between parents' and children's socioeconomic achievement. The GSS 2010 data set has information on the educational level of respondents and their mothers. Use the scatterplot and regression output below to interpret the relationship between mothers' level of education and one's own level of education (see Figures 11.19 and 11.20).

7. In this exercise, we will investigate the relationship between gross domestic product (GDP) per capita and birthrates in South America.

Country	GDP per Capita in 2008 (U.S. Dollars)	Birthrate in 2010 (Estimated)
Argentina	8,236	18
Bolivia	1,720	25
Brazil	8,205	18
Chile	10,084	15
Colombia	5,416	18
Ecuador	4,056	20
Paraguay	2,561	18
Peru	4,477	19
Uruguay	9,654	14
Venezuela	11,246	13

Sources: World Bank, *World Development Indicators*, 2010; *The World Factbook*, 2010.

Exercises

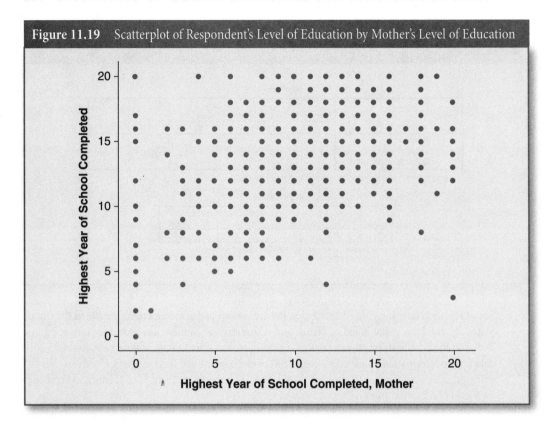

Figure 11.19 Scatterplot of Respondent's Level of Education by Mother's Level of Education

Figure 11.20 Linear Regression Output Specifying the Relationship Between Respondent's Level of Education by Mother's Level of Education

Coefficients[a]

Model		Unstandardized Coefficients		Standardized Coefficients		
		B	Std. Error	Beta	t	Sig.
1	(Constant)	9.988	.262		38.138	.000
	HIGHEST YEAR SCHOOL COMPLETED, MOTHER	.326	.022	.383	15.018	.000

a. Dependent Variable: HIGHEST YEAR OF SCHOOL COMPLETED

Model Summary

Model	R	R Square	Adjusted R Square	Std. Error of the Estimate
1	.383[a]	.146	.146	2.762

a. Predictors: (Constant), HIGHEST YEAR SCHOOL COMPLETED, MOTHER

a. Construct a scatterplot for GDP and birthrate. Do you think the scatterplot can be characterized by a linear relationship?
b. Calculate the coefficient of determination and correlation coefficient.
c. Use this information to describe the relationship between the variables.

8. In 2010, a Census Bureau report revealed that approximately 13% of all Americans were living below the poverty line in 2007. This figure is higher than in 2000, when the poverty rate was 12.2%. This translates into an increase of approximately 4.75 million Americans living below the poverty line. Individuals and families living below the poverty line face many obstacles, not the least of which is access to health care. In many cases, those living below the poverty line are without any form of health insurance. Using data from the Census Bureau, analyze the relationship between living below the poverty line and access to health care.

State	Percentage Below Poverty Line (2007)	Percentage Without Health Insurance (2007)
Alabama	16.9	12.0
California	12.4	18.2
Idaho	12.1	13.9
Louisiana	18.6	18.5
New Jersey	8.6	15.8
New York	13.7	13.2
Pennsylvania	11.6	9.5
Rhode Island	12.0	10.8
South Carolina	15.0	16.4
Texas	16.3	25.2
Washington	11.4	11.3
Wisconsin	10.8	8.2

Source: U.S. Census Bureau, *The 2010 Statistical Abstract*, 2010, Tables 693 and 150.

a. Construct a scatterplot, predicting the percentage without health insurance with the percentage living below the poverty level. Does it appear that a straight-line relationship will fit the data?
b. Calculate the regression equation with percentage of the population without health insurance as the dependent variable, and draw the regression line on the scatterplot. What is its slope? What is the intercept? Has your opinion changed about whether a straight line seems to fit the data? Are there any states that fall far from the regression line? Which one(s)?
c. What percentage of the population must be living below the poverty line to obtain a predicted value of 5% without health insurance?
d. Predicting a value that falls beyond the observed range of the two variables in a regression is problematic at best, so your answer in (c) isn't necessarily statistically believable. However, what is a nonstatistical, or substantive, reason? Why might making such a prediction be important?

Exercises

9. Let's examine the relationship between GNP per capita and the percentage of respondents willing to pay more in taxes.
   a. Calculate *a* and *b* and write out the regression equation (i.e., prediction equation).
   b. About what percentage of citizens are willing to pay higher taxes for a country with a GNP per capita of 3.0 (i.e., $3,000)? For a GNP per capita of 30.0 (i.e., $30,000)?

Country	GNP per Capita	Percentage Willing to Pay Higher Taxes
Canada	19.71	24.0
Chile	4.99	29.1
Finland	24.28	12.0
Ireland	18.71	34.3
Japan	32.35	37.2
Latvia	2.42	17.3
Mexico	3.84	34.7
Netherlands	24.78	51.9
Norway	14.60	22.8
Portugal	34.31	17.1
Russia	10.67	29.9
Spain	2.66	22.2
Sweden	14.10	19.5
Switzerland	25.58	33.5
United States	39.98	31.6

Sources: The World Bank Group, *Development Education Program; Learning Module: Economics, GNP per Capita,* 2004; International Social Survey Programme, 2000.

10. On completing this chapter, you should be able to correctly answer the following questions.
    a. True or false: It is possible, in fact it often is the case, that your slope, *b*, will be a positive value and your correlation coefficient, *r*, will be a negative value.
    b. Both *a* and *b* refer to changes in which variable, the independent or dependent?
    c. The coefficient of determination, r^2, is a PRE measure. What does this mean?
    d. True or false: All regression equations reflect *causal* relationships expressed as linear functions.

11. Does how many brothers and sisters one grows up with influence how many children he or she wants to have later in life? Researchers interested in this question produced the following SPSS output from the 2010 GSS. Use it to answer the following questions.

Model Summary

Model	R	R Square	Adjusted R Square	Std. Error of the Estimate
1	.221[a]	.049	.048	.915

a. Predictors: (Constant), NUMBER OF BROTHERS AND SISTERS

Coefficients[a]

Model		Unstandardized Coefficients		Standardized Coefficients	t	Sig.
		B	Std. Error	Beta		
1	(Constant)	2.301	.048		48.275	.000
	NUMBER OF BROTHERS AND SISTERS	.066	.010	.221	6.638	.000

a. Dependent Variable: IDEAL NUMBER OF CHILDREN

a. Report Pearson's correlation coefficient and use it to determine the strength of the relationship between number of siblings and one's ideal number of children.

b. Report and interpret the Y-intercept and the slope.

12. The following SPSS output from the 2010 GSS examines the effect of education on the amount of time respondents spend online per week. To determine whether there are differences between the sexes, separate output is provided for men and women. Use the output to answer the following questions.

Model Summary[a]

Model	R	R Square	Adjusted R Square	Std. Error of the Estimate
1	.219[b]	.048	.041	10.906

a. RESPONDENTS SEX = MALE

b. Predictors: (Constant), HIGHEST YEAR OF SCHOOL COMPLETED

Coefficients[a,b]

Model		Unstandardized Coefficients		Standardized Coefficients	t	Sig.
		B	Std. Error	Beta		
1	(Constant)	-4.301	5.377		-.800	.425
	HIGHEST YEAR OF SCHOOL COMPLETED	.978	.368	.219	2.661	.009

a. RESPONDENTS SEX = MALE

b. Dependent Variable: WWW HOURS PER WEEK

Model Summary[a]

Model	R	R Square	Adjusted R Square	Std. Error of the Estimate
1	.055[b]	.003	-.003	12.136

a. RESPONDENTS SEX = FEMALE

b. Predictors: (Constant), HIGHEST YEAR OF SCHOOL COMPLETED

Coefficients[a,b]

Model		Unstandardized Coefficients		Standardized Coefficients	t	Sig.
		B	Std. Error	Beta		
1	(Constant)	5.469	4.830		1.132	.259
	HIGHEST YEAR OF SCHOOL COMPLETED	.248	.337	.055	.737	.462

a. RESPONDENTS SEX = FEMALE

b. Dependent Variable: WWW HOURS PER WEEK

a. For which sex does education explain more of the variation in Internet habits? Cite and interpret the statistics you used to reach this conclusion.

b. Interpret and compare the Y-intercept and slope for each group. (Remember that it is problematic to read too much into a Y-intercept that extends beyond your observed cases of X.) Provide a brief summary statement on differences between the groups.

c. How many hours would we expect college graduates (16 years of education) to spend online weekly? Solve for men and women.

APPENDIX A
THE STANDARD NORMAL TABLE

The values in column A are Z scores. Column B lists the proportion of area between the mean and a given Z. Column C lists the proportion of area beyond a given Z. Only positive Z scores are listed. Because the normal curve is symmetrical, the areas for negative Z scores will be exactly the same as the areas for positive Z scores.

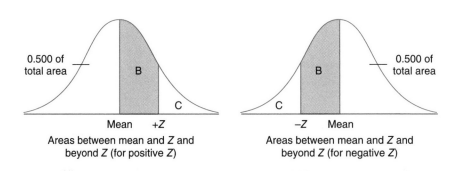

0.500 of total area — B — C — Mean — +Z
Areas between mean and Z and beyond Z (for positive Z)

C — −Z — Mean — B — 0.500 of total area
Areas between mean and Z and beyond Z (for negative Z)

A	B	C	A	B	C	A	B	C
	Area Between	Area Beyond		Area Between	Area Beyond		Area Between	Area Beyond
Z	Mean and Z	Z	Z	Mean and Z	Z	Z	Mean and Z	Z
0.00	0.0000	0.5000	0.11	0.0438	0.4562	0.21	0.0832	0.4168
0.01	0.0040	0.4960	0.12	0.0478	0.4522	0.22	0.0871	0.4129
0.02	0.0080	0.4920	0.13	0.0517	0.4483	0.23	0.0910	0.4090
0.03	0.0120	0.4880	0.14	0.0557	0.4443	0.24	0.0948	0.4052
0.04	0.0160	0.4840	0.15	0.0596	0.4404	0.25	0.0987	0.4013
0.05	0.0199	0.4801	0.16	0.0636	0.4364	0.26	0.1026	0.3974
0.06	0.0239	0.4761	0.17	0.0675	0.4325	0.27	0.1064	0.3936
0.07	0.0279	0.4721	0.18	0.0714	0.4286	0.28	0.1103	0.3897
0.08	0.0319	0.4681	0.19	0.0753	0.4247	0.29	0.1141	0.3859
0.09	0.0359	0.4641	0.20	0.0793	0.4207	0.30	0.1179	0.3821
0.10	0.0398	0.4602						

(Continued)

(Continued)

A Z	B Area Between Mean and Z	C Area Beyond Z	A Z	B Area Between Mean and Z	C Area Beyond Z	A Z	B Area Between Mean and Z	C Area Beyond Z
0.31	0.1217	0.3783	0.71	0.2611	0.2389	1.11	0.3665	0.1335
0.32	0.1255	0.3745	0.72	0.2642	0.2358	1.12	0.3686	0.1314
0.33	0.1293	0.3707	0.73	0.2673	0.2327	1.13	0.3708	0.1292
0.34	0.1331	0.3669	0.74	0.2703	0.2297	1.14	0.3729	0.1271
0.35	0.1368	0.3632	0.75	0.2734	0.2266	1.15	0.3749	0.1251
0.36	0.1406	0.3594	0.76	0.2764	0.2236	1.16	0.3770	0.1230
0.37	0.1443	0.3557	0.77	0.2794	0.2206	1.17	0.3790	0.1210
0.38	0.1480	0.3520	0.78	0.2823	0.2177	1.18	0.3810	0.1190
0.39	0.1517	0.3483	0.79	0.2852	0.2148	1.19	0.3830	0.1170
0.40	0.1554	0.3446	0.80	0.2881	0.2119	1.20	0.3849	0.1151
0.41	0.1591	0.3409	0.81	0.2910	0.2090	1.21	0.3869	0.1131
0.42	0.1628	0.3372	0.82	0.2939	0.2061	1.22	0.3888	0.1112
0.43	0.1664	0.3336	0.83	0.2967	0.2033	1.23	0.3907	0.1093
0.44	0.1700	0.3300	0.84	0.2995	0.2005	1.24	0.3925	0.1075
0.45	0.1736	0.3264	0.85	0.3023	0.1977	1.25	0.3944	0.1056
0.46	0.1772	0.3228	0.86	0.3051	0.1949	1.26	0.3962	0.1038
0.47	0.1808	0.3192	0.87	0.3078	0.1922	1.27	0.3980	0.1020
0.48	0.1844	0.3156	0.88	0.3106	0.1894	1.28	0.3997	0.1003
0.49	0.1879	0.3121	0.89	0.3133	0.1867	1.29	0.4015	0.0985
0.50	0.1915	0.3085	0.90	0.3159	0.1841	1.30	0.4032	0.0968
0.51	0.1950	0.3050	0.91	0.3186	0.1814	1.31	0.4049	0.0951
0.52	0.1985	0.3015	0.92	0.3212	0.1788	1.32	0.4066	0.0934
0.53	0.2019	0.2981	0.93	0.3238	0.1762	1.33	0.4082	0.0918
0.54	0.2054	0.2946	0.94	0.3264	0.1736	1.34	0.4099	0.0901
0.55	0.2088	0.2912	0.95	0.3289	0.1711	1.35	0.4115	0.0885
0.56	0.2123	0.2877	0.96	0.3315	0.1685	1.36	0.4131	0.0869
0.57	0.2157	0.2843	0.97	0.3340	0.1660	1.37	0.4147	0.0853
0.58	0.2190	0.2810	0.98	0.3365	0.1635	1.38	0.4162	0.0838
0.59	0.2224	0.2776	0.99	0.3389	0.1611	1.39	0.4177	0.0823
0.60	0.2257	0.2743	1.00	0.3413	0.1587	1.40	0.4192	0.0808
0.61	0.2291	0.2709	1.01	0.3438	0.1562	1.41	0.4207	0.0793
0.62	0.2324	0.2676	1.02	0.3461	0.1539	1.42	0.4222	0.0778
0.63	0.2357	0.2643	1.03	0.3485	0.1515	1.43	0.4236	0.0764
0.64	0.2389	0.2611	1.04	0.3508	0.1492	1.44	0.4251	0.0749
0.65	0.2422	0.2578	1.05	0.3531	0.1469	1.45	0.4265	0.0735
0.66	0.2454	0.2546	1.06	0.3554	0.1446	1.46	0.4279	0.0721
0.67	0.2486	0.2514	1.07	0.3577	0.1423	1.47	0.4292	0.0708
0.68	0.2517	0.2483	1.08	0.3599	0.1401	1.48	0.4306	0.0694
0.69	0.2549	0.2451	1.09	0.3621	0.1379	1.49	0.4319	0.0681
0.70	0.2580	0.2420	1.10	0.3643	0.1357	1.50	0.4332	0.0668

A	B	C	A	B	C	A	B	C
	Area	Area		Area	Area		Area	Area
	Between	Beyond		Between	Beyond		Between	Beyond
Z	Mean and Z	Z	Z	Mean and Z	Z	Z	Mean and Z	Z
1.51	0.4345	0.0655	1.91	0.4719	0.0281	2.31	0.4896	0.0104
1.52	0.4357	0.0643	1.92	0.4726	0.0274	2.32	0.4898	0.0102
1.53	0.4370	0.0630	1.93	0.4732	0.0268	2.33	0.4901	0.0099
1.54	0.4382	0.0618	1.94	0.4738	0.0262	2.34	0.4904	0.0096
1.55	0.4394	0.0606	1.95	0.4744	0.0256	2.35	0.4906	0.0094
1.56	0.4406	0.0594	1.96	0.4750	0.0250	2.36	0.4909	0.0091
1.57	0.4418	0.0582	1.97	0.4756	0.0244	2.37	0.4911	0.0089
1.58	0.4429	0.0571	1.98	0.4761	0.0239	2.38	0.4913	0.0087
1.59	0.4441	0.0559	1.99	0.4767	0.0233	2.39	0.4916	0.0084
1.60	0.4452	0.0548	2.00	0.4772	0.0228	2.40	0.4918	0.0082
1.61	0.4463	0.0537	2.01	0.4778	0.0222	2.41	0.4920	0.0080
1.62	0.4474	0.0526	2.02	0.4783	0.0217	2.42	0.4922	0.0078
1.63	0.4484	0.0516	2.03	0.4788	0.0212	2.43	0.4925	0.0075
1.64	0.4495	0.0505	2.04	0.4793	0.0207	2.44	0.4927	0.0073
1.65	0.4505	0.0495	2.05	0.4798	0.0202	2.45	0.4929	0.0071
1.66	0.4515	0.0485	2.06	0.4803	0.0197	2.46	0.4931	0.0069
1.67	0.4525	0.0475	2.07	0.4808	0.0192	2.47	0.4932	0.0068
1.68	0.4535	0.0465	2.08	0.4812	0.0188	2.48	0.4934	0.0066
1.69	0.4545	0.0455	2.09	0.4817	0.0183	2.49	0.4936	0.0064
1.70	0.4554	0.0466	2.10	0.4821	0.0179	2.50	0.4938	0.0062
1.71	0.4564	0.0436	2.11	0.4826	0.0174	2.51	0.4940	0.0060
1.72	0.4573	0.0427	2.12	0.4830	0.0170	2.52	0.4941	0.0059
1.73	0.4582	0.0418	2.13	0.4834	0.0166	2.53	0.4943	0.0057
1.74	0.4591	0.0409	2.14	0.4838	0.0162	2.54	0.4945	0.0055
1.75	0.4599	0.0401	2.15	0.4842	0.0158	2.55	0.4946	0.0054
1.76	0.4608	0.0392	2.16	0.4846	0.0154	2.56	0.4948	0.0052
1.77	0.4616	0.0384	2.17	0.4850	0.0150	2.57	0.4949	0.0051
1.78	0.4625	0.0375	2.18	0.4854	0.0146	2.58	0.4951	0.0049
1.79	0.4633	0.0367	2.19	0.4857	0.0143	2.59	0.4952	0.0048
1.80	0.4641	0.0359	2.20	0.4861	0.0139	2.60	0.4953	0.0047
1.81	0.4649	0.0351	2.21	0.4864	0.0136	2.61	0.4955	0.0045
1.82	0.4656	0.0344	2.22	0.4868	0.0132	2.62	0.4956	0.0044
1.83	0.4664	0.0336	2.23	0.4871	0.0129	2.63	0.4957	0.0043
1.84	0.4671	0.0329	2.24	0.4875	0.0125	2.64	0.4959	0.0041
1.85	0.4678	0.0322	2.25	0.4878	0.0122	2.65	0.4960	0.0040
1.86	0.4686	0.0314	2.26	0.4881	0.0119	2.66	0.4961	0.0039
1.87	0.4693	0.0307	2.27	0.4884	0.0116	2.67	0.4962	0.0038
1.88	0.4699	0.0301	2.28	0.4887	0.0113	2.68	0.4963	0.0037
1.89	0.4706	0.0294	2.29	0.4890	0.0110	2.69	0.4964	0.0036
1.90	0.4713	0.0287	2.30	0.4893	0.0107	2.70	0.4965	0.0035

(Continued)

(Continued)

A	B	C	A	B	C	A	B	C
Z	Area Between Mean and Z	Area Beyond Z	Z	Area Between Mean and Z	Area Beyond Z	Z	Area Between Mean and Z	Area Beyond Z
2.71	0.4966	0.0034	3.01	0.4987	0.0013	3.31	0.4995	0.0005
2.72	0.4967	0.0033	3.02	0.4987	0.0013	3.32	0.4995	0.0005
2.73	0.4968	0.0032	3.03	0.4988	0.0012	3.33	0.4996	0.0004
2.74	0.4969	0.0031	3.04	0.4988	0.0012	3.34	0.4996	0.0004
2.75	0.4970	0.0030	3.05	0.4989	0.0011	3.35	0.4996	0.0004
2.76	0.4971	0.0029	3.06	0.4989	0.0011	3.36	0.4996	0.0004
2.77	0.4972	0.0028	3.07	0.4989	0.0011	3.37	0.4996	0.0004
2.78	0.4973	0.0027	3.08	0.4990	0.0010	3.38	0.4996	0.0004
2.79	0.4974	0.0026	3.09	0.4990	0.0010	3.39	0.4997	0.0003
2.80	0.4974	0.0026	3.10	0.4990	0.0010	3.40	0.4997	0.0003
2.81	0.4975	0.0025	3.11	0.4991	0.0009	3.41	0.4997	0.0003
2.82	0.4976	0.0024	3.12	0.4991	0.0009	3.42	0.4997	0.0003
2.83	0.4977	0.0023	3.13	0.4991	0.0009	3.43	0.4997	0.0003
2.84	0.4977	0.0023	3.14	0.4992	0.0008	3.44	0.4997	0.0003
2.85	0.4978	0.0022	3.15	0.4992	0.0008	3.45	0.4997	0.0003
2.86	0.4979	0.0021	3.16	0.4992	0.0008	3.46	0.4997	0.0003
2.87	0.4979	0.0021	3.17	0.4992	0.0008	3.47	0.4997	0.0003
2.88	0.4980	0.0020	3.18	0.4993	0.0007	3.48	0.4997	0.0003
2.89	0.4981	0.0019	3.19	0.4993	0.0007	3.49	0.4998	0.0002
2.90	0.4981	0.0019	3.20	0.4993	0.0007	3.50	0.4998	0.0002
2.91	0.4982	0.0018	3.21	0.4993	0.0007	3.60	0.4998	0.0002
2.92	0.4982	0.0018	3.22	0.4994	0.0006	3.70	0.4999	0.0001
2.93	0.4983	0.0017	3.23	0.4994	0.0006	3.80	0.4999	0.0001
2.94	0.4984	0.0016	3.24	0.4994	0.0006	3.90	0.4999	<0.0001
2.95	0.4984	0.0016	3.25	0.4994	0.0006	4.00	0.4999	<0.0001
2.96	0.4985	0.0015	3.26	0.4994	0.0006			
2.97	0.4985	0.0015	3.27	0.4995	0.0005			
2.98	0.4986	0.0014	3.28	0.4995	0.0005			
2.99	0.4986	0.0014	3.29	0.4995	0.0005			
3.00	0.4986	0.0014	3.30	0.4995	0.0005			

APPENDIX B
DISTRIBUTION OF t

	Level of Significance for One-Tailed Test					
	.10	.05	.025	.01	.005	.0005
	Level of Significance for Two-Tailed Test					
df	.20	.10	.05	.02	.01	.001
1	3.078	6.314	12.706	31.821	63.657	636.619
2	1.886	2.920	4.303	6.965	9.925	31.598
3	1.638	2.353	3.182	4.541	5.841	12.941
4	1.533	2.132	2.776	3.747	4.604	8.610
5	1.476	2.015	2.571	3.365	4.032	6.859
6	1.440	1.943	2.447	3.143	3.707	5.959
7	1.415	1.895	2.365	2.998	3.499	5.405
8	1.397	1.860	2.306	2.896	3.355	5.041
9	1.383	1.833	2.262	2.821	3.250	4.781
10	1.372	1.812	2.228	2.764	3.169	4.587
11	1.363	1.796	2.201	2.718	3.106	4.437
12	1.356	1.782	2.179	2.681	3.055	4.318
13	1.350	1.771	2.160	2.650	3.012	4.221
14	1.345	1.761	2.145	2.624	2.977	4.140
15	1.341	1.753	2.131	2.602	2.947	4.073
16	1.337	1.746	2.120	2.583	2.921	4.015
17	1.333	1.740	2.110	2.567	2.898	3.965
18	1.330	1.734	2.101	2.552	2.878	3.922
19	1.328	1.729	2.093	2.539	2.861	3.883
20	1.325	1.725	2.086	2.528	2.845	3.850
21	1.323	1.721	2.080	2.518	2.831	3.819
22	1.321	1.717	2.074	2.508	2.819	3.792
23	1.319	1.714	2.069	2.500	2.807	3.767
24	1.318	1.711	2.064	2.492	2.797	3.745
25	1.316	1.708	2.060	2.485	2.787	3.725

(Continued)

(Continued)

df	Level of Significance for One-Tailed Test					
	.10	.05	.025	.01	.005	.0005
	Level of Significance for Two-Tailed Test					
	.20	.10	.05	.02	.01	.001
26	1.315	1.706	2.056	2.479	2.779	3.707
27	1.314	1.703	2.052	2.473	2.771	3.690
28	1.313	1.701	2.048	2.467	2.763	3.674
29	1.311	1.699	2.045	2.462	2.756	3.659
30	1.310	1.697	2.042	2.457	2.750	3.646
40	1.303	1.684	2.021	2.423	2.704	3.551
60	1.296	1.671	2.000	2.390	2.660	3.460
120	1.289	1.658	1.980	2.358	2.617	3.373
∞	1.282	1.645	1.960	2.326	2.576	3.291

Source: Abridged from R. A. Fisher and F. Yates, *Statistical Tables for Biological, Agricultural and Medical Research*, 6th ed. Copyright © R. A. Fisher and F. Yates 1963. Reprinted by permission of Pearson Education Limited.

APPENDIX C
DISTRIBUTION OF CHI-SQUARE

df	.99	.98	.95	.90	.80	.70	.50	.30	.20	.10	.05	.02	.01	.001
1	.03157	.03628	.00393	.0158	.0642	.148	.455	1.074	1.642	2.706	3.841	5.412	6.635	10.827
2	.0201	.0404	.103	.211	.446	.713	1.386	2.408	3.219	4.605	5.991	7.824	9.210	13.815
3	.115	.185	.352	.584	1.005	1.424	2.366	3.665	4.642	6.251	7.815	9.837	11.341	16.268
4	.297	.429	.711	1.064	1.649	2.195	3.357	4.878	5.989	7.779	9.488	11.668	13.277	18.465
5	.554	.752	1.145	1.610	2.343	3.000	4.351	6.064	7.289	9.236	11.070	13.388	15.086	20.517
6	.872	1.134	1.635	2.204	3.070	3.828	5.348	7.231	8.558	10.645	12.592	15.033	16.812	22.457
7	1.239	1.564	2.167	2.833	3.822	4.671	6.346	8.383	9.803	12.017	14.067	16.622	18.475	24.322
8	1.646	2.032	2.733	3.490	4.594	5.527	7.344	9.524	11.030	13.362	15.507	18.168	20.090	26.125
9	2.088	2.532	3.325	4.168	5.380	6.393	8.343	10.656	12.242	14.684	16.919	19.679	21.666	27.877
10	2.558	3.059	3.940	4.865	6.179	7.267	9.342	11.781	13.442	15.987	18.307	21.161	23.209	29.588
11	3.053	3.609	4.575	5.578	6.989	8.148	10.341	12.899	14.631	17.275	19.675	22.618	24.725	31.264
12	3.571	4.178	5.226	6.304	7.807	9.034	11.340	14.011	15.812	18.549	21.026	24.054	26.217	32.909
13	4.107	4.765	5.892	7.042	8.634	9.926	12.340	15.119	16.985	19.812	22.362	25.472	27.688	34.528
14	4.660	5.368	6.571	7.790	9.467	10.821	13.339	16.222	18.151	21.064	23.685	26.873	29.141	36.123
15	5.229	5.985	7.261	8.547	10.307	11.721	14.339	17.322	19.311	22.307	24.996	28.259	30.578	37.697
16	5.812	6.614	7.962	9.312	11.152	12.624	15.338	18.418	20.465	23.542	26.296	29.633	32.000	39.252
17	6.408	7.255	8.672	10.085	12.002	13.531	16.338	19.511	21.615	24.769	27.587	30.995	33.409	40.790
18	7.015	7.906	9.390	10.865	12.857	14.440	17.338	20.601	22.760	25.989	28.869	32.346	34.805	42.312
19	7.633	8.567	10.117	11.651	13.716	15.352	18.338	21.689	23.900	27.204	30.144	33.687	36.191	43.820
20	8.260	9.237	10.851	12.443	14.578	16.266	19.337	22.775	25.038	28.412	31.410	35.020	37.566	45.315
21	8.897	9.915	11.591	13.240	15.445	17.182	20.337	23.858	26.171	29.615	32.671	36.343	38.932	46.797
22	9.542	10.600	12.338	14.041	16.314	18.101	21.337	24.939	27.301	30.813	33.924	37.659	40.289	48.268
23	10.196	11.293	13.091	14.848	17.187	19.021	22.337	26.018	28.429	32.007	35.172	38.968	41.638	49.728
24	10.856	11.992	13.848	15.659	18.062	19.943	23.337	27.096	29.553	33.196	36.415	40.270	42.980	51.179
25	11.524	12.697	14.611	16.473	18.940	20.867	24.337	28.172	30.675	34.382	37.652	41.566	44.314	52.620
26	12.198	13.409	15.379	17.292	19.820	21.792	25.336	29.246	31.795	35.563	38.885	42.856	45.642	54.052
27	12.879	14.125	16.151	18.114	20.703	22.719	26.336	30.319	32.912	36.741	40.113	44.140	46.963	55.476
28	13.565	14.847	16.928	18.939	21.588	23.647	27.336	31.391	34.027	37.916	41.337	45.419	48.278	56.893
29	14.256	15.574	17.708	19.768	22.475	24.577	28.336	32.461	35.139	39.087	42.557	46.693	49.588	58.302
30	14.953	16.306	18.493	20.599	23.364	25.508	29.336	33.530	36.250	40.256	43.773	47.962	50.892	59.703

Source: R. A. Fisher & F. Yates, *Statistical Tables for Biological, Agricultural and Medical Research,* 6th ed. Copyright © R. A. Fisher and F. Yates 1963. Reprinted by permission of Pearson Education Limited.

APPENDIX D
DISTRIBUTION OF F

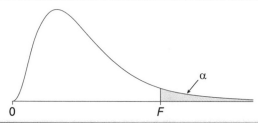

$\alpha = .05$

df$_2$	df$_1$									
	1	2	3	4	5	6	8	12	24	∞
1	161.4	199.5	215.7	224.6	230.2	234.0	238.9	243.9	249.0	254.3
2	18.51	19.00	19.16	19.25	19.30	19.33	19.37	19.41	19.45	19.50
3	10.13	9.55	9.28	9.12	9.01	8.94	8.84	8.74	8.64	8.53
4	7.71	6.94	6.59	6.39	6.26	6.16	6.04	5.91	5.77	5.63
5	6.61	5.79	5.41	5.19	5.05	4.95	4.82	4.68	4.53	4.36
6	5.99	5.14	4.76	4.53	4.39	4.28	4.15	4.00	3.84	3.67
7	5.59	4.74	4.35	4.12	3.97	3.87	3.73	3.57	3.41	3.23
8	5.32	4.46	4.07	3.84	3.69	3.58	3.44	3.28	3.12	2.93
9	5.12	4.26	3.86	3.63	3.48	3.37	3.23	3.07	2.90	2.71
10	4.96	4.10	3.71	3.48	3.33	3.22	3.07	2.91	2.74	2.54
11	4.84	3.98	3.59	3.36	3.20	3.09	2.95	2.79	2.61	2.40
12	4.75	3.88	3.49	3.26	3.11	3.00	2.85	2.69	2.50	2.30
13	4.67	3.80	3.41	3.18	3.02	2.92	2.77	2.60	2.42	2.21
14	4.60	3.74	3.34	3.11	2.96	2.85	2.70	2.53	2.35	2.13
15	4.54	3.68	3.29	3.06	2.90	2.79	2.64	2.48	2.29	2.07
16	4.49	3.63	3.24	3.01	2.85	2.74	2.59	2.42	2.24	2.01
17	4.45	3.59	3.20	2.96	2.81	2.70	2.55	2.38	2.19	1.96
18	4.41	3.55	3.16	2.93	2.77	2.66	2.51	2.34	2.15	1.92
19	4.38	3.52	3.13	2.90	2.74	2.63	2.48	2.31	2.11	1.88
20	4.35	3.49	3.10	2.87	2.71	2.60	2.45	2.28	2.08	1.84
21	4.32	3.47	3.07	2.84	2.68	2.57	2.42	2.25	2.05	1.81
22	4.30	3.44	3.05	2.82	2.66	2.55	2.40	2.23	2.03	1.78
23	4.28	3.42	3.03	2.80	2.64	2.53	2.38	2.20	2.00	1.76
24	4.26	3.40	3.01	2.78	2.62	2.51	2.36	2.18	1.98	1.73
25	4.24	3.38	2.99	2.76	2.60	2.49	2.34	2.16	1.96	1.71
26	4.22	3.37	2.98	2.74	2.59	2.47	2.32	2.15	1.95	1.69
27	4.21	3.35	2.96	2.73	2.57	2.46	2.30	2.13	1.93	1.67
28	4.20	3.34	2.95	2.71	2.56	2.44	2.29	2.12	1.91	1.65
29	4.18	3.33	2.93	2.70	2.54	2.43	2.28	2.10	1.90	1.64
30	4.17	3.32	2.92	2.69	2.53	2.42	2.27	2.09	1.89	1.62
40	4.08	3.23	2.84	2.61	2.45	2.34	2.18	2.00	1.79	1.51
60	4.00	3.15	2.76	2.52	2.37	2.25	2.10	1.92	1.70	1.39
120	3.92	3.07	2.68	2.45	2.29	2.17	2.02	1.83	1.61	1.25
∞	3.84	2.99	2.60	2.37	2.21	2.09	1.94	1.75	1.52	1.00

α= .01									
					df$_1$				
df$_2$	1	2	3	4	5	6	8	12	24	∞
1	4052	4999	5403	5625	5764	5859	5981	6106	6234	6366
2	98.49	99.01	99.17	99.25	99.30	99.33	99.36	99.42	99.46	99.50
3	34.12	30.81	29.46	28.71	28.24	27.91	27.49	27.05	26.60	26.12
4	21.20	18.00	16.69	15.98	15.52	15.21	14.80	14.37	13.93	13.46
5	16.26	13.27	12.06	11.39	10.97	10.67	10.27	9.89	9.47	9.02
6	13.74	10.92	9.78	9.15	8.75	8.47	8.10	7.72	7.31	6.88
7	12.25	9.55	8.45	7.85	7.46	7.19	6.84	6.47	6.07	5.65
8	11.26	8.65	7.59	7.01	6.63	6.37	6.03	5.67	5.28	4.86
9	10.56	8.02	6.99	6.42	6.06	5.80	5.47	5.11	4.73	4.31
10	10.04	7.56	6.55	5.99	5.64	5.39	5.06	4.71	4.33	3.91
11	9.65	7.20	6.22	5.67	5.32	5.07	4.74	4.40	4.02	3.60
12	9.33	6.93	5.95	5.41	5.06	4.82	4.50	4.16	3.78	3.36
13	9.07	6.70	5.74	5.20	4.86	4.62	4.30	3.96	3.59	3.16
14	8.86	6.51	5.56	5.03	4.69	4.46	4.14	3.80	3.43	3.00
15	8.68	6.36	5.42	4.89	4.56	4.32	4.00	3.67	3.29	2.87
16	8.53	6.23	5.29	4.77	4.44	4.20	3.89	3.55	3.18	2.75
17	8.40	6.11	5.18	4.67	4.34	4.10	3.79	3.45	3.08	2.65
18	8.28	6.01	5.09	4.58	4.25	4.01	3.71	3.37	3.00	2.57
19	8.18	5.93	5.01	4.50	4.17	3.94	3.63	3.30	2.92	2.49
20	8.10	5.85	4.94	4.43	4.10	3.87	3.56	3.23	2.86	2.42
21	8.02	5.78	4.87	4.37	4.04	3.81	3.51	3.17	2.80	2.36
22	7.94	5.72	4.82	4.31	3.99	3.76	3.45	3.12	2.75	2.31
23	7.88	5.66	4.76	4.23	3.94	3.71	3.41	3.07	2.70	2.26
24	7.82	5.61	4.72	4.22	3.90	3.67	3.36	3.03	2.66	2.21
25	7.77	5.57	4.68	4.18	3.86	3.63	3.32	2.99	2.62	2.17
26	7.72	5.53	4.64	4.14	3.82	3.59	3.29	2.96	2.58	2.13
27	7.68	5.49	4.60	4.11	3.78	3.56	3.26	2.93	2.55	2.10
28	7.64	5.45	4.57	4.07	3.75	3.53	3.23	2.90	2.52	2.06
29	7.60	5.42	4.54	4.04	3.73	3.50	3.20	2.87	2.49	2.03
30	7.56	5.39	4.51	4.02	3.70	3.47	3.17	2.84	2.47	2.01
40	7.31	5.18	4.31	3.83	3.51	3.29	2.99	2.66	2.29	1.80
60	7.08	4.98	4.13	3.65	3.34	3.12	2.82	2.50	2.12	1.60
120	6.85	4.79	3.95	3.48	3.17	2.96	2.66	2.34	1.95	1.38
∞	6.64	4.60	3.78	3.32	3.02	2.80	2.51	2.18	1.79	1.00

Appendix E
A Basic Math Review

by James Harris

Y ou have probably already heard that there is a lot of math in statistics, and for this reason you are somewhat anxious about taking a statistics course. Although it is true that courses in statistics can involve a great deal of mathematics, you should be relieved to hear that this course will stress interpretation rather than the ability to solve complex mathematical problems. With that said, however, you will still need to know how to perform some basic mathematical operations as well as understand the meanings of certain symbols used in statistics. Following is a review of the symbols and math you will need to know to successfully complete this course.

◉ SYMBOLS AND EXPRESSIONS USED IN STATISTICS

Statistics provides us with a set of tools for describing and analyzing *variables.* A variable is an attribute that can vary in some way. For example, a person's age is a variable because it can range from just born to more than one hundred years old. "Race" and "gender" are also variables, though with fewer categories than the variable "age." In statistics, variables you are interested in measuring are often given a symbol. For example, if we wanted to know something about the age of students in our statistics class, we would use the symbol Y to represent the variable "age." Now let's say for simplicity we asked only the students sitting in the first row their ages—19, 21, 23, and 32. These four ages would be scores of the Y variable.

Another symbol that you will frequently encounter in statistics is Σ, or uppercase sigma. Sigma is a Greek letter that stands for summation in statistics. In other words, when you see the symbol Σ, it means you should sum all of the scores. An example will make this clear. Using our sample of students' ages represented by Y, the use of sigma as in the expression ΣY (read as the sum of Y) tells us to sum all the scores of the variable Y. Using our example, we would find the sum of the set of scores from the variable "age" by adding the scores together:

$$19 + 21 + 23 + 32 = 95$$

So, for the variable "age," $\Sigma Y = 95$.

Sigma is also often used in expressions with an exponent, as in the expression ΣY^2 (read as the sum of squared scores). This means that we should first square all the scores of the Y variable and then sum the squared products. So using the same set of scores, we would solve the expression by squaring each score first and then adding them together:

$$19^2 + 21^2 + 23^2 + 32^2 = 361 + 441 + 529 + 1{,}024 = 2{,}355$$

So, for the variable "age," $\Sigma Y^2 = 2{,}355$.

A similar, but slightly different, expression, which illustrates the function of parentheses, is $(\Sigma Y)^2$ (read as the sum of scores, squared). In this expression, the parentheses tell us to first sum all the scores and then square this summed total. Parentheses are often used in expressions in statistics, and they always tell us to perform the expression within the parentheses first and then the part of the problem that is outside of the parentheses. To solve this expression, we need to sum all the scores first. However, we already found that $\Sigma Y = 95$, so to solve the expression $(\Sigma Y)^2$, we simply square this summed total,

$$95^2 = 9{,}025$$

So, for the variable "age," $(\Sigma Y)^2 = 9{,}025$.

You should also be familiar with the different symbols that denote multiplication and division. Most students are familiar with the times sign ($\times$); however, there are several other ways to express multiplication. For example,

$$3(4) \quad (5)6 \quad (4)(2) \quad 7 \cdot 8 \quad 9 * 6$$

all symbolize the operation of multiplication. In this text, the first three are most often used to denote multiplication. There are also several ways division can be expressed. You are probably familiar with the conventional division sign ($\div$), but division can also be expressed in these other ways:

$$4/6 \quad \frac{6}{3}$$

This text uses the latter two forms to express division.

In statistics, you are likely to encounter greater than and less than signs ($>$, $<$), greater than or equal to and less than or equal to signs ($\geq$, $\leq$), and not equal to signs ($\neq$). It is important you understand what each sign means, though admittedly it is easy to confuse them. Use the following expressions for review. Notice that numerals and symbols are often used together:

$4 > 2$ means 4 is greater than 2

$H_1 > 10$ means H_1 is greater than 10

$7 < 9$ means 7 is less than 9

$a < b$ means a is less than b

$Y \geq 10$ means that the value for Y is a value greater than or equal to 10

$a \leq b$ means that the value for a is less than or equal to the value for b

$8 \neq 10$ means 8 does not equal 10

$H_1 \neq H_2$ means H_1 does not equal H_2

▣ PROPORTIONS AND PERCENTAGES

Proportions and percentages are commonly used in statistics and provide a quick way to express information about the relative frequency of some value. You should know how to find proportions and percentages.

Proportions are identified by P; to find a proportion, apply this formula:

$$P = \frac{f}{N}$$

where f stands for the frequency of cases in a category and N the total number of cases in all categories. So, in our sample of four students, if we wanted to know the proportion of men in the front row, there would be a total of two categories, female and male. Because there are 3 women and 1 man in our sample, our N is 4; and the number of cases in our category "male" is 1. To get the proportion, divide 1 by 4:

$$P = \frac{f}{N} \qquad P = \frac{1}{4} = .25$$

So, the proportion of men in the front row is .25. To convert this to a percentage, simply multiply the proportion by 100 or use the formula for percentaging:

$$\% = \frac{f}{N} \times 100 \qquad \% = \frac{1}{4} \times 100 = 25\%$$

▣ WORKING WITH NEGATIVES

Addition, subtraction, multiplication, division, and squared numbers are not difficult for most people; however, there are some important rules to know when working with negatives that you may need to review.

1. When adding a number that is negative, it is the same as subtracting:

    $$5 + (-2) = 5 - 2 = 3$$

2. When subtracting a negative number, the sign changes:

    $$8 - (-4) = 8 + 4 = 12$$

3. When multiplying or dividing a negative number, the product or quotient is always negative:

    $$6 \times -4 = -24 \qquad -10 \div 5 = -2$$

4. When multiplying or dividing two negative numbers, the product or quotient is always positive:

$$-3 \times -7 = 21 \qquad -12 \div -4 = 3$$

5. Squaring a number that is negative always gives a positive product because it is the same as multiplying two negative numbers:

$$-5^2 = 25 \text{ is the same as } -5 \times -5 = 25$$

▣ ORDER OF OPERATIONS AND COMPLEX EXPRESSIONS

In statistics you are likely to encounter some fairly lengthy equations that require several steps to solve. To know what part of the equation to work out first, follow two basic rules. The first is called the rules of precedence. They state that you should solve all squares and square roots first, then multiplication and division, and finally, all addition and subtraction from left to right. The second rule is to solve expressions in parentheses first. If there are brackets in the equation, solve the expression within parentheses first and then the expression within the brackets. This means that parentheses and brackets can override the rules of precedence. In statistics, it is common for parentheses to control the order of calculations. These rules may seem somewhat abstract here, but a brief review of their application should make them more clear.

To solve the problem

$$4 + 6 \cdot 8 = 4 + 48 = 52$$

do the multiplication first and then the addition. Not following the rules of precedence will lead to a substantially different answer:

$$4 + 6 \cdot 8 = 10 \cdot 8 = 80$$

which is incorrect.

To solve the problem

$$6 - 4(6)/3^2$$

first, find the square of 3,

$$6 - 4(6)/9$$

then do the multiplication and division from left to right,

$$6 - \frac{24}{9} = 6 - 2.67$$

and finally, work out the subtraction,

$$6 - 2.67 = 3.33$$

To work out the following equation, do the expressions within parentheses first:

$$(4 + 3) - 6(2)/(3 - 1)^2$$

First, solve the addition and subtraction in the parentheses,

$$(7) - 6(2)/(2)^2$$

Now that you have solved the expressions within parentheses, work out the rest of the equation based on the rules of precedence, first squaring the 2,

$$(7) - 6(2)/4$$

Then do the multiplication and division next:

$$(7) - \frac{12}{4} = (7) - 3$$

Finally, work out the subtraction to solve the equation:

$$7 - 3 = 4$$

The following equation may seem intimidating at first, but by solving it in steps and following the rules, even these complex equations should become manageable:

$$\sqrt{\left(8(4-2)^2\right)\Big/\left(12/4\right)^2}$$

For this equation, work out the expressions within parentheses first; note that there are parentheses within parentheses. In this case, work out the inner parentheses first,

$$\sqrt{\left(8(2)^2\right)\Big/3^2}$$

Now do the outer parentheses, making sure to follow the rules of precedence within the parentheses—square first and then multiply:

$$\sqrt{\frac{32}{3^2}}$$

Now, work out the square of 3 first and then divide:

$$\sqrt{\frac{32}{9}} = \sqrt{3.55}$$

Last, take the square root:

$$1.88$$

LEARNING CHECK SOLUTIONS

▣ CHAPTER 1

(p. 7)

Learning Check. Review the definitions of exhaustive and mutually exclusive. Now look at Figure 1.2. What other categories could be added to the variable religion *to make it exhaustive and mutually exclusive? What other categories could be added to social class? To income?*

Answer:

To the variable *social class,* we could add "lower class" as a category. Monthly *income* requires many additional categories and could be recoded as an ordinal measure: $0 to $25,000, $25,001 to $50,000, $50,001 to 75,000, $75,001 and higher. For *religion,* we can also include "Protestant" and those without a religion.

(p. 10)

Learning Check. Identify the independent and dependent variables in the following hypotheses:

- *Younger Americans are more likely to support stricter gun control laws than older Americans.*
- *People who attend church regularly are more likely to oppose abortion than people who do not attend church regularly.*
- *Elderly women are more likely to live alone than elderly men.*
- *Individuals with postgraduate education are likely to have fewer children than those with less education.*

What are the independent and dependent variables in your hypothesis?

Answer:

Independent	Dependent
Age	Support for stricter gun control
Church attendance	Opposition to abortion
Gender	Living arrangement
Educational attainment	Number of children

▣ CHAPTER 2

(p. 29)

Learning Check. Examine Table 2.4 and answer the following questions: What is the percentage of white non-Hispanics who are employed? What is the base (N) for this percentage? What is the percentage of Hispanics who are not in the labor force? What is the base (N) for this percentage?

Answer:

For white non-Hispanics: 56.2%, $N = 7,363$
For Hispanics: 29.2%, $N = 17,162$

(p. 43)

Learning Check. Inspect Table 2.14 and answer the following questions:

- *What is the source of this table?*
- *How many variables are presented? What are their names?*
- *What is represented by the numbers presented in the second column? In the last row of the table?*

Answer:

The source for the data is noted at the bottom of the table.

There are nine variables, listed in the first column of the table. The first variable name is "Prenatal care in first three months of pregnancy."

The second column corresponds to mothers who are Mexican immigrants. The numbers correspond to the percentage of these mothers who utilized each health and public assistance program.

The last row corresponds to WIC utilization.

▣ C HAPTER 3

(p. 78)

Learning Check. Calculate the mean of the following distribution: 22, 15, 18, 33, 17, 5, 11, 28, 40, 19, 8, 20. Is it the same as the median, or is it different?

Answer:

$$\text{Mean} = \frac{22+15+18+33+17+5+11+28+40+19+8+20}{12} = 19.67$$

So the mean, 19.67, is larger than the median, 18.5.

▣ CHAPTER 4

(p. 99)

Learning Check. Why can't we use the range to describe diversity in nominal variables? The range can be used to describe diversity in ordinal variables (e.g., we can say that responses to a question ranged from "somewhat satisfied" to "very dissatisfied"), but it has no quantitative meaning. Why not?

Answer:

In nominal variables, the numbers are used only to represent the different categories of a variable without implying anything about the magnitude or quantitative difference between these categories. Therefore, the range, being a measure of variability that gives the quantitative difference between two values that a variable takes, is not an appropriate measure for nominal variables. Similarly, in ordinal variables, numbers corresponding with the categories of a variable are only used to rank-order these categories without having any meaning in terms of the quantitative difference between these categories. Therefore, the range does not convey any quantitative meaning when used to describe the diversity in ordinal variables.

(p. 102)

Learning Check. Why is the IQR better than the range as a measure of variability, especially when there are extreme scores in the distribution? To answer this question, you may want to examine Figure 4.2.

Answer:

Extreme scores directly influence the range, which is by definition the difference between the highest and the lowest scores. Therefore, if a distribution has extreme (very high and/or very low) scores, the range does not provide an accurate description of the distribution. The IQR, on the other hand, is not affected by extreme scores. Thus, it is a better measure of variability than the range when there are extreme scores in the distribution.

(p. 106)

Learning Check. Examine Table 4.4 again and note the disproportionate contribution of the western region to the sum of the squared deviations from the mean (it actually accounts for about 45% of the sum of squares). Can you explain why? (Hint: It has something to do with the sensitivity of the mean to extreme values.)

Answer:

The western region has the highest projected percentage change in the elderly population between 2008 and 2015, which is 27%. Therefore, it deviates more from the mean than the other regions. The more a category of a variable deviates from the mean, the larger the square of the deviation gets, and hence the more this category contributes to the sum of the squared deviations from the mean.

▣ CHAPTER 5

(p. 126)

Learning Check. Transform the Z scores in Table 5.2 back into raw scores. Your answers should agree with the raw scores listed in the table.

Answer:

Z Score	Raw Score
−2.93	$Y = 70.07 − 2.93(10.27) = 40$
−0.98	$Y = 70.07 − 0.98(10.27) = 60$
0.97	$Y = 70.07 + 0.97(10.27) = 80$
2.91	$Y = 70.07 + 2.91(10.27) = 100$

▣ CHAPTER 6

(p. 151)

Learning Check. Suppose a population distribution has a mean $\mu_Y = 150$ and a standard deviation $\sigma_Y = 30$, and you draw a simple random sample of N = 100 cases. What is the probability that the mean is between 147 and 153? What is the probability that the sample mean exceeds 153? Would you be surprised to find a mean score of 159? Why? (Hint: To answer these questions, you need to apply what you learned in Chapter 5 about Z scores and areas under the normal curve [Appendix A].) Remember, to translate a raw score into a Z score we used this formula:

$$Z = \frac{Y - \overline{Y}}{S_Y}$$

However, because here we are dealing with a sampling distribution, replace Y with the sample mean $\overline{Y}$, $\overline{Y}$ with the sampling distribution's mean $\mu_{\overline{Y}}$, and S_Y with the standard error of the mean $\sigma_Y / \sqrt{N}$

$$Z = \frac{\overline{Y} - \mu_{\overline{Y}}}{\sigma_Y / \sqrt{N}}$$

Answer:

The Z score equivalent of 147 is

$$Z = \frac{\bar{Y} - \mu_{\bar{Y}}}{\sigma_Y / \sqrt{N}} = \frac{147 - 150}{30 / \sqrt{100}} = \frac{-3}{3} = -1$$

The Z score equivalent of 153 is

$$Z = \frac{\bar{Y} - \mu_{\bar{Y}}}{\sigma_Y / \sqrt{N}} = \frac{153 - 150}{30 / \sqrt{100}} = \frac{3}{3} = 1$$

Using the standard normal table (Appendix A), we can see that the probability of the area between the mean and a score 1 standard deviation above or below the mean is 0.3413. So the probability that the mean is between 147 and 153, both of which deviate from the mean by 1 standard deviation, is 0.6826 (0.3413 + 0.3413), or 68.26%.

The probability of the area beyond 1 standard deviation from the mean is 0.1587. So the probability that the mean exceeds 153 is 0.1587, or 15.87%.

The Z score equivalent of 159 is

$$Z = \frac{\bar{Y} - \mu_{\bar{Y}}}{\sigma_Y / \sqrt{N}} = \frac{159 - 150}{30 / \sqrt{100}} = \frac{9}{3} = 3$$

The probability of the area beyond 3 standard deviations from the mean, according to the standard normal table, is 0.0014. Therefore, it would be surprising to find a mean score of 159, as the probability is very low (0.14%).

▣ CHAPTER 7

(p. 160)

Learning Check. What is the difference between a point estimate and a confidence interval?

Answer:

When the estimate of a population parameter is a single number, it is called a point estimate. When the estimate is a range of scores, it is called an interval estimate. Confidence intervals are used for interval estimates.

(p. 161)

Learning Check. To understand the relationship between the confidence level and Z, review the material in Chapter 5. What would be the appropriate Z value for a 98% confidence interval?

Answer:

The appropriate Z value for a 98% confidence interval is 2.33.

(p. 163)

Learning Check. What is the 90% confidence interval for the mean commuting time? (Hint: First, find the Z value associated with a 90% confidence level.)

Answer:

$$90\% \text{ CI} = 7.5 \pm 1.65(0.07)$$

$$= 7.5 \pm 0.12$$

$$= 7.38 \text{ to } 7.62$$

CHAPTER 8

(p. 195)

Learning Check. For the following research situations, state your research and null hypotheses:

- There is a difference between the mean statistics grades of social science majors and the mean statistics grades of business majors.
- The average number of children in two-parent black families is lower than the average number of children in two-parent nonblack families.
- Grade point averages are higher among girls who participate in organized sports than among girls who do not.

Answer:

Null Hypothesis	Research Hypothesis
Means are presumed equal for all statements.	Two-tailed test. No direction is stated.
	One-tailed test, left.
	One-tailed test, right.

(p. 201)

Learning Check. Would you change your decision in the previous example if alpha were .01? Why or why not?

Answer:

The significance of 4.266 is .000, which is less than .01. Our decision to reject the null hypothesis would not change.

(p. 204)

Learning Check. Review the information provided in Table 8.4. What would be the critical t statistic at the .05 level for the first indicator, EC Index? Assume a two-tailed test.

Answer:

The N's are reported as a note at the bottom of the table. The df calculation would be $(78 + 113) - 2 = 189$. Based on Appendix B, $df = \infty$, the critical t value is 1.960.

▣ CHAPTER 9

(p. 216)

Learning Check. Examine Table 9.2. Make sure that you can identify all the parts just described and that you understand how the numbers were obtained. Can you identify the independent and dependent variables in the table? You will need to know this to convert the frequencies to percentages.

Answer:

The independent variable is race, and home ownership is the dependent variable.

(p. 228)

Learning Check. Construct a bivariate table (in percentages) showing no association between age and first-generation college status.

Answer:

Age and First-Generation College Status

	19 Years or Younger	*20 Years or Older*	
Firsts	41.9%	41.9%	41.9% (1,934)
Nonfirsts	58.1%	58.1%	58.1% (2,683)
	100.0%	100.0%	4,617

(p. 229)

Learning Check. Refer to the data in the previous Learning Check. Are the variables age and first-generation college status statistically independent? Write out the research and the null hypotheses for your practice data.

Answer:

Null hypothesis: There is no association between age and first-generation college status.

Research hypothesis: Age and first-generation college status are statistically dependent.

(p. 231)

Learning Check. Refer to the data in the Learning Check on page 228. Calculate the expected frequencies for age and first-generation college status and construct a bivariate table. Are your column and row marginals the same as in the original table?

Answer:

Using the format of Table 9.14, construct a table to calculate chi-square for age and educational attainment.

	f_o	f_e	$f_o - f_e$	$(f_o - f_e)^2$	$(f_o - f_e)^2/f_e$
19/Firsts	916	1,138.53	−222.53	49,519.60	43.49
19/Nonfirsts	1,802	1,579.47	222.53	49,519.60	31.35
20/Firsts	1,018	795.47	222.53	49,519.60	62.25
20/Nonfirsts	−881	1,103.53	−222.53	49,519.60	44.87

Chi-square = 181.96, with Yates's correction = 181.15.

(p. 235)

Learning Check. Based on Appendix C, identify the probability for each chi-square value (df in parentheses)

Answer:

- *12.307 (15)* Between .70 and .50
- *20.337 (21)* Exactly .50
- *54.052 (24)* Less than .001

(p. 235)

Learning Check. What decision can you make about the association between age and first-generation college status? Should you reject the null hypothesis at the .05 alpha level or at the .01 level?

Answer:

We would reject the null hypothesis of no difference. Our calculated chi-square is significant at the .05 and the .01 levels. We have evidence that age is related to first-generation college status—older students are more likely to be first-generation students than younger students. Fifty-four percent of students 20 years or older are first-generation students versus 33.7% of students 19 years or younger.

(p. 238)

Learning Check. For the bivariate table with age and first-generation college status, the value of the obtained chi-square is 181.15 with 1 degree of freedom. Based on Appendix C, we determine that its probability is less than .001. This probability is less than our alpha level of .05. We reject the null hypothesis of no relationship between age and first-generation college status. If we reduce our sample size by half, the obtained chi-square is 90.58. Determine the P value for 90.58. What decision can you make about the null hypothesis?

Answer:

Even if we reduce the chi-square by half, we would still reject the null hypothesis.

▣ CHAPTER 11

(p. 287)

Learning Check. Use Figure 11.3 to predict the percentage of residents with bachelor's degrees in a state with a median household income of $47,500 and one with a median household income of $50,000.

Answer:

The percentage of residents with bachelor's degrees in a state with a median household income of $47,500 is about 26%. The comparable percentage in a state with a median household income of $50,000 is about 27%.

(p. 290)

Learning Check. Use the linear equation describing the relationship between seniority and salary of teachers to obtain the predicted salary of a teacher with 12 years of seniority.

Answer:

The predicted salary of a teacher with 12 years of seniority is $36,000 ($Y = 12,000 + 2,000[12]$).

(p. 290)

Learning Check. For each of these four lines, as X goes up by 1 unit, what does Y do? Be sure you can answer this question using both the equation and the line.

Answer:

For the line $Y = 1X$, as X goes up by 1 unit, Y also goes up by 1 unit. In the second line, $Y = 2 + 0.5X$, Y increases by 0.5 units as a result of a 1-unit increase in X. The line $Y = 6 − 2X$ tells that every 1-unit increase in X results in a 2-unit decrease in Y. Finally, in the fourth line, Y decreases by 0.33 units as a result of a 1-unit increase in X.

(p. 295)

Learning Check. Use the prediction equation to calculate the predicted values of Y for New York, Georgia, and Ohio. Verify that the regression line in Figure 11.6 passes through these points.

Answer:

$$\text{New York: } \hat{Y} = 7.41 + 0.0004(54{,}659) = 29.27\%$$

$$\text{Georgia: } \hat{Y} = 7.41 + 0.0004(47{,}590) = 26.45\%$$

$$\text{Ohio: } \hat{Y} = 7.41 + 0.0004(45{,}395) = 25.57\%$$

ANSWERS TO ODD-NUMBERED EXERCISES

▣ CHAPTER 1

1. Once our research question, the hypothesis, and the study variables have been selected, we move on to the next stage of the research process—measuring and collecting the data. The choice of a particular data collection method or instrument depends on our study objective. After our data have been collected, we have to find a systematic way to organize and analyze our data and set up some set of procedures to decide what we mean.

3. a. Interval-ratio
   b. Nominal
   c. Interval-ratio
   d. Ordinal
   e. Nominal
   f. Interval-ratio
   g. Interval-ratio
   h. Nominal

5. There are many possible variables from which to choose. Some of the most common selections by students will probably be type of occupation or industry, work experience, and educational training or expertise. Students should first address the relationship between these variables and gender. For example, men have more years of work experience than women in the same occupation. Students may also consider measuring structural bias or discrimination.

7. In general, the difficulty with studying criminal acts is that the criminal act needs to be reported first. It is estimated that the majority of crimes are not reported to authorities. Data on reported crimes are routinely collected by the Federal Bureau of Investigation and the Bureau of Justice.

9. Individual age: This variable could be measured as an interval-ratio variable, with actual age in years reported. As discussed in the chapter, interval-ratio variables are the highest level of measurement and can also be measured at ordinal or nominal levels.

   Annual income: This variable could be measured as an interval-ratio variable, with actual dollar earnings reported.

   Religiosity: This variable could be measured in several ways. For example, as church attendance, the variable could be ordinal (number of times attended church in a month: every week, at least twice a month, less than two times a month, none at all).

   Student performance: This could be measured as an interval-ratio variable as GPA or test score.

Social class: This variable is an ordinal variable, with categories low, working, middle, and upper.

Attitude toward gun control: This variable is an ordinal variable, with categories strongly disagree, disagree, neutral, agree, and strongly agree.

▣ CHAPTER 2

1. a. Race is a nominal variable. Class is an ordinal variable, since the categories can be ordered from lower to higher status.
    b.

Frequency Table for Race

Race	Frequency (f)
White	17
Nonwhite	13
Total (N)	30

Frequency Table for Class

Class	Frequency (f)
Lower	3
Working	15
Middle	11
Upper	1
Total (N)	30

3.

Number of Traumas	Frequency (f)
0	15
1	11
2	4
Total (N)	30

  a. Trauma is an interval or ratio-level variable, since it has a real zero point and a meaningful numeric scale.
  b. People in this survey are more likely to have experienced no traumas last year (50% of the group).
  c. The proportion who experienced one or more traumas is calculated by first adding 36.7% and 13.3% = 50%. Then divide that number by 100 to obtain the proportion, 0.50, or half the group.

5. Ranking them from highest to lowest level of support: strong Democrats, strong Republicans, and Independents. Support does vary by group: however, the majority of strong Democrats (56.8%) and strong

Republicans (50%) agree or strongly agree with the statement. The group with the lowest level of support is Independents with 42.3%.

7. a.
    For whites.

Education	f	%	C%
Less than high school	72	12.3	12.3
High school graduate	272	46.5	58.8
Junior college	46	7.9	66.7
Bachelor	118	20.2	86.9
Graduate	77	13.2	100.1
TOTAL	585		

For blacks.

Education	f	%	C%
Less than high school	26	22.0	22.0
High school graduate	59	50	72.0
Junior college	10	8.5	80.5
Bachelor	16	13.6	94.1
Graduate	7	5.9	100.0
TOTAL	118		

For men.

Education	f	%	C%
Less than high school	46	14.0	14.0
High school graduate	151	45.9	59.9
Junior college	24	7.3	67.2
Bachelor	65	19.8	87
Graduate	43	13.1	100.1
TOTAL	329		

For women.

Education	f	%	C%
Less than high school	67	15.0	15.0
High school graduate	214	47.8	62.8
Junior college	37	8.2	71.0
Bachelor	81	18.1	89.1
Graduate	49	10.9	100.0
TOTAL	448		

b. 40.2% of men attended school beyond high school. A lower percentage of women (37.2%) did the same.

c. 58.8% for whites and 72.0% for blacks.

d. Cumulative percentages are more similar for men and women than for whites and blacks. Inequality appears to be larger between racial groups. A larger percentage of whites complete bachelor or graduate degrees than do blacks.

9. The group with the largest increase in voting rates is blacks, from 53% in 1996 to 66.2% in 2012. Blacks are the only group that did not experience a decline in voting rates for the years presented. Hispanic voting rates exceeded the voting rates for Asians in 2000 and remained higher than Asians through 2012. Hispanics and Asians have the lowest voting rates for all groups. As noted in the exercise, in the 2012 presidential election, blacks had the highest voting rates for all groups, followed by non-Hispanic whites, Hispanics, and Asians. White voting rates declined by 2% from 2008 to 2012. The highest voting rate for whites was in 2004 (67.2%), 2008 for Hispanics (49.9%) and for Asians (47.6%).

11. a. Victimization rates are highest for those 12 to 17 years of age.

b. Victimization rates have been declining since 1994–1998. In the last time period, 2005–2010, all rates are below 5 per 1,000 females. Across the three time periods, victimization rates are highest for females ages 12 to 17 (11.3–4.1). Second highest rates are among females ages 18 to 34 years (7.0–3.7).

13. The group with the highest level of support for stricter gun laws is graduate degree (84.4%), followed by the high school degree group (77.02%). The lowest level of support is reported for the junior college group (71.43%).

◉ CHAPTER 3

1. a. The mode can be found two ways: looking for either the highest frequency (470) or the highest percentage (48.3). The mode is the category that corresponds to these values, "exciting."

b. The median can be found two ways: by using either the frequencies column or the cumulative percentages.

Using Frequencies	Using Cumulative Percentages
$$\frac{N+1}{2} = \frac{974+1}{2} = 487.5 \text{th case}$$	Notice that 48.3% of the observations fall in the "exciting" percentage category; 93.9% fall in or before the "routine" category.
Starting with the frequency in the first category (470), add up the frequencies until you find where the 487th and 488th cases fall. Both of these cases correspond to the category "routine," which is the median.	The 50% mark, or the median, is located somewhere within the "routine" category. So the median is "routine."

   c. The mode is simply the category with the highest frequency (or percentage) in the distribution. The median divides the distribution into two equal parts so that half the cases are below it and half are above it.

   d. Because this variable is an ordinal-level variable.

3. a. Interval-ratio. The mode can be found two ways: by looking either for the highest frequency (14) or the highest percentage (43.8). The mode is the category that corresponds to the value "40 hours worked last week." The median can be found two ways: by using either the frequencies column or the cumulative percentages.

Using Frequencies	*Using Cumulative Percentages*
$$\frac{N+1}{2} = \frac{32+1}{2} = 16.5\text{th case}$$	Notice that 34.4% of the observations fall in or below the "32 hours worked last week" category; 78.1% fall in or below the "40 hours worked last week" category.
Starting with the frequency in the first category (1), add up the frequencies until you find where the 16th and 17th cases fall. Both of these cases correspond to the category "40 hours worked last week," which is the median.	The 50% mark, or the median, is located somewhere within the "40 hours worked last week" category. So the median is "40 hours worked last week."

   b. Since the median is merely a synonym for the 50th percentile, we already know that its value is 40 hours worked last week.

     25th percentile = (32 × 0.25) = 8th case = 30 hours worked last week
     75th percentile = (32 × 0.75) = 24th case = 40 hours worked last week

5. a. The mode can be found by looking for the highest frequency in each column; the mode for each group is listed below:

     18–29: Good
     30–39: Good
     40–49: Good
     50–59: Good

   The median can be found two ways: by using either the frequencies column or the cumulative percentages. However, since the problem only gives the frequencies, we'll use those to solve for the median.

Age-Group			
18–29	*30–39*	*40–49*	*50–59*
$$\frac{N+1}{2} = \frac{164+1}{2} = 82.5\text{th case}$$	$$\frac{N+1}{2} = \frac{169+1}{2} = 85\text{th case}$$	$$\frac{N+1}{2} = \frac{168+1}{2} = 84.5\text{th case}$$	$$\frac{N+1}{2} = \frac{173+1}{2} = 87\text{th case}$$
Starting with the frequency in the first category (56), add up the frequencies until you find where the 82nd and 83rd cases fall. Both cases correspond to "good," which is the median.	Starting with the frequency in the first category (55), add up the frequencies until you find where the 85th case falls. This case corresponds to "good," which is the median.	Starting with the frequency in the first category (41), add up the frequencies until you find where the 84th and 85th cases fall. Both cases correspond to "good," which is the median.	Starting with the frequency in the first category (38), add up the frequencies until you find where the 87th case falls. It corresponds to "good," which is the median.

b. Since the mode and median for all four age-groups was "good," it has to do with how respondents interpreted the question. For instance, it is possible that one's health status was assessed relative to his or her age. Neither the median nor the mode provides a better description of the data since they provide the same information.

7. We begin by multiplying each household size by its frequency.

Household Size	Frequency	Frequency × Y (fY)
1	381	381
2	526	1,052
3	227	681
4	200	800
5	96	480
6	42	252
7	19	133
8	5	40
9	2	18
10	2	20
Total	$N = 1,500$	$\sum fY = 3,857$

$$\bar{Y} = \frac{\sum fY}{N} = \frac{3,857}{1,500} = 2.57$$

So, the mean number of people per household is 2.57.

9. a. There appear to be a few outliers (i.e., extremely high values); this leads us to believe that the distribution is skewed in the positive direction.

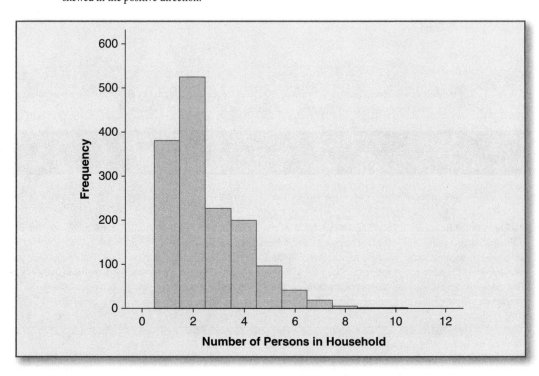

b. The median can be found two ways: by using either the frequencies column or the cumulative percentages. The data are in frequencies; we'll use those to solve the median. Since the median (2) is less than the mean (2.57), we can conclude that the distribution is skewed in a positive direction. Our answer to question 9a is further supported.

Using Frequencies

$$\frac{N+1}{2} = \frac{1,500+1}{2} = 750.5 \text{th case}$$

Starting with the frequency in the first category (381), add up the frequencies until you find where the 750th and 751st cases fall. Both of these cases correspond to the category "2," which is the median.

---

11. Yes, both politicians can be correct, at least in a technical sense. One politician can be referring to the mean; the other could be using the median. It would be unusual if these two statistics were exactly equal. The average or mean income of Americans can be greater than the median if the distribution of income is positively skewed, which is certainly true.

13. The mode can be found by looking for the highest frequency in each column; the mode for each group is listed below:

Men: Working
Women: Working

The median can be found in two ways, either by using the frequencies column or by using the cumulative percentages. However, because the problem gives only the frequencies, we'll use those to solve for the median.

Men	*Women*
$$\frac{N+1}{2} = \frac{474+1}{2} = 237.5 \text{th case}$$	$$\frac{N+1}{2} = \frac{646+1}{2} = 328.5 \text{th case}$$
Starting with the frequency in the first category (390), add up the frequencies until you find where the 237th and 238th cases fall. We actually don't need to do any adding, as both these cases correspond with the first category, "working."	Starting with the frequency in the first category (441), add up the frequencies until you find where the 328th and 329th cases fall. We actually don't need to do any adding, as both these cases correspond with the first category, "working."

When using both the mode and median to estimate participation in the labor force, it appears that there are no substantial differences between men and women. This might be due to the emphasis on being able to provide for oneself as well as the necessity for households to have two sources of income.

15. a. The data are measured at the ordinal level.
    b. Whites: mode = somewhat agree and somewhat disagree (bimodal), median = somewhat agree.
       Blacks: mode = somewhat disagree, median = somewhat disagree.
    c. The data suggest a difference between whites and blacks for whether racism is a thing of the past. The most common category for blacks is "somewhat disagree," whereas for whites it is a tie between "somewhat agree" and "somewhat disagree." The median—the most useful measure of central tendency for

ordinal variables—is "somewhat disagree" for blacks and "somewhat agree" for whites. Whites are more likely to believe that racism is in the past, while blacks are more likely to see it as ongoing.

▣ CHAPTER 4

1. a. The range of convictions in 1990 is (583 – 79) = 504. The range of convictions in 2009 is (426 – 102) = 324. The range of convictions is larger in 1990 than 2009.
    b. The mean number of convictions is 295.67 in 1990 and 261.67 in 2009.
    c.

1990

Government Level	No. of Convictions	$(Y - \bar{Y})$	$(Y - \bar{Y})^2$
Federal	583	287.33	82,558.53
State	79	−216.67	46,945.89
Local	225	−70.67	4,994.25
Total	887	−0.01	134,498.67

$$\bar{Y} = 295.67$$

$$S_Y = \sqrt{S_Y^2} = \sqrt{\frac{\sum(Y - \bar{Y})^2}{N - 1}} = \sqrt{\frac{134,498.67}{2}} = 259.32$$

2009

Government Level	No. of Convictions	$(Y - \bar{Y})$	$(Y - \bar{Y})^2$
Federal	426	164.33	27,004.35
State	102	−159.67	25,494.51
Local	257	−4.67	21.81
Total	785	−0.01	52,520.67

$$\bar{Y} = 261.67$$

$$S_Y = \sqrt{S_Y^2} = \sqrt{\frac{\sum(Y - \bar{Y})^2}{N - 1}} = \sqrt{\frac{52,520.67}{2}} = 162.05$$

    d. The standard deviation is larger in 1990 than in 2009, thus indicating more variability in number of convictions in 1990 than 2009. This supports our results from 3a.

3. a. Because the mean and median values for occupational prestige are relatively similar (mean = 40.59 and median = 40.00 for high school degree, mean = 50.95 and median = 51.00 for bachelor's degree), these statistics do not suggest any significant skew between either of the types of graduates.

    b. Using the variance, standard deviation, range, and interquartile range to compare the variability for the two groups, those with bachelor's degrees have variance (167.185), standard deviation (12.930), and interquartile range (23) values that exceed those of the high school group (130.393, 11.419, and 17, respectively). Only in terms of range does the high school group exceed the college group (58, compared with 52). Overall, this suggests that there is more variability of prestige in the college graduate group.

5. a. For those diagnosed with cancer:

$$Variance = \frac{\Sigma(Y - \bar{Y})^2}{N - 1} = \frac{3{,}059.14}{187 - 1} = 16.45$$

Thus, the standard deviation is $\sqrt{16.45}$, or 4.06

For those not diagnosed with cancer:

$$Variance = \frac{\Sigma(Y - \bar{Y})^2}{N - 1} = \frac{25{,}180.20}{1{,}200 - 1} = 21$$

Thus, the standard deviation is $\sqrt{21}$, or 4.58

    b. Although there is slightly more variability in the psychological distress score among those not diagnosed with cancer compared with those diagnosed with cancer, the standard deviations of the two groups are very close (4.58 vs. 4.06). An interesting finding is that the respondents diagnosed with cancer are not only slightly less diverse in terms of distress level: their average psychological distress score is also less than for those not diagnosed with cancer (3.9 vs. 4.87).

    c. No, we didn't need the mean of Y to calculate the variance or standard deviation.

7. Because the variable is interval-ratio, we should use variance (or standard deviation, range, or IQR). Among these three measures, variance and/or standard deviation is preferred. For measurements of central tendency, as discussed in Chapter 3, if we are looking for the average life expectancy for these 10 countries, we should rely on the mean.

On average, non-European countries have a slightly higher life expectancy at birth. Both the mean and median are higher for non-European countries than for European countries. Also, the distribution of European countries tended to exhibit more variability; the standard deviation for European countries is 3.5 years, while for non-European countries, it is 2.6 years. The IQRs also attest to more variability in the distribution of life expectancy for European countries (IQR = 4.2) compared with non-European countries (IQR = 2.5).

These differences might be explained by access and availability of health care and/or diet. However, the difference might simply be random, due to the small number of countries presented in this example. Perhaps we would find different results if more countries were incorporated into the analyses.

A table of results is shown on the next page:

Life Expectancy		Statistic
European countries	Mean	82.2
	Median	82.7
	Variance	12.1
	Standard deviation	3.5
	Minimum	76.3
	Maximum	84.8
	Range	8.5
	Interquartile range	4.2
Non-European countries	Mean	83.3
	Median	83.7
	Variance	6.7
	Standard deviation	2.6
	Minimum	79.2
	Maximum	86.4
	Range	7.2
	Interquartile range	2.5

9. We should be cautious when making generalized statements about the relationship between education and ideal number of children because we have statistics for only two groups. We would need more data from a number of groups in order to make specific statements about this relationship. Therefore, we must restrict our discussion to Chinese Americans and Filipino Americans. On average, Chinese Americans are more educated than Filipino Americans (15.55 years versus 13.42 years), and both groups have about the same standard deviation (3.643 for Chinese Americans and 3.704 for Filipino Americans). Additionally, Chinese Americans report a lower number of ideal children (2.88) than Filipino Americans (4.00). Again, for this variable, both groups have about the same standard deviation (2.167 for Chinese Americans and 2.098 for Filipino Americans). Based on these findings, we might suggest that as level of education increases, the ideal number of children decreases (but remember: we can't be certain this is the case for all Americans!).

▣ CHAPTER 5

1. a. The Z score for a person who watches more than 8 hr/day:

$$Z = \frac{8 - 3.01}{2.65} = 1.88$$

   b. We first need to calculate the Z score for a person who watches 5 hr/day:

$$Z = \frac{5 - 3.01}{2.65} = 0.75$$

The area between Z and the mean is 0.2734. We then need to add 0.50 to 0.2734 to find the proportion of people who watch television less than 5 hr/day. Thus, we conclude that the proportion of people who watch television less than 5 hr/day is 0.7734. This corresponds to 783.45 (0.7734 × 1,013).

c. 5.66 television hours per day corresponds to a Z score of +1.

$$Y = \overline{Y} + Z(S_Y) = 3.01 + 1(2.65) = 5.66$$

d. The Z score for a person who watches 1 hour of television per day is

$$Z = \frac{1 - 3.01}{2.65} = -0.76$$

The area between the mean and Z is 0.2764.
The Z score for a person who watches 6 hours of television per day is

$$Z = \frac{6 - 3.01}{2.65} = 1.13$$

The area between the mean and Z is 0.3708.
Therefore, the percentage of people who watch between 1 and 6 hours of television per day is 64.72% (0.2764 + 0.3708 = 0.6472 × 100).

3. a. For an individual with 13.47 years of education, his or her Z score would be

$$Z = \frac{13.47 - 13.47}{3.1} = 0.0$$

b. Since our friend's number of years of education completed is associated with the 60th percentile, we need to solve for Y. However, we must first use the logic of the normal distribution to find Z. For any normal distribution, 50% of all cases will fall above the mean. Since our friend is in the 60th percentile, we know that the area between the mean and our friend's score is 0.10. Similarly, the area beyond our friend's score is 0.40. We can now look in Appendix A, column B for 0.10 or in column C for 0.40. We find that the Z associated with these values is 0.25. Now, we can solve for Y:

$$Y = \overline{Y} + Z(S_Y) = 13.47 + 0.25(3.1) = 14.25$$

5. a. The mean and standard deviation are

$$\overline{Y} = \frac{11,877.3}{4} = 2,969.33 \quad S_Y = 821.35$$

b. One standard deviation above the mean is 2,969.33 + 1.0(821.35) = 3,790.68. One region has unemployment above this value. We would expect, from Appendix A, about 15.9% of the distribution to lie above a Z score of 1.0. Then 15.9% of 4 regions is less than one region, which means that the number of regions with scores greater than 1 standard deviation above the mean is more than what we would expect from a normal distribution.

c. Because we have only four regions and each has a different unemployment number, the distribution is obviously not normal. Because the South has such a large number of unemployed persons, the distribution is technically positively skewed. See the figure on the next page.

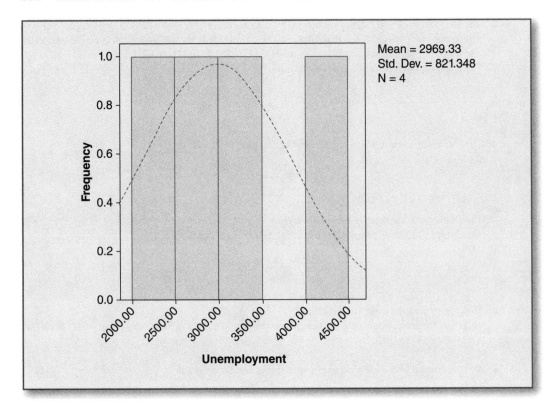

7. a. The area beyond the Z score is about 0.1056, so 10.56% of students should score above 625.

$$Z = \frac{625 - 500}{100} = 1.25$$

b. The area between this score and the mean is 0.3413.

$$Z = \frac{400 - 500}{100} = -1.00$$

$$Z = \frac{600 - 500}{100} = 1.00$$

The area between this score and the mean is also 0.3413, so 68.26% of all students should score between 400 and 600 (or ±1 standard deviation from the mean of 500).

9. a. The area between the value and the upper tail of the distribution is 0.1020. So the probability that someone will work more than 60 hours per week is 0.1020. This translates into approximately 85 (838 × 0.1020) respondents in the sample.

$$Z_{60} = \frac{60 - 40.62}{15.26} = 1.27$$

b. The area between the value and the lower tail of the distribution is 0.2420. So the probability that someone will work fewer than 30 hours per week is 0.2420. This translates into approximately 203 (838 × 0.2420) respondents in the sample.

$$Z = \frac{30 - 40.62}{15.26} = -0.70$$

11. a. For the eligibility criterion, Team A has a Z score of

$$Z = \frac{917 - 974.3}{40.4} = -1.42$$

For the retention criterion,

$$Z = \frac{913 - 970.2}{38.8} = -1.47$$

Team A has a higher Z score on the eligibility criterion and therefore is better on that.

b. For the eligibility criterion, Team B has a Z score of

$$Z = \frac{962 - 974.3}{40.4} = -0.30$$

For the retention criterion,

$$Z = \frac{962 - 970.2}{38.8} = -0.21$$

Team B is better on the retention criterion.

c. Team B's retention Z score was –0.21, below the mean. We simply look to column C and find that the proportion of teams that did worse than Team B on the retention criterion is 0.4168. This is the area between Team B's retention Z score and the tail end of the distribution.

13. a. The 95th percentile corresponds to a Z score of about 1.65. Translating this into a raw score for the number of women needing shelter yields

$$Y = \bar{Y} + Z(S_Y) = 250 + 1.65(75) = 373.75 \text{ women.}$$

Unfortunately, a capacity of 350 is below this value, so there will not be enough space for all abused women on 95% of all nights. Obviously, the city needs at least 374 beds.

b. The area below this value is 0.3446, so the area exceeding this Z is $1 - 0.3446 = 0.6554$. Thus, on 65.54% of all nights, the number of women seeking shelter will exceed the capacity of 220.

$$Z = \frac{220 - 250}{75} = -0.40$$

15. a. The Z score with only .01 of the population above it (in column C) is 2.33.

   $65,628.80 + 2.33(59,373.92) = 203,970.03$

   So, a household would have to have made \$203,970.03 in the past year to have had more income than 99% of other houses.

   b. $Z = \dfrac{0-65,628.80}{59,373.92} = \dfrac{-65,628.80}{59,373.92} = -1.11$

   The area below this Z score is .1335. Our classmate has a point: it is problematic to assume that the distribution is normal, when that would entail that 13.35% of people come from households making less than \$0 in the past year.

▣ CHAPTER 6

1. The relationship between the standard error and the standard deviation is $\sigma_{\bar{Y}} = \sigma_Y/\sqrt{N}$ where $\sigma_{\bar{Y}} = \sigma_Y/\sqrt{N}$ is the standard error of the mean and σ_Y is the standard deviation. Since σ_Y is divided by $\sqrt{N}$, $\sigma_{\bar{Y}} = \sigma_Y/\sqrt{N}$ must always be smaller than σ_Y, except in the trivial case where $N = 1$. Theoretically, the dispersion of the mean must be less than the dispersion of the raw scores. This implies that the standard error of the mean is less than the standard deviation.

3. a. The mean number of active military personnel per region in 2009 was

   $$\bar{Y} = \frac{1,082,228}{9} = 120,247.6$$

   with a standard deviation of 119,819.

   b. Ten sample means calculated from samples of size 3:

   223,410.7
   52,547.0
   84,785.0
   165,595.3
   53,129.3
   142,018.0
   88,199.0
   193,251.3
   41,593.0
   89,428.0

   c. The mean of these 10 means is 113,395.66. Right away we notice that the population mean and the mean of the sampling distribution are somewhat close, a feature we should come to expect given the fact that $\mu_Y = \mu_{\bar{Y}}$.

   d. The standard deviation for all regions is 119,819. The standard error is calculated using the following equation:

   $$\sigma_{\bar{Y}} = \frac{\sigma_Y}{\sqrt{N}} = \frac{119,819}{\sqrt{3}} = 69,177.53$$

   We know that the value of the standard error will always be less than the value of the standard deviation in the population.

e. The population distribution is positively skewed and not close to normal. Because a very small sample size is used in this problem, the histogram for the samples of size 3 does not look normal. The distributions appear unimodal. The fact that the sample distribution of the means tends toward normality because of the central limit theorem would become even more apparent if we took samples of size 5 or 6.

f. We treated the distribution for all regions as the population distribution.

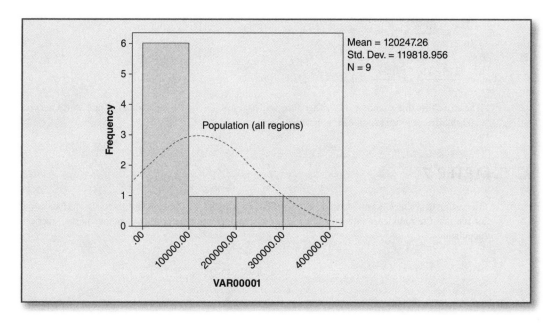

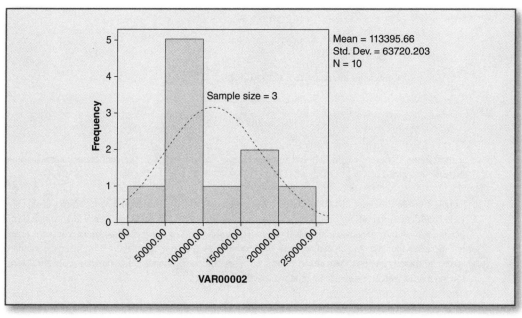

5. a. Mean = 5.3; standard deviation = 3.27.
   b. Here are 10 means from random samples of size 3: 6.33, 5.67, 3.33, 5.00, 7.33, 2.33, 6.00, 6.33, 7.00, 3.00.
   c. The mean of these 10 sample means is 5.23. The standard deviation is 1.76. The mean of the sample means is very close to the mean for the population. The standard deviation of the sample means is much less than the standard deviation for the population. The standard deviation of the means from the samples is an estimate of the standard error of the mean we would find from one random sample of size 3.

7.
a. $\dfrac{1.734}{\sqrt{100}} = \dfrac{1.734}{10} = .1734$

b. $Z = \dfrac{2.00 - 1.97}{.1734} = \dfrac{.03}{.1734} = .17$

The area above this Z score is .4325. Therefore, we'd have a .4325 chance of drawing a sample with a mean of 2.00 children or greater from this "population."

◉ CHAPTER 7

1. a. The estimate at the 90% confidence level is 16.035% to 16.365%. This means that there are 90 chances out of 100 that the confidence interval will contain the true population percentage of victims in the American population.

$$\text{Standard error} = \sqrt{\dfrac{(16.2)(100 - 16.2)}{143,120}} = 0.10$$

$$\text{Confidence interval} = 16.2 \pm 1.65(0.10)$$

$$= 16.2 \pm 0.165$$

$$= 16.035 \text{ to } 16.365$$

b.
$$\text{Confidence interval} = 16.2 \pm 2.58(0.10)$$

$$= 16.2 \pm 0.258$$

$$= 15.942 \text{ to } 16.458$$

   c. If the sample size were cut in half and the percentage remained at 16.2, the confidence intervals would increase by a factor of $1/\sqrt{\dfrac{1}{2}}$, or about 41%.

   d. Sample sizes on the order of 1,000 to 1,500 are a trade-off between precision and cost. Samples of that size yield confidence intervals for proportions with errors around ±3% at the 95% confidence level, which is sufficient for most purposes in applied social research. Doubling the sample size to 3,000 does reduce the errors but increases the cost. As you can see with this example, such a large sample size minimizes the standard error and thereby narrows the confidence interval. Therefore, our estimates of victimization are fairly accurate no matter which confidence interval we select.

3. a.

$$S_p = \sqrt{\frac{(0.39)(1-0.39)}{1,511}} = 0.013$$

Confidence interval $= 0.39 \pm 1.96(0.013)$

$$= 0.39 \pm 0.025$$

$$= 0.365 \text{ to } 0.415$$

b.

Confidence interval $= 0.39 \pm 2.58(0.013)$

$$= 0.39 \pm 0.034$$

$$= 0.356 \text{ to } 0.424$$

c. There is very little difference between the 95% and 99% confidence intervals here because the sample size is reasonably large. The former interval is only one-half of a percentage point wide, the latter nearly two-thirds of a percentage point. Most large survey organizations use the 95% confidence interval routinely, and that seems like the best choice here. Our conclusions about Americans' opinions about global warming will be the same in either case. The intent of this problem is to get students to recognize that they always have a choice as to what confidence interval they choose for a particular problem.

5.

$$\text{Standard error} = \sqrt{\frac{(51)(100-51)}{5,490}} = 0.67$$

Confidence interval $= 51 \pm 1.96(0.67)$

$$= 49.69\% \text{ to } 52.31\%$$

We set the interval at the 95% confidence level. However, no matter whether the 90%, 95%, or 99% confidence level is chosen, the calculated interval includes values below 50% for the vote for a Republican candidate. Therefore, you should tell your supervisors that it would not be possible to declare a Republican candidate the likely winner of the votes coming from men if there were an election today because it seems quite possible that less than a majority of male voters would support her or him.

7. a.

$$S_p = \sqrt{\frac{(0.727)(1-0.727)}{1,500}} = 0.012$$

Confidence interval $= 0.727 \pm 1.96(0.012)$

$$= 0.727 \pm 0.024$$

$$= 0.703 \text{ to } 0.751 \text{ or } 70.3\% \text{ to } 75.1\%$$

b. Based on our answer to 7a, we know that a 90% confidence interval will be more precise than a 95% confidence interval that has a lower bound of 70.3% and an upper bound of 75.1%. Accordingly, a 90% confidence interval will have a lower bound that is greater than 70.3% and an upper bound that is less than 75.1%. Additionally, we know that a 99% confidence interval will be less precise than what we calculated in 7a. Thus, the lower bound for a 99% confidence interval will be less than 70.3% and the upper bound will be greater than 75.1%.

9.

$$S_{\bar{Y}} = \frac{S_Y}{\sqrt{N}} = \frac{1.73}{\sqrt{1,496}} = 0.045$$

$$\text{Confidence interval} = 1.97 \pm 1.65(0.045)$$

$$= 1.97 \pm 0.074$$

$$= 1.896 \text{ to } 2.044$$

11.

$$S_p = \sqrt{\frac{(21)(100-21)}{2,257}} = 0.86$$

$$\text{Confidence interval} = 21 \pm 1.96(0.86)$$

$$= 21 \pm 1.69$$

$$= 19.31\% \text{ to } 22.69\%$$

13. a. For those who thought that homosexual relations are always wrong:

$$S_p = \sqrt{\frac{(50.2)(100-50.2)}{930}} = 1.64$$

$$\text{Confidence interval} = 50.2 \pm 1.96(1.64)$$

$$= 50.2 \pm 3.21$$

$$= 46.99\% \text{ to } 53.41\%$$

For those who thought that homosexual relations are not wrong at all:

$$S_p = \sqrt{\frac{(37.2)(100-37.2)}{930}} = 1.58$$

$$\text{Confidence interval} = 37.2 \pm 1.96(1.58)$$

$$= 37.2 \pm 3.10$$

$$= 34.10\% \text{ to } 40.30\%$$

b.

$$S_p = \sqrt{\frac{(13)(100-13)}{930}} = 1.10$$

$$\text{Confidence interval} = 13 \pm 1.96(1.10)$$

$$= 13 \pm 2.16$$

$$= 10.84\% \text{ to } 15.16\%$$

c. Because the 95% confidence interval for those who think that homosexual relations are always wrong does include a value less than 50%, we cannot have a definite conclusion that the majority of the American public thinks that homosexual relations are always wrong.

15. a. The 95% confidence interval is 2.47 to 2.60. It does not contain 2.3, so we can safely say that the population parameter is not exactly 2.3 children at this confidence level.

b. First, we must find the estimated standard error.

$$\frac{.940}{\sqrt{865}} = \frac{.940}{29.41} = .032$$

$$254 \pm (2.58).032 = 2.54 \pm .083 = 2.457 \text{ to } 2.623$$

This interval still does not contain 2.3 children. Once again, we can safely say that the true population parameter for the ideal number of children Americans want is not 2.3.

▣ CHAPTER 8

Please note that in this chapter, small differences in calculations may occur between student results and those listed below because of rounding.

1. a. $H_0: \mu_Y = 13.5$ years; $H_1: \mu_Y < 13.5$ years.
   b. The Z value obtained is -4.19. The P value for a Z of -4.19 is less than .001 for a one-tailed test. This is less than the alpha of .01, so we reject the null hypothesis and conclude that the doctors at the HMO do have less experience than the population of doctors at all HMOs.

3. a. Two-tailed test, $\mu_1 \neq \$50,054$; null hypothesis, $\mu_1 = \$50,054$
   b. One-tailed test, $\mu_1 > 3.2$; null hypothesis, $\mu_1 = 3.2$
   c. One-tailed test, $\mu_1 < \mu_2$; null hypothesis, $\mu_1 = \mu_2$
   d. Two-tailed test, $\mu_1 \neq \mu_2$; null hypothesis, $\mu_1 = \mu_2$
   e. One-tailed test, $\mu_1 > \mu_2$; null hypothesis, $\mu_1 = \mu_2$
   f. One-tailed test, $\mu_1 < \mu_2$; null hypothesis, $\mu_1 = \mu_2$

5. a. $H_0: \mu_1 = 37.2; H_1: \mu_1 \neq 37.2$

   b. The t value obtained is 48.32, and its P value is <.001.

$$t = \frac{49.28 - 37.2}{17.21 \big/ \sqrt{4857}} = \frac{12.08}{.25} = 48.32$$

   c. We conclude that we can reject the null hypothesis in favor of the research hypothesis. There is a difference between the mean age of the GSS sample and the mean age of all American adults. Relative to age, the GSS sample is not representative of all American adults (the GSS sample is significantly older).

7. a. The appropriate test statistic is Z for proportions.
   b. Z obtained is -5.00, $P <.0001$. Since P (.0001) $< .05$, we reject the null hypothesis. This indicates that there is a statistical difference between conservatives and liberals in their views on affirmative action. Liberals are more likely to support affirmative action policies in the workplace than conservatives.

$$S_{p_1 - p_2} = \sqrt{\frac{.12(1-.12)}{336} = + \frac{.27(1-.27)}{267}} = .03$$

$$Z = \frac{.12 - .27}{.03} = -5.00$$

    c. If alpha were .01, there would be no change to our final decision (.0002 < .01).

9.   a. "Less than" indicates a one-tailed test.

    b. $Z = -3.00$ with a significance of .0014. We can reject the null hypothesis and conclude that the proportion of men who support President Obama is significantly less than proportion of female voters who support the president (.49 − .58 = .09).

    c. The significance of −3.00 is less than .01 (.0014 < .01). The decision to reject the null hypothesis does not change.

11. a. The t obtained = −1.17. We fail to reject the null hypothesis. Based on 123 degrees of freedom, the t obtained is less than the t critical of 1.658.

$$S_{\bar{Y}_1 - \bar{Y}_2} = 7.25(.19) = 1.377 = 1.38$$

$$t = \frac{5.71 - 7.32}{1.38} = -1.17$$

    b. The t obtained = 3.23. We reject the null hypothesis. Based on 355 degrees of freedom, the t obtained is larger than the t critical of 1.960. High school graduates spend more hours per week watching television than college graduates. The difference of .84 hours (3.25 − 2.41) is significant.

$$S_{\bar{Y}_1 - \bar{Y}_2} = 2.37(.11) = .2607 = .26$$

$$t = \frac{3.25 - 2.41}{.26} = 3.23$$

13. Based on the t obtained of −8.593 (equal variances assumed), we reject the null hypothesis. The probability of obtaining this t statistic is .000 (less than our alpha of .05). Respondents with high school degrees have their first children at a younger age than respondents with bachelor's degrees. The age difference between the two groups is 4.09 years (22.66 − 26.75).

回 CHAPTER 9

1.   a. The independent variable is race; the dependent variable is fear of walking alone at night.

	Race	
Fear of Walking Alone at Night	Black	White
Yes	3	4
No	5	9

b. Approximately 69% of whites (69.2%) are not afraid to walk alone in their neighborhoods at night, whereas approximately 63% of blacks (62.5%) are not afraid to walk alone. This amounts to about a 7% difference (69.2% − 62.5%) between whites and blacks who are not afraid to walk alone at night, indicating a weak relationship. Also, although we went ahead and compared percentage differences in this exercise, it is important to keep in mind that our sample size inhibits our ability to make any meaningful comparisons.

	Race	
Fear of Walking Alone at Night	*Black*	*White*
Yes	37.5%	30.8%
No	62.5%	69.2%

c. There is some difference in fears between home owners and renters. A total of 25.0% of home owners and 38.5% of renters are afraid to walk in their neighborhoods at night. The difference between the two groups is 13.5%. Thus, there is a weak to moderate relationship between home ownership and fear of walking in one's neighborhood at night.

	Home Ownership	
Fear of Walking Alone at Night	*Yes*	*No*
Yes	2	5
Percentage	25.0%	38.5%
No	6	8
Percentage	75.0%	61.5%

3. a. Based on the student's argument the independent variable is attitude toward homosexual relations and the dependent variable is political views.
   b. $451/792 = 56.9\%$
   c. Those who believe that homosexuality is always wrong are more likely to be conservative (50.8%) than moderate or liberal. On the other hand, those who believe that homosexuality is not wrong at all are more likely to indicate liberal political views (45.4%) than moderate or conservative.
   d.

	f_o	f_e	$f_o - f_e$	$(f_o - f_e)^2$	$\dfrac{(f_0 - f_e)^2}{f_e}$
Always wrong/liberal	82	134.96	52.96	2,804.76	20.78
Always wrong/moderate	140	148.05	8.05	64.8025	0.44
Always wrong/conservative	229	167.99	61.01	3,722.2201	22.16
Not wrong at all/liberal	155	102.04	52.96	2,804.76	27.49
Not wrong at all/moderate	120	111.94	8.06	64.9636	0.58
Not wrong at all/ conservative	66	127.01	61.01	3,722.2201	29.31
$\chi^2 = 100.76$					

Chi-square obtained, 100.76, is larger than chi-square critical of 9.210 (2 degrees of freedom, alpha = .01). The null hypothesis is rejected. There is a relationship between attitudes about homosexuality and general political views.

5. In contrast with black and white male students, Hispanic male students are less likely to report being moderately or very drunk. The majority of white men and black men report being moderately or very drunk. For Hispanic men, the total percentage in these two categories is 44.5%, which is lower than the totals for the other groups (51.4% of white students and 51.7% of black students). The relationship between race and getting drunk while drinking alcohol is weak.

alchhowdrunk When you drink, how drunk do you get? * race Respondent's race (trichotomized B/W/H) Crosstabulation[a]

			race Respondent's race (trichotomized B/W/H)			
			1 BLACK:(1)	2 WHITE:(2)	3 HISPANIC: (3)	Total
alchhowdrunk When you drink, how drunk do you get?	1 NOT @ALL:(1)	Count	10	71	14	95
		% within race Respondent's race (trichotomized B/W/H)	34.5%	25.5%	25.9%	26.3%
	2 A LITTLE:(2)	Count	4	64	16	84
		% within race Respondent's race (trichotomized B/W/H)	13.8%	23.0%	29.6%	23.3%
	3 MODERATE:(3)	Count	10	108	19	137
		% within race Respondent's race (trichotomized B/W/H)	34.5%	38.8%	35.2%	38.0%
	4 VERY:(4)	Count	5	35	5	45
		% within race Respondent's race (trichotomized B/W/H)	17.2%	12.6%	9.3%	12.5%
Total		Count	29	278	54	361
		% within race Respondent's race (trichotomized B/W/H)	100.0%	100.0%	100.0%	100.0%

a. sex Respondent's sex = 1 MALE:(1)

7. Female seniors have higher educational expectations than male seniors. For example, 73.9% (32.6 + 41.3) of female students expected to complete a bachelor's degree or higher. This is higher than the combined percentage for male students, 63.3% (34.4 + 28.9).

9. Based on the SPSS output, we would fail to reject the null hypothesis. The obtained chi-square is 4.872, significant at the .771 level. Teen residence and marijuana access are not associated.

11. a. Ignoring the sex of the offender, we would make 1,730 errors. $E_1 = 5,940 - 4,210 = 1,730$.
    b. Considering the sex of the offender to predict the sex of the victim, we would make 1,730 errors. For male offenders, we would make 1,590 errors, and for female offenders, we would make 140 errors.
    c. Lambda = (1,730 - 1,730)/1,730 = 0.0% Information about the sex the of the offender reduces our error in predicting the sex of the victim by (0.0 × 100).

13. SPANKING and SEX: gamma = .118. There's a very weak, positive relationship between being male and opinion of spanking.

FAVOR SPANKING TO DISCIPLINE CHILD * RESPONDENTS SEX Crosstabulation

Count

		RESPONDENTS SEX		
		MALE	FEMALE	Total
FAVOR SPANKING TO DISCIPLINE CHILD	STRONGLY AGREE	120	121	241
	AGREE	217	260	477
	DISAGREE	104	126	230
	STRONGLY DISAGREE	15	44	59
Total		456	551	1007

SPANKING and CLASS: gamma = −.178. There's a weak, negative relationship between class and opinion of spanking.

FAVOR SPANKING TO DISCIPLINE CHILD * SUBJECTIVE CLASS IDENTIFICATION Crosstabulation

Count

		SUBJECTIVE CLASS IDENTIFICATION				
		LOWER CLASS	WORKING CLASS	MIDDLE CLASS	UPPER CLASS	Total
FAVOR SPANKING TO DISCIPLINE CHILD	STRONGLY AGREE	23	128	85	3	239
	AGREE	39	224	195	15	473
	DISAGREE	19	84	121	4	228
	STRONGLY DISAGREE	5	19	33	2	59
Total		86	455	434	24	999

SPANKING and MARITAL STATUS: lambda = 0.0. There's no relationship between marital status and opinion of spanking.

FAVOR SPANKING TO DISCIPLINE CHILD * MARITAL STATUS Crosstabulation

Count

		MARITAL STATUS					
		MARRIED	WIDOWED	DIVORCED	SEPARATED	NEVER MARRIED	Total
FAVOR SPANKING TO DISCIPLINE CHILD	STRONGLY AGREE	120	20	29	8	64	241
	AGREE	245	33	70	12	117	477
	DISAGREE	99	23	43	9	56	230
	STRONGLY DISAGREE	31	6	6	2	14	59
Total		495	82	148	31	251	1007

▣ CHAPTER 10

1.

$\bar{Y}_1 = 2.875$	$\bar{Y}_2 = 2.250$	$\bar{Y}_3 = 2.00$	$\bar{Y}_4 = 1.375$
$\Sigma Y_1 = 23$	$\Sigma Y_2 = 18$	$\Sigma Y_3 = 16$	$\Sigma Y_4 = 11$
$\Sigma Y_1^2 = 71$	$\Sigma Y_2^2 = 44$	$\Sigma Y_3^2 = 38$	$\Sigma Y_4^2 = 17$
$n_1 = 8$	$n_2 = 8$	$n_3 = 8$	$n_4 = 8$
$\bar{Y} = 2.125$			
$N = 32$			

$$\begin{aligned} SSB &= 8(2.875 - 2.125)^2 + 8(2.250 - 2.125)^2 + 8(2.00 - 2.125)^2 + \\ &\quad 8(1.375 - 2.125)^2 \\ &= 8(0.5625) + 8(.015625) + 8(.015625) + 8(.5625) \\ &= 4.5 + .125 + .125 + 4.5 \end{aligned}$$

$$SSB = 9.25$$

$$dfb = 4 - 1$$
$$dfb = 3$$

Mean square between $= 9.25/3 = 3.08$

$$\begin{aligned} SSW &= (71 + 44 + 38 + 17) - [(23^2/8) + (18^2/8) + (16^2/8) + (11^2/8)] \\ &= 170 - (66.125 + 40.5 + 32 + 15.125) \\ &= 170 - 153.75 \end{aligned}$$

$$SSW = 16.25$$

$$dfw = 32 - 4$$
$$dfw = 28$$

Mean square within $= 16.25/28 = 0.58$

$$F = 3.08/0.58$$
$$F = 5.31$$

Decision: If we set alpha at 0.05, F critical would be 2.95 ($df_1 = 3$ and $df_2 = 28$). Based on our F obtained of 5.31, we would reject the null hypothesis and conclude that at least one of the means is significantly different than the others. Upper-class respondents rate their health the highest (1.375), followed by middle- and working-class respondents (2.00 and 2.25, respectively) and lower-class respondents (2.875) on a scale where 1 = excellent and 4 = poor.

3. a.

$\bar{Y}_1 = 1.6$	$\bar{Y}_2 = 1.4$	$\bar{Y}_3 = 0.6$
$\Sigma Y_1 = 16$	$\Sigma Y_2 = 14$	$\Sigma Y_3 = 6$
$\Sigma Y_1^2 = 30$	$\Sigma Y_2^2 = 24$	$\Sigma Y_3^2 = 8$
$n_1 = 10$	$n_2 = 10$	$n_3 = 10$
$\bar{Y} = 1.2$		
$N = 30$		

$$SSB = 10(1.6 - 1.2)^2 + 10(1.4 - 1.2)^2 + 10(0.6 - 1.2)^2$$
$$= 10(0.16) + 10(0.04) + 10(0.36)$$
$$= 1.6 + 0.4 + 3.6$$

$$SSB = 5.6$$
$$dfb = 3 - 1$$
$$\mathbf{dfb = 2}$$

Mean square between $= 5.6/2 = \mathbf{2.8}$

$$SSW = (30 + 24 + 8) - [(16^2/10) + (14^2/10) + (6^2/10)]$$
$$= 62 - (25.6 + 19.6 + 3.6)$$
$$= 62 - 48.8$$
$$SSW = 13.2$$
$$dfw = 30 - 3$$
$$\mathbf{dfw = 27}$$

Mean square within $= 13.2/27 = \mathbf{0.488889}$

$$F = 2.8/0.49$$
$$\mathbf{F = 5.71}$$

Decision: If we set alpha at .05, F critical would be 3.35. Based on our F obtained of 5.71, we would reject the null hypothesis and conclude that at least one of the means is significantly different than the others. Respondents with no degrees rate their church attendance highest (1.6), followed by respondents with secondary degrees (1.4), and then respondents with university degrees (0.6).

b. If we set alpha to .01, we would still reject the null hypothesis. F critical would be 5.49, less than the F obtained of 5.71.

5. The calculated F ratio is .070, significant at the .991 level. We would fail to reject the null hypothesis of no difference between the group means.

7. a.

$\bar{Y}_1 = 4.29$ $\bar{Y}_2 = 2.29$ $\bar{Y}_3 = 3.14$

$\sum Y_1 = 30$ $\sum Y_2 = 16$ $\sum Y_3 = 22$

$\sum Y_1^2 = 134$ $\sum Y_2^2 = 44$ $\sum Y_3^2 = 84$

$n_1 = 7$ $n_2 = 7$ $n_3 = 7$

$\bar{Y} = 3.24$

$N = 21$

$$SSB = 7(4.29 - 3.24)^2 + 7(2.29 - 3.24)^2 + 7(3.14 - 3.24)^2$$

$$= 7(1.10) + 7(0.90) + 7(0.01)$$

$$= 7.70 + 6.30 + 0.07$$

$$SSB = 14.07$$

$$df_b = 3 - 1$$

$$df_b = 2$$

$$\text{Mean square between} = 14.07/2 = 7.035$$

$$SSW = (134 + 44 + 84) - [(30^2/7) + (16^2/7) + (22^2/7)]$$

$$= 262 - (128.57 + 36.57 + 69.14)$$

$$= 262 - 234.28$$

$$SSW = 27.72$$

$$df_w = 21 - 3$$

$$df_w = 18$$

$$\text{Mean square within} = 27.72/18 = 1.54$$

$$F = 7.035/1.54$$

$$F = 4.57$$

Decision: If we set alpha at .05, F critical would be 3.55 ($df_1 = 2$ and $df_2 = 18$). Based on our F obtained of 4.57, we would reject the null hypothesis and conclude that at least one of the means is significantly different from the others. On average, white respondents have the highest number of school days missed in the past 4 weeks (4.29), followed by Hispanic respondents (3.14), and then black respondents (2.29).

b. If alpha were changed to .01, F critical would be 6.01. We would fail to reject the null hypothesis at this alpha level.

9. For each sociocultural resource we would reject the null hypothesis. For social support, the obtained F ratio is 12.17, $P < .001$. Whites report the highest level of social support (2.85) while non-Cuban Hispanics have the lowest (2.58). For religious attendance, the obtained F ratio is 56.43, $P < .001$. Church attendance is highest for African Americans and non-Cuban Hispanics in the sample (3.94 and 3.37 on the five-point scale).

11. Based on the F obtained of 7.318, we reject the null hypothesis of no difference and conclude that there is a significant difference in GPA among these three student groups. High school GPA is highest for white students, followed by Hispanic and black students.

▣ CHAPTER 11

1. a. On the scatterplot below, the regression line has been plotted to make it easier to see the relationship between the two variables.

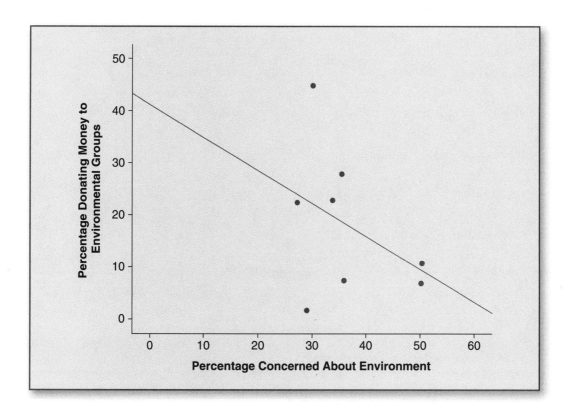

b. The scatterplot shows that there is a general linear relationship between the two variables. There is not a lot of scatter about the straight line describing the relationship. As the percentage of respondents concerned about the environment increases, the percentage of respondents donating money to environmental groups decreases.

c. The Pearson correlation coefficient between the two variables is –.40. This is consistent with the scatterplot, which indicates a negative relationship between being concerned about the environment and actually donating money to environmental groups.

	(1)	(2)	(3)	(4)	(5)	(6)	(7)
	Percentage Concerned	Percentage Donating					
Country	X	Y	$(X-\bar{X})$	$(X-\bar{X})^2$	$(Y-\bar{Y})$	$(Y-\bar{Y})^2$	$(X-\bar{X})(Y-\bar{Y})$
United States	33.8	22.8	−2.69	7.22	4.77	22.80	−12.83
Austria	35.5	27.8	−0.99	0.98	9.77	95.55	−9.67
Netherlands	30.1	44.8	−6.39	40.80	26.77	716.90	−171.06
Slovenia	50.3	10.7	13.81	190.79	−7.33	53.66	−101.23
Russia	29.0	1.6	−7.49	56.06	−16.43	269.78	123.06
Philippines	50.1	6.8	13.61	185.30	−11.23	126.00	−152.84
Spain	35.9	7.4	−0.59	0.35	−10.63	112.89	6.27
Denmark	27.2	22.3	−9.29	86.26	4.27	18.28	−39.67
	$\sum X = 291.9$	$\sum Y = 144.2$	−0.02[a]	567.76	0.04[a]	1,415.85	−357.97

$$\text{Mean } X = \bar{X} = \frac{\sum X}{N} = \frac{291.9}{8} = 36.49$$

$$\text{Mean } Y = \bar{Y} = \frac{\sum Y}{N} = \frac{144.2}{8} = 18.03$$

$$\text{Variance}(X) = S_X^2 = \frac{\Sigma(X-\bar{X})^2}{N-1} = \frac{567.8}{7} = 81.11$$

$$\text{Standard deviation}(X) = S_X = \sqrt{81.11} = 9.01$$

$$\text{Variance}(Y) = S_Y^2 = \frac{\Sigma(Y-\bar{Y})^2}{N-1} = \frac{1,415.9}{7} = 202.3$$

$$\text{Standard deviation}(Y) = S_Y = \sqrt{202.3} = 14.22$$

$$\text{Covariance}(X,Y) = S_{XY} = \frac{\Sigma(X-\bar{X})(Y-\bar{Y})}{N-1} = \frac{-357.97}{7} = -51.14$$

$$r = \frac{S_{XY}}{S_X S_Y} = \frac{-51.14}{(9.01)(14.22)} = -0.40[a]$$

a. Answers may differ slightly because of rounding.

3. a. The correlation coefficient is −0.45.

	(1)	(2)	(3)	(4)	(5)	(6)	(7)
	GNP per Capita	Percentage Willing to Pay					
Country	X	Y	$(X-\bar{X})$	$(X-\bar{X})^2$	$(Y-\bar{Y})$	$(Y-\bar{Y})^2$	$(X-\bar{X})(Y-\bar{Y})$
United States	29.24	44.9	2.72	7.40	−1.64	2.69	−4.46
Ireland	18.71	53.3	−7.81	61.00	6.76	45.70	−52.80
Netherlands	24.78	61.2	−1.74	3.03	14.66	214.92	−25.51
Norway	34.31	40.7	7.79	60.68	−5.84	34.11	−45.49
Sweden	25.58	32.6	−0.94	0.88	−13.94	194.32	13.10
	$\sum X = 132.62$	$\sum Y = 232.7$	−0.02[a]	132.99	0.04[a]	491.74	−115.16

$$\text{Mean } X = \bar{X} = \frac{\sum X}{N} = \frac{132.62}{5} = 26.52$$

$$\text{Mean } Y = \bar{Y} = \frac{\sum Y}{N} = \frac{232.7}{5} = 46.54$$

$$\text{Variance}(X) = S_X^2 = \frac{\sum(X-\bar{X})^2}{N-1} = \frac{132.99}{4} = 33.25$$

$$\text{Standard deviation}(X) = S_X = \sqrt{33.25} = 5.77$$

$$\text{Variance}(Y) = S_Y^2 = \frac{\sum(Y-\bar{Y})^2}{N-1} = \frac{491.74}{4} = 122.94$$

$$\text{Standard deviation}(Y) = S_Y = \sqrt{122.94} = 11.09$$

$$\text{Covariance}(X,Y) = S_{XY} = \frac{\sum(X-\bar{X})(Y-\bar{Y})}{N-1} = \frac{-115.16}{4} = -28.79$$

$$r = \frac{S_{XY}}{S_X S_Y} = \frac{-28.79}{(5.77)(11.09)} = -0.45^{a}$$

a. Answers may differ slightly because of rounding.

   b. A correlation coefficient of −0.45 means that relatively high values of GNP are moderately negatively associated with low values of percentage of residents willing to pay higher prices to protect the environment.

5. Although somewhat difficult to visually determine, there is evidence within the scatterplot that indicates a negative relationship between education and hours per day spent watching TV. Looking to the regression equation output, this is confirmed by the negative coefficient (−.231). This means that for each year of education a person has, the number of hours he or she spends watching TV per day decreases by 0.231 hours. The value of the intercept (6.130) indicates that a person with 0 years of education spends 6.130 hr/day watching television. The value for r (.268) suggests a moderate relationship between education and hours per day spent watching television. The value of r^2 is .072; thus, 7.2 percent of the variation in hours per day spent watching TV can be explained by taking into account a person's level of education.

7. a. The scatterplot for GDP per capita and birthrate can be summarized with a straight line or linear relationship.

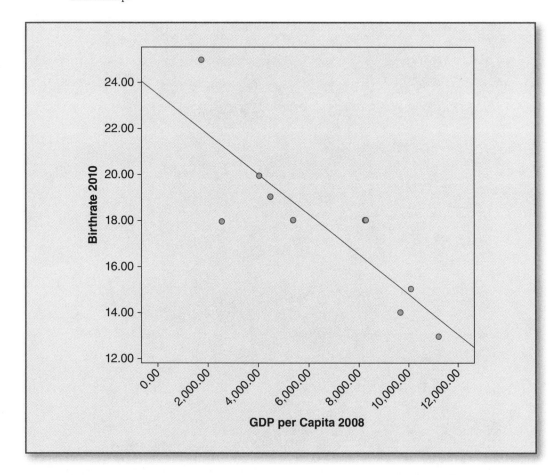

   b. Students must be careful and not calculate the coefficient of determination first and then take the square root to get r, because they won't get the negative sign for r.

   c. The relationship between GDP per capita and birthrate is strong and is negative, with GDP explaining about 73.1% of the variation in birthrate. As a country becomes wealthier, couples have fewer children.

	(1)	(2)	(3)	(4)	(5)	(6)	(7)
	GDP per Capita	Birthrate					
Country	X	Y	$(X-\bar{X})$	$(X-\bar{X})^2$	$(Y-\bar{Y})$	$(Y-\bar{Y})^2$	$(X-\bar{X})(Y-\bar{Y})$
Argentina	8,236	18	1,670.5	2,790,570.25	0.2	0.04	334.1
Bolivia	1,720	25	−4,845.5	23,478,870.25	7.2	51.84	−34,887.6
Brazil	8,205	18	1,639.5	2,687,960.25	0.2	0.04	327.9
Chile	10,084	15	3,518.5	12,379,842.25	−2.8	7.84	−9,851.8
Colombia	5,416	18	−1,149.5	1,321,350.25	0.2	0.04	−229.9
Ecuador	4,056	20	−2,509.5	6,297,590.25	2.2	4.84	−5,520.9
Paraguay	2,561	18	−4,004.5	16,036,020.25	0.2	0.04	−800.9
Peru	4,477	19	−2,088.5	4,361,832.25	1.2	1.44	−2,506.2
Uruguay	9,654	14	3,088.5	9,538,832.25	−3.8	14.44	−11,736.3
Venezuela	11,246	13	4,680.5	21,907,080.25	−4.8	23.04	−22,466.4
	$\Sigma X = 65{,}655$	$\Sigma Y = 178$	0.0^a	100,799,948.50	0.0^a	103.60	−87,338

$$\text{Mean } X = \bar{X} = \frac{\Sigma X}{N} = \frac{65{,}655}{10} = 6{,}565.5$$

$$\text{Mean } Y = \bar{Y} = \frac{\Sigma Y}{N} = \frac{178}{10} = 17.8$$

$$\text{Variance}(X) = S_Y^2 = \frac{\Sigma(Y-\bar{Y})^2}{N-1} = \frac{100{,}799{,}948.5}{9} = 11{,}199{,}994.28$$

$$\text{Standard deviation}(X) = S_X = \sqrt{11{,}199{,}994.28} = 3{,}346.64$$

$$\text{Variance}(Y) = S_X^2 = \frac{\Sigma(X-\bar{X})^2}{N-1} = \frac{103.6}{9} = 11.51$$

$$\text{Standard deviation}(Y) = S_Y = \sqrt{11.51} = 3.39$$

$$\text{Covariance}(X,Y) = S_{XY} = \frac{\Sigma(X-\bar{X})(Y-\bar{Y})}{N-1} = \frac{-87{,}338}{9} = -9{,}704.22$$

$$r^2 = \frac{S_{XY}^2}{S_X^2 S_Y^2} = \frac{(-9{,}704.22)^2}{(11{,}199{,}994.28)(11.51)} = .731^*$$

$$r = \sqrt{r^2} = \sqrt{.731} = .855 \ (\textit{Note}: r \text{ is actually } -.855, \text{ but students might leave it as positive.})$$

a. Answers may differ slightly because of rounding.

9. a. The equation is $\hat{Y} = 24.95 + .157X$, indicating a positive relationship between GNP per capita and willingness to pay more in taxes.

$$b = \frac{S_{XY}}{S^2_X} = \frac{22.73}{144.93} = 0.157\,*$$

$$a = \bar{Y} - b\bar{X} = 27.81 - (.157)(18.20) = 24.95\,^*$$

$$\hat{Y} = 24.95 + .157X$$

*Answers may differ because of rounding.

b. $\hat{Y} = 24.95 + .157(3) = 25.42\%$ are willing to pay more in taxes.

$\hat{Y} = 24.95 + .157(30) = 29.66\%$ are willing to pay more in taxes.

	(1) GNP per capita	(2) % Willing to Pay Higher Taxes	(3)	(4)	(5)	(6)	(7)
Country	X	Y	$(X-\bar{X})$	$(X-\bar{X})^2$	$(Y-\bar{Y})$	$(Y-\bar{Y})^2$	$(X-\bar{X})(Y-\bar{Y})$
Canada	19.71	24.0	1.51	2.28	−3.81	14.52	−5.75
Chile	4.99	29.1	−13.21	174.50	1.29	1.66	−17.04
Finland	24.28	12.0	6.08	36.97	−15.81	249.96	−96.12
Ireland	18.71	34.3	.51	.26	6.49	42.12	3.31
Japan	32.35	37.2	14.15	200.22	9.39	88.17	132.87
Latvia	2.42	17.3	−15.78	249.01	−10.51	110.46	165.85
Mexico	3.84	34.7	−14.36	206.21	6.89	47.47	−98.94
Netherlands	24.78	51.9	6.58	43.30	24.09	580.33	158.51
Norway	14.60	22.8	−3.60	12.96	−5.01	25.10	18.04
Portugal	34.31	17.1	16.11	259.53	−10.71	114.70	−172.54
Russia	10.67	29.9	−7.53	56.70	2.09	4.37	−15.74
Spain	2.66	22.2	−15.54	241.49	−5.61	31.47	87.18
Sweden	14.10	19.5	−4.10	16.81	−8.31	69.06	34.07
Switzerland	25.58	33.5	7.38	54.46	5.69	32.38	41.99
United States	39.98	31.6	21.78	474.37	3.79	14.36	82.55
	$\Sigma x = 272.98$	$\Sigma y = 417.10$	0.0[a]	2029.08	0.0[a]	1426.13	318.23

$$\text{Mean } X = \frac{\Sigma X}{N} = \frac{272.98}{15} = 18.20$$

$$\text{Mean } Y = \frac{\Sigma Y}{N} = \frac{417.10}{15} = 27.81$$

$$\text{Variance } X = S^2_x = \frac{\Sigma(X-\bar{X})^2}{(N-1)} = \frac{2,029.08}{14} = 144.93$$

$$\text{Standard Deviation } X = S_X = \sqrt{144.93} = 12.04$$

$$\text{Variance } Y = S^2_Y = \frac{\Sigma(Y-\bar{Y})^2}{(N-1)} = \frac{1,426.13}{14} = 101.87$$

$$\text{Standard Deviation } Y = S_Y = \sqrt{101.87} = 10.09$$

$$\text{Covariance } X,Y = S_{XY} = \frac{\Sigma(X-\bar{X})(Y-\bar{Y})}{(N-1)} = \frac{318.23}{14} = 22.73$$

a. Answers may differ slightly due to rounding.

11. a. The Pearson correlation coefficient of .221 indicates a weak, positive relationship between one's number of siblings and the number of children one wishes to have.

b. The Y-intercept of 2.301 means that an only child will be predicted to think that 2.301 children is the ideal number to have. The slope of 0.066 indicates that for each additional brother or sister a respondent has, his or her ideal number of children is predicted to rise by 0.066.

Glossary

Alpha (α) The level of probability at which the null hypothesis is rejected. It is customary to set alpha at the .05, .01, or .001 level.

Analysis of variance (ANOVA) An inferential statistics technique designed to test for a significant relationship between two variables in two or more groups or samples.

Asymmetrical measure of association A measure of association whose value may vary depending on which variable is considered the independent variable and which the dependent variable.

Bar graph A graph showing the differences in frequencies or percentages among the categories of a nominal or an ordinal variable. The categories are displayed as rectangles of equal width with their heights proportional to the frequency or percentage of the category.

Between-group sum of squares (SSB) The sum of squared deviations between each sample mean and the overall mean score.

Bivariate analysis A statistical method designed to detect and describe the relationship between two nominal or ordinal variables.

Bivariate table A table that displays the distribution of one variable across the categories of another variable.

Cell The intersection of a row and a column in a bivariate table.

Central limit theorem If all possible random samples of size N are drawn from a population with a mean μ_Y and a standard deviation σ_Y, then as N becomes larger, the sampling distribution of sample means becomes approximately normal, with mean $\mu_{\bar{Y}}$ equal to the population mean and standard deviation equal to $\sigma_{\bar{Y}} = \sigma_Y / \sqrt{N}$.

Chi-square (obtained) The test statistic that summarizes the differences between the observed (f_o) and the expected (f_e) frequencies in a bivariate table.

Chi-square test An inferential statistics technique designed to test for a significant relationship between two nominal or ordinal variables organized in a bivariate table.

Coefficient of determination (r^2) A PRE measure reflecting the proportional reduction of error that results from using the linear regression model. It reflects the proportion of the total variation in the dependent variable, Y, explained by the independent variable, X.

Column variable A variable whose categories are the columns of a bivariate table.

Confidence interval (CI) A range of values defined by the confidence level within which the population parameter is estimated to fall. Sometimes confidence intervals are referred to as margin of error.

Confidence level The likelihood, expressed as a percentage or a probability, that a specified interval will contain the population parameter.

Control variable An additional variable considered in a bivariate relationship. The variable is controlled for when we take into account its effect on the variables in the bivariate relationship.

Cross-tabulation A technique for analyzing the relationship between two nominal or ordinal variables that have been organized in a table.

Cumulative frequency distribution A distribution showing the frequency at or below each category (class interval or score) of the variable.

Cumulative percentage distribution A distribution showing the percentage at or below each category (class interval or score) of the variable.

Data Information represented by numbers, which can be the subject of statistical analysis.

Degrees of freedom (*df*) The number of scores that are free to vary in calculating a statistic.

Dependent variable The variable to be explained (the "effect").

Descriptive statistics Procedures that help us organize and describe data collected from either a sample or a population.

Deterministic (perfect) linear relationship A relationship between two interval-ratio variables in which all the observations (the dots) fall along a straight line. The line provides a predicted value of Y (the vertical axis) for any value of X (the horizontal axis).

Dichotomous variable A variable that has only two values.

Elaboration A process designed to further explore a bivariate relationship; it involves the introduction of control variables.

Empirical research Research based on evidence that can be verified by using our direct experience.

Estimation A process whereby we select a random sample from a population and use a sample statistic to estimate a population parameter.

Expected frequencies (f_e) The cell frequencies that would be expected in a bivariate table if the two variables were statistically independent.

F critical The F-test statistic that corresponds to the alpha level, df_w, and df_b (as in Appendix D).

F obtained The F-test statistic that is calculated.

F ratio or F statistic The test statistic for ANOVA, calculated as the ratio of mean square to mean square within.

Frequency distribution A table reporting the number of observations falling into each category of the variable.

Gamma A symmetrical measure of association suitable for use with ordinal variables or with dichotomous nominal variables. It can vary from 0.0 to ±1.0 and provides us with an indication of the strength and direction of the association between the variables.

Histogram A graph showing the differences in frequencies or percentages among categories of an interval-ratio variable. The categories are displayed as contiguous bars, each with width proportional to the width of the category and height proportional to the frequency or percentage of that category.

Hypothesis A tentative answer to a research problem.

Independent variable The variable expected to account for (the "cause" of) the dependent variable.

Inferential statistics The logic and procedures concerned with making predictions or inferences about a population from observations and analyses of a sample.

Interquartile range (IQR) The width of the middle 50% of the distribution. It is defined as the difference between the lower and upper quartiles (Q_1 and Q_3).

Interval-ratio measurement Measurements for all cases are expressed in the same units.

Kendall's tau-*b* A symmetrical measure of association suitable for use with ordinal variables. It can vary from 0.0 to ±1.0. It provides an indication of the strength and direction of the association between the variables. Kendall's tau-*b* will always be lower than gamma.

Lambda An asymmetrical measure of association, lambda is suitable for use with nominal variables and may range from 0.0 to 1.0. It provides us with an indication of the strength of an association between the independent and dependent variables.

Least squares line (best-fitting line) A line where the residual sum of squares, or Σe^2, is at a minimum.

Least squares method The technique that produces the least squares line.

Left-tailed test A one-tailed test in which the sample outcome is hypothesized to be at the left tail of the sampling distribution.

Linear relationship A relationship between two interval-ratio variables in which the observations displayed in a scatter diagram can be approximated with a straight line.

Line graph A graph showing the differences in frequencies or percentages among categories of an interval-ratio variable. Points representing the frequencies of each category are placed above the midpoints of the categories and are joined by a straight line.

Marginals The row and column totals in a bivariate table.

Margin of error The radius of a confidence interval.

Mean A measure of central tendency that is obtained by adding up all the scores and dividing by the total number of scores. It is the arithmetic average.

Mean square between Sum of squares between divided by its corresponding degrees of freedom.

Mean square within Sum of squares within divided by its corresponding degrees of freedom.

Measure of association A single summarizing number that reflects the strength of a relationship, indicates the usefulness of predicting the dependent variable from the independent variable, and often shows the direction of the relationship.

Measures of central tendency Categories or scores that describe what is average or typical of the distribution.

Measures of variability Numbers that describe diversity or variability in the distribution of a variable.

Median The score that divides the distribution into two equal parts so that half the cases are above it and half are below it.

Mode The category or score with the highest frequency (or percentage) in the distribution.

Negatively skewed distribution A distribution with a few extremely low values.

Negative relationship A bivariate relationship between two variables measured at the ordinal level or higher in which the variables vary in opposite directions.

Nominal measurement Numbers or other symbols are assigned to a set of categories for the purpose of naming, labeling, or classifying the observations.

Normal distribution A bell-shaped and symmetrical theoretical distribution with the mean, the median, and the mode all coinciding at its peak and with the frequencies gradually decreasing at both ends of the curve.

Null hypothesis (H_0) A statement of "no difference" that contradicts the research hypothesis and is always expressed in terms of population parameters.

Observed frequencies (f_o) The cell frequencies actually observed in a bivariate table.

One-tailed test A type of hypothesis test that involves a directional hypothesis. It specifies that the values of one group are either larger or smaller than some specified population value.

One-way ANOVA Analysis of variance application with one dependent variable and one independent variable.

Ordinal measurement Numbers are assigned to rank-ordered categories ranging from low to high.

Parameter A measure (e.g., mean or standard deviation) used to describe the population distribution.

Pearson's correlation coefficient (r) The square root of r^2; it is a measure of association for interval-ratio variables, reflecting the strength of the linear association between two interval-ratio variables. It can be positive or negative in sign.

Percentage A relative frequency obtained by dividing the frequency in each category by the total number of cases and multiplying by 100.

Percentage distribution A table showing the percentage of observations falling into each category of the variable.

Percentile A score below which a specific percentage of the distribution falls.

Pie chart A graph showing the differences in frequencies or percentages among the categories of a nominal or an ordinal variable. The categories are displayed as segments of a circle whose pieces add up to 100% of the total frequencies.

Point estimate A sample statistic used to estimate the exact value of a population parameter.

Population A group that includes all the cases (individuals, objects, or groups) in which the researcher is interested.

Positively skewed distribution A distribution with a few extremely high values.

Positive relationship A bivariate relationship between two variables measured at the ordinal level or higher in which the variables vary in the same direction.

Proportion A relative frequency obtained by dividing the frequency in each category by the total number of cases.

Proportional reduction of error (PRE) The concept that underlies the definition and interpretation of several measures of association. PRE measures are derived by comparing the errors made in predicting the dependent variable while ignoring the independent variable with errors made when making predictions that use information about the independent variable.

P **value** The probability associated with the obtained value of *Z*.

Range A measure of variation in interval-ratio variables. It is the difference between the highest (maximum) and the lowest (minimum) scores in the distribution.

Rate A number obtained by dividing the number of actual occurrences in a given time period by the number of possible occurrences.

Research hypothesis (H_1) A statement reflecting the substantive hypothesis. It is always expressed in terms of population parameters, but its specific form varies from test to test.

Research process A set of activities in which social scientists engage to answer questions, examine ideas, or test theories.

Right-tailed test A one-tailed test in which the sample outcome is hypothesized to be at the right tail of the sampling distribution.

Row variable A variable whose categories are the rows of a bivariate table.

Sample A subset of cases selected from a population.

Sampling distribution A theoretical probability distribution of all possible sample values for the statistics in which we are interested.

Sampling distribution of the difference between means A theoretical probability distribution that would be obtained by calculating all the possible mean differences that would be obtained by drawing all the possible independent random samples of size N_1 and N_2 from two populations where N_1 and N_2 are both greater than 50.

Sampling distribution of the mean A theoretical probability distribution of sample means that would be obtained by drawing from the population all possible samples of the same size.

Sampling error The discrepancy between a sample estimate of a population parameter and the real population parameter.

Scatter diagram (scatterplot) A visual method used to display a relationship between two interval-ratio variables.

Skewed distribution A distribution with a few extreme values on one side of the distribution.

Slope (*b*) The amount of change in a dependent variable per unit change in an independent variable.

Standard deviation A measure of variation for interval-ratio variables; it is equal to the square root of the variance.

Standard error of the mean The standard deviation of the sampling distribution of the mean. It describes how much dispersion there is in the sampling distribution of the mean.

Standard normal distribution A normal distribution represented in standard (*Z*) scores, with mean = 0 and standard deviation = 1.

Standard normal table A table showing the area (as a proportion, which can be translated into a percentage) under the standard normal curve corresponding to any *Z* score or its fraction.

Standard (*Z*) **score** The number of standard deviations that a given raw score is above or below the mean.

Statistic A measure (e.g., mean or standard deviation) used to describe the sample distribution.

Statistical hypothesis testing A procedure that allows us to evaluate hypotheses about population parameters based on sample statistics.

Statistical independence The absence of association between two cross-tabulated variables. The percentage distributions of the dependent variable within each category of the independent variable are identical.

Statistics A set of procedures used by social scientists to organize, summarize, and communicate information.

Symmetrical distribution The frequencies at the right and left tails of the distribution are identical; each half of the distribution is the mirror image of the other.

Symmetrical measure of association A measure of association whose value will be the same when either variable is considered the independent variable or the dependent variable.

t **distribution** A family of curves, each determined by its degrees of freedom (*df*). It is used when the population standard deviation is unknown and the standard error is estimated from the sample standard deviation.

Theory An elaborate explanation of the relationship between two or more observable attributes of individuals or groups.

Time-series chart A graph displaying changes in a variable at different points in time. It shows time (measured in units such as years or months) on the horizontal axis and the frequencies (percentages or rates) of another variable on the vertical axis.

Total sum of squares (*SST*) The total variation in scores, calculated by adding *SSB* and *SSW*.

t **statistic (obtained)** The test statistic computed to test the null hypothesis about a population mean when the population standard deviation is unknown and is estimated using the sample standard deviation.

Two-tailed test A type of hypothesis test that involves a non-directional research hypothesis. We are equally interested in whether the values are less than or greater than one another. The sample outcome may be located at both the low and high ends of the sampling distribution.

Type I error The probability associated with rejecting a null hypothesis when it is true.

Type II error The probability associated with failing to reject a null hypothesis when it is false.

Unit of analysis The level of social life on which social scientists focus. Examples of different levels are individuals and groups.

Variable A property of people or objects that takes on two or more values.

Variance A measure of variation for interval-ratio variables; it is the average of the squared deviations from the mean.

Within-group sum of squares (*SSW*) Sum of squared deviations within each group, calculated between each individual score and the sample mean.

Y-**intercept** (*a*) The point where the regression line crosses the *Y*-axis, or the value of *Y* when *X* is 0.

Z **statistic (obtained)** The test statistic computed by converting a sample statistic (such as the mean) to a *Z* score. The formula for obtaining *Z* varies from test to test.

Notes

Chapter 1

1. U.S. Bureau of Labor Statistics, *Economic News Release: Usual Weekly Earnings Summary,* January 18, 2013.

2. U.S. Census Bureau, *Statistical Abstract of the United States: 2012,* Table 616.

3. Rampell, Catherine. "Women Now a Majority in American Workplaces," *The New York Times,* February 5, 2010. Retrieved from http://www.nytimes.com/2010/02/06/business/economy/06women.html.

4. Chava Frankfort-Nachmias and David Nachmias, *Research Methods in the Social Sciences* (New York: Worth Publishers, 2000), p. 56.

5. Barbara Reskin and Irene Padavic, *Women and Men at Work* (Thousand Oaks, CA: Pine Forge Press, 2002), pp. 65, 144.

6. Frankfort-Nachmias and Nachmias, 2000, p. 50.

7. Ibid., p. 52.

Chapter 2

1. Jennifer Medina, "New Suburban Dream Born of Asia and Southern California" (*The New York Times,* April 29, 2012), p. A9.

2. Gary Hytrek and Kristine Zentgraf, *America Transformed: Globalization, Inequality and Power* (New York: Oxford University Press, 2007).

3. Elizabeth Grieco, Yesenia Acosta, C. Patricia de la Cruz, Christine Gambino, Thomas Gryn, Luke Larsen, Edward Trevelyan, and Nathan Walters, *The Foreign-Born Population in the United States: 2010* (ACS-19; Washington, DC: U.S. Census Bureau, 2012). Retrieved from http://www.census.gov/prod/2012pubs/acs-19.pdf.

4. Greico et al., 2012.

5. David Knoke and George W. Bohrnstedt, *Basic Social Statistics* (New York: Peacock, 1991), p. 25.

6. Ibid., p. 41.

7. The idea of the "Reading the Research Literature" sections that appear in most chapters was inspired by Joseph F. Healey's *Statistics: A Tool for Social Research.*

8. Yolanda Padilla, Melissa Dalton Radey, Robert Hummer, and Eunjeong Kim, "The Living Conditions of U.S.-Born Children of Mexican Immigrants in Unmarried Families," *Hispanic Journal of Behavioral Sciences* 28, no. 3 (2006), pp. 343–344.

9. Harry Moody, *Aging: Concepts and Controversies* (Thousand Oaks, CA: SAGE, 2010), p. xxiii.

10. The Census Bureau notes that persons of Hispanic origin may be of any race.

11. U.S. Census Bureau, *Marital Status and Living Arrangements: March 1996,* Current Population Reports, P20-496, 1998, p. 5.

12. U.S. Census Bureau, *65+ in America,* Current Population Reports, Special Studies, P23-190, 1996, pp. 2–3.

13. Ibid., p. 6-2.

14. Edward R. Tufte, *The Visual Display of Quantitative Information* (Cheshire, CT: Graphics Press, 1983), p. 53.

Chapter 3

1. U.S. Bureau of Labor Statistics, Current Population Survey 2012, *Household Data Annual Averages,* Table 39.

2. This rule was adapted from David Knoke and George W. Bohrnstedt, *Basic Statistics* (New York: Peacock Publishers, 1991), pp. 56–57.

3. The rates presented in Table 3.4 are computed for aggregate units (countries) of different sizes. The mean of 13.3 is therefore called an unweighted mean. It is not the same as the gun ownership rate for the population in the combined countries.

4. Three variables, TVHOURS, SIBS, and EDUC, were taken from a GSS sample; EDUC was then recoded into another variable including only the respondents without high school diplomas.

Chapter 4

1. Johnneta B. Cole, "Commonalities and Differences," in *Race, Class, and Gender,* eds. Margaret L. Andersen and Patricia Hill Collins (Belmont, CA: Wadsworth, 1998), pp. 128–129.

2. Ibid., pp. 129–130.

3. Recent census data reveal that the recession of 2008–2009 has halted this dominant migration trend.

4. The percentage increase in the population 65 years and older for each state and region was obtained by the following formula:

Percentage increase = [(2015 population − 2008 population) / 2008 population] × 100

5. U.S. Census Bureau, *The Older Population: 2010*, p. 3.

6. $N − 1$ is used in the formula for computing variance because usually we are computing from a sample with the intention of generalizing to a larger population. $N − 1$ in the formula gives a better estimate and is also the formula used in SPSS.

7. A good discussion of the relationship between the standard deviation and the mean can be found in Stephen Gould's "The Median Isn't the Message," *Discover Magazine*, June 1985.

8. Herman J. Loether and Donald G. McTavish, *Descriptive and Inferential Statistics: An Introduction* (Boston: Allyn and Bacon, 1980), pp. 160–161.

9. Stephanie A. Bohon, Monica Kirkpatrick Johnson, and Bridget K. Gorman, "College Aspirations and Expectations Among Latino Adolescents in the United States," *Social Problems* 53, no. 2 (2006): 207–225.

10. Ibid, p. 210.

11. Ibid., p. 213.

12. Ibid.

Chapter 6

This discussion is based on C. Stephen Layman's *The Power of Logic* (Mountain View, CA: Mayfield, 2004).

1. This discussion has benefited from a more extensive presentation on the aims of sampling in Richard Maisel and Caroline Hodges Persell, *How Sampling Works* (Thousand Oaks, CA: Pine Forge Press, 1996).

2. The population of the 20 individuals presented in Table 6.2 is considered a finite population. A finite population consists of a finite (countable) number of elements (observations). Other examples of finite populations include all women in the labor force in 2008 and all public hospitals in New York City. A population is considered infinite when there is no limit to the number of elements it can include. Examples of infinite populations include all women in the labor force, in the past or the future. Most samples studied by social scientists come from finite populations. However, it is also possible to form a sample from an infinite population.

3. *Time Magazine*, U.S. edition, Vol. 180, No. 18, October 29, 2012.

4. *The New York Times*, May 2, 2013.

5. U.S. Census Bureau, American Fact Finder (http://www.factfinder2.census.gov/).

Chapter 7

1. The relationship between sample size and interval width when estimating means also holds true for sample proportions. When the sample size increases, the standard error of the proportion decreases, and therefore, the width of the confidence interval decreases as well.

2. "In U.S., Record-High Say Gay, Lesbian Relations Morally OK," Gallup Poll, May 20, 2013.

3. "Partisan Interest, Reactions to IRS and AP Controversies," Pew Research Center, May 20, 2013.

4. "Most Say Immigration Policy Needs Big Changes," Pew Research Center, May 9, 2013.

5. Data from "More Say There Is Solid Evidence of Global Warming," Pew Research Center, October 15, 2012.

6. Data from "'Enthusiastic' Voters Prefer GOP by 20 Points in 2010 Vote," *Gallup*, April 27, 2010.

7. Data from "22% of Americans Used Social Networking or Twitter for Politics in 2010 Campaign," Pew Research Center, January 27, 2011.

8. Data from "The Millennials: Confident. Connected. Open to Change," Pew Research Center, February 24, 2010.

Chapter 8

1. Jeff Somer, "Numbers That Sway Markets and Voters," 2012. Retrieved from http://www.nytimes.com/2012/03/04/your-money/rising-gasoline-prices-could-soon-have-economic-effects.html?pagewanted=all&_r=0.

2. Steve Hargreaves, "Gas Prices Hit Working Class," 2007. Retrieved from http://money.cnn.com/2007/11/13/news/economy/gas_burden/index.htm.

3. American Automobile Association, *Daily Fuel Gauge Report*, May 7, 2013. Retrieved from www.fuelgaugereport.com/.

4. To compute the sample variance for any particular sample, we must first compute the sample mean. Since the sum of the deviations about the mean must equal 0, only $N − 1$ of the deviation scores are free to vary with each variance estimate.

5. U.S. Census Bureau, *Statistical Abstract of the United States: 2012*, Table 232.

6. Degrees of freedom formula based on Dennis Hinkle, William Wiersma, and Stephen Jurs, *Applied Statistics for the Behavioral Sciences* (Boston: Houghton Mifflin, 1998), p. 268.

7. Lloyd D. Johnson, Patrick M. O'Malley, Jerald G. Bachman, and John E. Schulenberg, *Monitoring the Future National Results on Adolescent Drug Use: Overview of Key Findings, 2008* (Bethesda, MD: National Institute on Drug Abuse, 2009).

8. Pew Research Center, *Second-Generation Americans: A Portrait of the Adult Children of Immigrants*, February 7, 2013.

9. The sample proportions are unbiased estimates of the corresponding population proportions. Therefore, we can use the Z statistic, although our standard error is estimated from the sample proportions.

10. Paula Y. Goodwin, William D. Mosher, and Anjani Chandra, "Marriage and Cohabitation in the United States: A Statistical Portrait Based on Cycle 6 (2002) of the National Survey of Family Growth," *Vital Health Statistics* 23, no. 28 (2010): 1–45.

11. Robert E. Jones and Shirley A. Rainey, "Examining Linkages Between Race, Environmental Concern, Health and Justice in a Highly Polluted Community of Color," *Journal of Black Studies* 36, no. 4 (2006): 473–496.

12. Pew Research Center, *The Gender Gap: Three Decades Old, as Wide as Ever,* March 29, 2012.

That is, if first-generation status and gender were statistically independent, we would also expect to see the distribution of first-generation status identical in each gender category.

10. Although this general formula provides a framework for all PRE measures of association, only lambda is illustrated with this formula. Gamma, which is discussed in the next section, is calculated with a different formula. Both are interpreted as PRE measures.

11. "NSA Bugged European Union Offices, Computer Networks: Report," *Reuters*, June 29, 2013.

12. For 2×2 tables, a measure identical to gamma—Yule's Q—was first introduced by the statistician Udny Yule. However, whereas gamma is suitable for any size table, Yule's Q is appropriate only for 2×2 tables. When gamma is calculated for 2×2 tables, it is sometimes referred to as Yule's Q.

13. Because the two variables are nominal (not ordered), we can ignore the negative sign of gamma.

Chapter 9

1. Pew Research Center, "Immigration: Key Data Points From Pew Research," May 16, 2013. Retrieved June 24, 2013, from http://www.pewresearch.org/key-data-points/immigration-tip-sheet-on-u-s-public-opinion/.

2. *USA Today*, October 9, 1992.

3. Full consideration of the question of detecting the presence of a bivariate relationship requires the use of inferential statistics. Inferential statistics are discussed in Chapters 8 through 11.

4. Note that this group is but a small sample taken from the GSS national sample. The relationship between home ownership and race noted here may not necessarily hold true in other (larger) samples.

5. Another way in which percentages are sometimes expressed is with the total number of cases (N) used as the base. These overall percentages express the proportion of the sample who share two properties. For example, 7 of 89 respondents (7.9%) support abortion and have job security. Overall percentages do not have as much research utility as row and column percentages and are used less frequently.

6. The same three properties are also discussed by Joseph F. Healey in *Statistics: A Tool for Social Research* (Belmont, CA: Cengage, 2012), pp. 308–337.

7. Church attendance has been recoded into three categories.

8. For purposes of illustration, only selected categories of educational level and attendance of religious services are shown.

9. Because statistical independence is a symmetrical property, the distribution of the independent variable within each category of the dependent variable will also be identical.

Chapter 10

1. U.S. Census Bureau, *Statistical Abstract of the United States: 2012*, Table 229.

2. Since the N in our computational example is small ($N = 21$), the assumptions of normality and homogeneity of variance are required. We've selected a small N to demonstrate the calculations for F and have proceeded with Assumptions 3 and 4. If a researcher is not comfortable making these assumptions for a small sample, she or he can increase the size of N. In general, the F test is known to be robust with respect to moderate violations of these assumptions. A larger N increases the F test's robustness to severe departures from the normality and homogeneity of variance assumptions.

3. Carol Musil, Camille Warner, Jaclene Zauszniewski, May Wykle, and Theresa Standing, "Grandmother Caregiving, Family Stress and Strain, and Depressive Symptoms," *Western Journal of Nursing Research* 31, no. 3 (2009): 389–408.

4. Ibid., p. 395.

5. Ibid., p. 399.

Chapter 11

1. Refer to Paul Allison's *Multiple Regression: A Primer* (Thousand Oaks, CA: Pine Forge Press, 1999) for a complete discussion of multiple regression—statistical methods and techniques that consider the relationship between one dependent variable and one or more independent variables.

2. Center for Disease Control, "Births: Final Data for 2010," *National Vital Statistics Report*: 61(1), August 28, 2012.

3. J. M. Greene and C. L. Ringwalt, "Pregnancy Among Three National Samples of Runaway Homeless Youth," *Journal of Adolescent Health* 23 (1998): 370–377.

4. William J. Wilson, *The Truly Disadvantaged: The Inner City, the Underclass, and Public Policy* (Chicago: University of Chicago Press, 1987).

5. Stephanie Coontz, "The Welfare Discussion We Really Need," *Christian Science Monitor* (December 29, 1994): 19.

6. Preliminary analysis revealed several outliers. The scatterplots in Figures 13.12 and 13.13 depict state-level data with the outliers omitted.

7. Analysis is limited to women ages 40 years and older and those who identified working 40 hours per week or more the week prior to taking the GSS.

Index

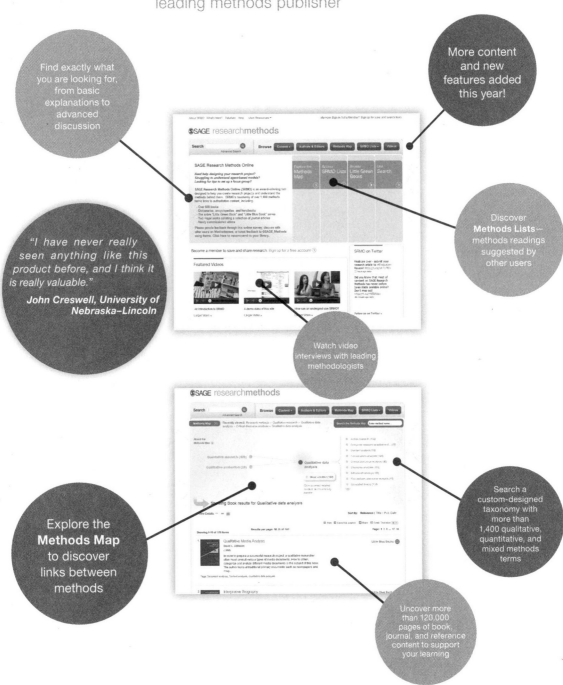

⊗SAGE researchmethods

The essential online tool for researchers from the world's leading methods publisher

Find exactly what you are looking for, from basic explanations to advanced discussion

More content and new features added this year!

"I have never really seen anything like this product before, and I think it is really valuable."

John Creswell, University of Nebraska–Lincoln

Discover **Methods Lists**— methods readings suggested by other users

Watch video interviews with leading methodologists

Explore the **Methods Map** to discover links between methods

Search a custom-designed taxonomy with more than 1,400 qualitative, quantitative, and mixed methods terms

Uncover more than 120,000 pages of book, journal, and reference content to support your learning

Find out more at
www.sageresearchmethods.com